T0254153

# Construction Equipment Management for Engineers, Estimators, and Owners

# Construction Equipment Management for Engineers, Estimators, and Owners

2nd Edition

Douglas D. Gransberg
Jorge A. Rueda-Benavides

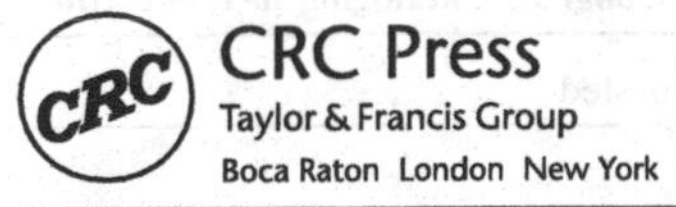

CRC Press is an imprint of the
Taylor & Francis Group, an Informa business

CRC Press
Taylor & Francis Group
6000 Broken Sound Parkway NW, Suite 300
Boca Raton, FL 33487-2742



CRC Press is an imprint of Taylor & Francis Group, an Informa business



Printed on acid-free paper

International Standard Book Number-13: 978-1-4987-8848-9 (Hardback)

International Standard Book Number-13: 978-0-367-48884-0 (Paperback)



**Library of Congress Cataloging-in-Publication Data**

A catalog record for this title has been requested

**Visit the Taylor & Francis Web site at**
**www.taylorandfrancis.com**

**and the CRC Press Web site at**
**www.crcpress.com**

# Contents

# Preface

The book is intended to be both a reference source for construction project managers, estimators, construction equipment fleet managers, professional engineers, and a textbook for use in university construction engineering and management courses. It is directed at both the public and private sectors. It contains a great deal of "hands-on, how-to" information about equipment management gleaned from the authors' personal construction experiences throughout the world. It was written to provide a guide for those individuals who find they need to estimate the cost of equipment on a given project who don't have that data at their fingertips because their routine business does not involve a lot of equipment-related construction. The authors also hope that it will be useful to the public agency equipment fleet manager whose need is to minimize equipment costs rather than to maximize the profit earned by the equipment.

The book is useful to all parties in the architect/engineer/construction industry as well as to project owners. The first chapter describes the evolution of construction equipment and serves to set the stage for the following chapters that provide specific up-to-date information on the state-of-art in the area. The chapters on estimating equipment ownership and operating costs, and determining economic life and replacement policy will be of great value to construction estimators. The chapters on determining the optimum mix of equipment and estimating the equipment productivity show the estimator how to maximize the profit of an equipment-intensive construction project. The chapter on scheduling demonstrates how to convert a linear schedule into a precedence diagram for use in a project that has a mandated scheduling methodology. This information shows the equipment fleet manager how to ensure that a production-driven, equipment-intensive project can be scheduled to achieve target production rates and hence target equipment-related unit costs and profits.

The book also shows managers and engineers how to avoid making costly common mistakes during project equipment selection. It contains a matrix that will help the novice equipment manager select the proper piece of equipment based on the requirements of the project. It is full of detailed examples of the types of calculations made to allow both public and private equipment owning organizations to determine an optimum equipment utilization plan for any project regardless of their level of experience. Finally, it includes material on the sustainability issues associated with equipment fleet management as well as brief descriptions of current and emerging technologies like automated machine guidance and intelligent compaction that will impact the fleet management field in the years to come.

The authors hope that the combination of both the analytical and practical will result in a reference document that will be of value to a wide range of individuals and organizations within the architect/engineer/construction industry.

*Douglas D. Gransberg*
*Jorge A. Rueda-Benavides*

# Preface

# Acknowledgments

The First Edition of the book was the brainchild of Dr. Calin Popescu of the University of Texas and flows out of an early work that he did to support his graduate civil engineering class in construction equipment management. Dr. Popescu's focus was on equipment used for heavy-civil projects. Professor Richard Ryan of the University of Oklahoma blended much of his work on managing construction equipment in building construction for his construction science undergraduate class on construction equipment in Dr. Popescu's outline to produce a reference that, for the first time treats both horizontal and vertical construction projects. Both original authors have retired, from academia, and while the 2nd edition has undergone a major revision, it still owes its focus and pragmatic tone to Professors Popescu and Ryan and for that the authors of this 2nd Edition are sincerely grateful.

# Author Biographies

**Douglas D. Gransberg, PhD, PE | President – Gransberg & Associates, Inc.**
Contact Information: Mailing address: Gransberg & Associates, 903 Flaming Oaks Dr., Suite 1, Norman, OK 73026; 405-503-3393; dgransberg@gransberg.com

Douglas D. Gransberg, PhD, PE, is the president of Gransberg & Associates, Inc., a construction management/project delivery consulting firm headquartered in Norman, Oklahoma. The firm was founded in 1996 and provides complex project management and cost engineering services to public agencies, consultants, and construction companies. G&A, Inc. has been called on to assist with projects throughout the United States and Canada, as well as in New Zealand, Okinawa, Latin America, Europe, and the Middle East. The firm specializes in the development of project management services for complex mega-projects. He received both his B.S. and M.S. degrees in Civil Engineering from Oregon State University and his Ph.D. in Civil Engineering from the University of Colorado at Boulder. He is a licensed Professional Engineer in Oklahoma, Texas and Oregon, a Certified Cost Engineer, a Fellow of the American Society of Civil Engineers a Fellow of the American Society of Civil Engineers, and a Fellow of the Royal Institution of Chartered Surveyors in the United Kingdom.

Dr. Gransberg retired in 2017 as a professor of construction engineering at Iowa State University, where he held an endowed research chair. He is also an Emeritus Professor of Architecture and Engineering at the University of Oklahoma, retiring from that institution in 2010. Before that he held a tenured position at Texas Tech University.

Before moving to academia in 1994, he spent over 20 years in the US Army Corps of Engineers, retiring at the rank of lieutenant colonel. His research is centered in the delivery of complex infrastructure/transportation projects. He is the author of four books on construction management topics and over 200 articles, conference papers, and other publications.

**Jorge A. Rueda-Benavides, PhD | Assistant Professor, Auburn University.**
Contact Information: Mailing address: Dept. of Civil Engineering, 238 Harbert Engineering Center, Auburn, AL, 36849-5337; 614-745-9422; jrueda@auburn.edu

Dr. Jorge Rueda-Benavides is an assistant professor of civil engineering at Auburn University in Auburn, Alabama. He specializes in transportation asset management, risk management, data analytics, qualitative and quantitative modeling, decision-making modeling, alternative contracting methods, and big data. Dr. Rueda taught construction equipment management and automated machine guidance courses at Iowa State University.

He received his BS degree in Civil Engineering at the Universidad Industrial de Santander in Bucaramanga, Columbia and his MS and PhD degrees in Construction Engineering and Management at Iowa State. Dr. Rueda been involved in various studies sponsored by the National Cooperative Highway Research Program (NCHRP) as well as studies for the Iowa, Montana, and Minnesota Departments of Transportation.

# Author Biographies

# 1 The Fundamentals of Construction Equipment Management

## 1.0 INTRODUCTION

Construction projects are built using three primary resources: materials, labor, and equipment. Since the competing contractors are bidding to build the same project, they will be required to purchase the same set of materials, and those materials will usually be purchased from the nearest supplier. So, there is very little price competition found in this required resource. Additionally, all the competitors will generally have to pay the prevailing wages for the location in which the project is being constructed, and again price competition is limited in the labor resource. Therefore, the one resource that is not standard amongst competing contractors is the equipment they choose to use in the project. Equipment choice is a function of the means and methods selected to accomplish the work. Hence, in a low-bid contract, the contractor who has the most efficient equipment fleet and the most cost-effective approach to construction equipment management will gain the necessary edge over its competitor.

Cost-effective construction equipment fleet management is the subject of this book. The book contains a comprehensive collection of construction equipment management practices that represent the state-of-the-practice on this topic. Construction equipment fleet management practices are important to project owners, both private and public, who own and operate their own fleet of equipment to provide resources to repair and maintain the capital facility infrastructure that is built for them by construction contractors. While the intensity of the profit motive is different, the fundamental principles of equipment management remain the same for both owners and construction contractors. Therefore, the remainder of the book will detail those principles and then demonstrate them in both contexts to account for both perspectives.

## 1.1 CONSTRUCTION EQUIPMENT MANAGEMENT DEFINED

Construction equipment is a critical resource for a construction company or public/private owner whose core business involves the construction, operations, and/or maintenance of equipment-intensive projects, such as highways, utilities, or industrial facilities. Reliable construction equipment production is essential to the timely completion of all types of construction projects. In fact, equipment purchases are often the most capital intensive, long-term investments made by a construction company or public/private owner. The expenses related to construction equipment are key contributors to the private construction company's profitability and a public/private owner's ability to operate within its authorized annual budget. Hence, managing construction equipment and its related expenses are crucial to the organization's success, making it a priority undertaking.

> A problem which frequently confronts a contractor …[planning] to construct a project is the selection of the most suitable equipment. [The contractor] should consider the money spent for the equipment as an investment which [it] can expect to recover, with a profit, during the useful life of the equipment.
>
> **A contractor does not pay for construction equipment; the equipment must pay for itself by earning the contractor more money that it cost.**
>
> R.L. Peurifoy 1956 [1]

The type of equipment managed is a function of the type of construction project on which they will be used. Table 1.1 is a synopsis of the level of equipment usage listed by type of construction project and shows the typical kinds of work activities that are generally performed by standard pieces of construction equipment on the project's site. One can see that the level of use increases as the size and type of project get bigger and more equipment intensive. Residential projects are labor-intensive with equipment use providing supporting work rather than production work. On the other end of the spectrum are heavy civil projects where the cardinal production rates are equipment related, and the construction equipment fleet will drive project progress.

The term "construction equipment management" refers to the continuing assessment of the organization's construction equipment fleet and its cost within the context of the portfolio of projects on which the equipment is actively employed. Thus, construction equipment management, in its most essential form, embodies optimizing the construction equipment expense and usage with the project schedules and cash flow. A construction equipment fleet manager must be able to make pragmatic decisions regarding the management of the construction equipment to ensure maximum benefits while minimizing the expenses of the fleet.

Ideally, a piece of construction equipment should pay for itself by earning more than it costs to own, operate, maintain, and store. Equipment operating costs are recurring expenses based

**TABLE 1.1**
**Equipment Use by Construction Sector**

| Sector | Project Types | Usage | Work Activities |
|---|---|---|---|
| Residential | Single and multi-family housing | Light | Finish site-work, foundation excavation, ground material moving, up to three-story lifting, pneumatic assembly tools |
| Commercial | Hotels, restaurants, warehouses, etc. | Moderate | Rough and finish site-work, stabilizing and compacting, multiple story material and personnel lifting, ground and on-structure material moving, miscellaneous types of assembly and support equipment |
| Industrial | Process plants, energy, water/waste-water | Heavy | Large volume rough and finish site-work, stabilizing and compacting, ground and on-structure material moving, multiple story heavy lifting and precision placing, numerous miscellaneous special types of equipment for assembly and support |
| Heavy Civil | Highways, bridges, dams, locks, canals, aviation | Intense | Mass dirt and material excavating and moving, stabilizing and compacting, ground material moving and hoisting, concrete and asphalt paving and finishing, miscellaneous special types of equipment for support |
| Specialty | Utilities, oil production, marine | Intense | Pipeline, power, transmission line, steel erection, railroad, offshore, pile driving, logging, concrete pumping, boring and sawing, many others |

on the hours of productive usage, but the financing and storage costs are incurred regardless of whether the equipment is working. Thus, idle machines produce no income while continuing to incur expenses to finance, maintain, and store. The equipment fleet manager must participate in a constant balancing act to ensure that the fleet remains productive and that idle time is minimized. Therefore, being able to forecast future equipment requirements based on the company's long-term workload is a necessary part of good construction equipment management. Much of the fleet manager's activities involve making the following types of decisions:

- Should new equipment be added to the fleet, and if yes, when?
- Which pieces of equipment need to be replaced?
- When should under-utilized or poorly running machines be eliminated from the fleet?
- What cost parameters indicate a piece of equipment has exceeded its economic life?
- Should project-specific or specialty equipment be purchased or leased?

Project equipment fleet composition is another key factor in ensuring timely completion of the construction or maintenance project within the available budget. Construction equipment fleet managers must thoroughly understand the construction activities to be completed on each project type, along with the appropriate machines that can be employed to meet both project budget and time requirements.

Construction equipment management generally operates on two levels. The first is management of the organization's entire fleet. The second level is the assignment of individual pieces of equipment to specific projects. Since this is often a dynamic environment, cost-effectiveness is the primary objective of the construction equipment manager. To achieve this objective, the fleet manager often must conduct the necessary analyses to determine the following:

- Selecting the number and type of construction equipment required for each project.
- Determining whether to own, lease, or rent the equipment and its effect on the bottom line.
- Planning the period in which each piece of equipment will be assigned to a given job site.
- Calculating transportation and operating costs when dealing with multiple projects and schedules.
- Maximizing the productivity of the construction equipment in the fleet.
- Ensuring that preventive maintenance is scheduled and accomplished to minimize the cost of downtime and reduce equipment-related delays to the project.

Given the above discussion, this book is intended to provide the necessary information as well as the tools to assist the construction equipment fleet manager in making the necessary analyses required to inform the full range of construction equipment management decisions required for current construction and maintenance projects.

## 1.2 EQUIPMENT FLEET MANAGEMENT

Managing a construction equipment fleet is rife with conflicting priorities. One of the most significant is the economic trade-off between the capital cost of replacing a piece of equipment and the ownership costs of operating and maintaining if the machine in question is retained for another year. Additionally, the fleet manager's funding typically comes from two sources: funds authorized to make capital equipment purchases and funding allocated to equipment operations and maintenance (O&M). Often the fleet manager must forecast the need to purchase replacement equipment as much as 5 years in advance of the actual purchase. It is crucial that the construction company or public/private owners collect the necessary performance data to permit it to make the replacement decision with a reasonable degree of confidence. Hence, one key aspect of successful equipment fleet management is to maintain the necessary cost, usage, and production data to inform the decisions made by the fleet manager. The data provide the objective input to determine when a piece of equipment has reached a point where it is no longer cost-effective to retain. This principle is called equipment life and represents one of the major aspects of fleet management.

Equipment life can be mathematically defined in three different ways: physical life, profit life, and economic life. [2] While profit life is not directly applicable to public agencies, physical and economic life are, and both must be defined and calculated when considering equipment life because they furnish two important means to approach replacement analysis and to ultimately make an equipment replacement decision. [3] The concepts of depreciation, inflation, investment, maintenance and repairs, downtime, and obsolescence are all integral to the calculation of equipment costs.

Vorster and Sears [4] proposed the "Cumulative Cost Model" shown in Figure 1.1, and Mitchell [2] extended it to accommodate repair costs in a manner that accounts for direct comparison with potential replacement models.

> With the cumulative cost model, it is possible to depict and understand changes in total costs, average costs, and marginal costs. The cumulative cost model is the only one of the economic replacement models that incorporates both classic economic replacement theory and repair limit theory.
>
> Mitchell. [2]

The cumulative cost model can be used to minimize costs or to maximize profits – it is not implicitly tied to one method or the other. It is also possible to explicitly show the three basic steps of buy, operate, and sell at any point in the machine's life. The cumulative cost model allows for more than one definition of economic life for heavy construction equipment [2].

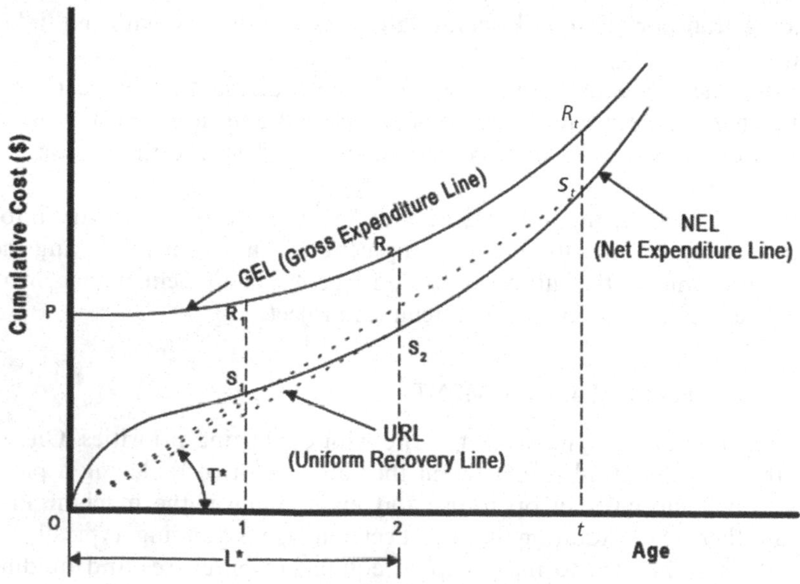


**FIGURE 1.1** Cumulative Cost Model after Mitchell [2].

Thus, the cumulative cost model furnishes an excellent theoretical foundation for equipment fleet management decision-making. This and other key concepts will be covered in detail in Chapter 3.

## 1.3 UNDERSTANDING EQUIPMENT PRODUCTION IMPACT DURING THE PROJECT PLANNING, SCHEDULING, AND COST ESTIMATING PROCESS

Many advancements have been made in the development of larger, faster, and more productive construction machinery. Increased machine productivity has resulted in an increase in the overall project size. These two factors have combined to produce a capital-intensive risk environment in which the construction contractor must operate. As a result, the construction industry has been forced to search for methods to reduce the high level of risk. Historically, the lowest cost method for reducing risk has been to provide detailed estimating and planning prior to submitting a bid and solid management throughout the course of the project. Estimating and planning involve the judicious selection of equipment, the careful scheduling of time and resources, and the accurate determination of system productivity and costs. Management involves putting the plan into action. The key management ingredient is having predetermined standards by which actual system outputs can be measured and upon which future decisions can be based.

Even a seemingly straightforward operation such as earthmoving is a highly dynamic system. A hauling operation contains several components that interact in a very complex manner. Analytical methods, based on engineering fundamentals, have been developed to solve the problem of bringing these components together in a logical manner. These methods mathematically model hauling systems. Their solutions are numerical results that may be used in the decision-making process of estimating, planning, and managing an earthmoving project.

Early construction equipment management methods made the somewhat naive assumption that optimizing productivity based on physical constraints of the environment would, in turn, minimize the overall production cost. Therefore, no effort was made to include cost or profit variables in those mathematical models. The models developed by Gates and Scarpa [5] were the first to recognize the importance of the cost function in overall system optimization. Many methods currently in use do not adequately model physical conditions. They rely on the judgment and experience of the user that may be very good or very bad with corresponding outputs from the models.

The modeling of the physical parameters that impact an equipment fleet's production is an essential aspect to developing accurate cost estimates during the contract bidding process. Parameters such as rolling resistance, grade, engine horsepower, and altitude all impact the efficiency of the equipment production process. Modeling methods range from simple algebraic formulas to sophisticated stochastic simulations. Each of the major methods will be covered in Chapters 5 and 6.

The modeling output includes estimated equipment system production rates. These data are adjusted from theoretical maximums, termed "instantaneous production," to account for delays, breaks, and other factors that will arise during a normal working day to determine a "sustained production" rate that is the average expected production for the typical daily shift. That information is then used as input for estimating durations for various equipment-related activities in the project's schedule. A comparison of the contractual schedule constraints with the estimated duration of production-driven activities is then used to determine the number and types of construction equipment needed to complete the project on time. These are then grouped into activity-specific crews, that is, excavation, material hauling, crane lift, and so on, which then give the estimator the number and types of operators needed for the project to estimate the project labor costs.

Maximizing project equipment fleet productivity is crucial to accurately estimate a project's cost. In the low-bid environment, optimizing the construction equipment fleet selected for a project can provide the necessary competitive advantage to win the contract while maintaining a competitive target profit on the project.

## 1.4 CONSTRUCTION EQUIPMENT MANAGEMENT IN THE PAST AND TODAY

Figure 1.2 chronicles the development of earthmoving equipment from the 18th century to the present. Essentially, the construction industry transitioned from building with hand tools to building with machines. The machines were specialized adaptations of the tools they replaced. The tracked excavator replaced the shovel, and the dump truck replaced the wheelbarrow. As a result, equipment management transitioned from a master builder who ensured that all the project's craftworkers were supplied with the appropriate tools to the equipment fleet manager who optimizes the production, operation, and maintenance of the machinery assigned to the job. The advent of computers created a new step in the equipment evolution by providing a means to electronically control various operating systems on the equipment itself and giving the machine a "smart" capability to adjust its engine's performance as its operating parameters change. The result was seen in enhanced fuel efficiency, increased productivity, and reduced emissions.

Construction fleet managers had to add a new skill set to their toolbelts when the Global Positioning System (GPS) was launched, making it possible to maintain line and grade remotely without the need for surveyors' staking. Prior to the advent of GPS, the project's equipment production was often constrained by the ability of survey crews to establish horizontal and vertical control for the project. As the precision of GPS improved, automated machine guidance (AMG) systems were added to construction machinery, permitting the onboard computer to automatically make the adjustments of cutting blades and other machine-mounted tools, leaving the equipment operator to concentrate on merely driving the piece of equipment.

AMG-enabled equipment added another level of equipment O&M cost to pay for the periodic calibration of the computerized apparatus and to update the software, as well as the cost of training for AMG staff technicians. Additionally, two-dimensional construction plans must be converted to three-dimensional terra-models to permit the efficient use of AMG on large earthmoving projects. At this writing, the design conversion issue is being supplanted by the implementation of three-dimensional design software. In the vertical construction sector, Building Information Modeling (BIM) permits the designer to produce virtual construction documents that can digitally communicate with other systems that have the capability to translate geospatial coordinate data. In the heavy highway construction sector, Civil Integrated Management (CIM) performs a similar function regarding 3-dimensional design products and adds a full suite of other key data in a machine-readable format.

The upshot is that construction equipment management is now a computer and data-driven activity. Where the equipment fleet managers of the 20th century were often master mechanics and skilled operators themselves, the fleet managers of the 21st century must be computer literate, technology - savvy, and masters of analytics. Advancements in equipment technology will continue to occur, and fleet managers must be prepared to exploit them as they arrive.

## 1.5 THE FUTURE OF CONSTRUCTION EQUIPMENT MANAGEMENT

The most recent entry into the construction equipment arena is the appearance of fully autonomous construction machinery, which does not require a human operator to guide the machine. Essentially, these are robots in every sense of the word. Thus, with GPD, AMG, BIM, CIM, and autonomous machinery, the construction industry now has the means to enhance its potential maximum production by eliminating the weakest link in the production chain: the human

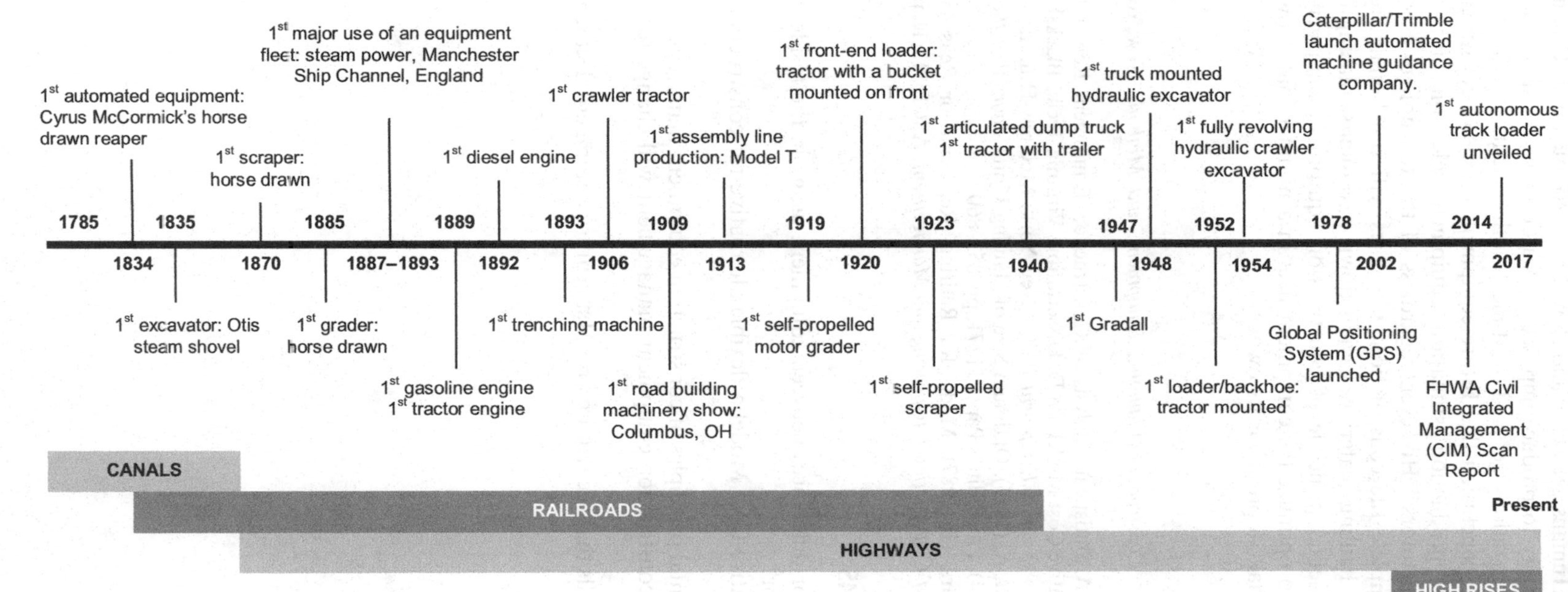


**FIGURE 1.2** Earthmoving and Excavating Equipment Development Timeline.

equipment operators. Humans get weary and slow down. Robots don't. Humans are subject to constraints on their ability to produce imposed by statutory work rules and union contracts. Robots don't have to join unions, get paid overtime, health benefits, or take vacations.

All this leads to the conclusion that future equipment-intensive construction projects will require new models for production estimation, equipment selection, and a different skillset for the equipment "operators." However, regardless of the technology used to perform the tasks required to complete the construction, the tasks will functionally remain the same: lifting loads, digging, hauling materials, laying down pavements, placing concrete, and so on. Thus, the fundamental principle of construction equipment management will remain essentially the same: to optimize the equipment fleet in a manner that allows the machinery to pay for itself and make a profit for its owner.

## REFERENCES

[1] Peurifoy, R.L. (1956). *Construction Planning, Equipment, and Methods.* New York: McGraw-Hill, Inc. p. 66.

[2] Mitchell, Z. (1998). A Statistical Analysis of Construction Equipment Repair Costs Using Field Data and the Cumulative Cost Model, Ph.D. Dissertation, Virginia Tech., Blacksburg, VA.

[3] Douglas, J. (1975). *Construction Equipment Policy.* New York: McGraw-Hill, Inc.

[4] Gates, M. and Scarpa, A. (1975). Optimum Size of Hauling Units. *Journal of the Construction Division*, ASCE, 101(CO4), Proceedings Paper 11771, pp. 853–860.

[5] Vorster, M. and Sears, G. (1987). Model for Retiring, Replacing or Reassigning Construction Equipment. *Journal of Construction Engineering and Management*, ASCE 113(1), pp. 125–137.

## CHAPTER PROBLEMS

1. What are two major differences between equipment usage in the residential sector versus the heavy civil sector?
2. Why is the Cumulative Cost Model a flexible alternative for construction equipment fleet decision-making?
3. List three fundamental principles of construction equipment management.
4. Speculate on how construction equipment management will change to construct projects on the moon.
5. List five specific skills that are required by 21st century equipment fleet managers.

# 2 Cost of Owning and Operating Construction Equipment

## 2.0 INTRODUCTION

A thorough understanding of both estimated and actual costs of operating and owning equipment drives profitable equipment management. This chapter will develop that understanding in detail and help the reader to understand the calculations that go into determining these fundamental costs for an equipment-intensive project.

Plant, equipment, and tools used in construction operations are priced in the following three categories in the estimate.

1. *Small tools and consumables*: Hand tools up to a certain value together with blades, drill bits, and other consumables used in the work are priced as a percentage of the total labor price of the estimate.
2. *Equipment that is usually shared by several work activities*: These kinds of equipment items are kept at the site over a period of time and used on the work in progress.
3. *Equipment that is used for specific tasks*: These are capital items and used on the project such a digging trench or hoisting material into place. This equipment is priced directly against the take-off quantities for the work it is to be used on. The equipment is not kept on site for extended periods like those in the previous classification, but the equipment is shipped to the site, used for its task, and then immediately shipped back to its source. Excavation equipment, cranes, hoisting equipment and highly specialized, items such as concrete saws fall into this category.

This chapter's focus is on estimating the cost of owning and operating construction equipment of the third category. For contractors in the heavy civil construction industry, the cost of owning and operating equipment is a key part of doing business in a profitable manner. Without knowing the actual equipment ownership costs, contractors might report higher-than-justified paper profits due to inaccurate accounting practices that do not factor the cost of idle equipment into the company's overall profit picture.

Total equipment costs comprise two separate components: ownership costs and operating costs. Except for the one-time initial capital cost of purchasing the machine, ownership costs are fixed costs that are incurred each year, regardless of whether the equipment is operated or idle. Operating costs are the costs incurred only when the equipment is used. All the above costs have different characteristics of their own and are calculated using different methods.

## 2.1 OWNERSHIP COST

Ownership costs are fixed costs. Almost all these costs are annual in nature and include:

- Initial capital cost
- Depreciation
- Investment (or interest) cost
- Insurance cost
- Taxes
- Storage cost

### 2.1.1 Initial Cost

On average, the initial cost makes up about 25% of the total cost that will be invested during the equipment's useful life [1]. This cost is paid for getting equipment into the contractor's yard, or construction site, and getting them ready for operation. Many kinds of ownership and operating costs are calculated using initial cost as a basis, and normally this cost can be calculated accurately. Initial cost consists of the following items:

- Price at factory + extra equipment + sales tax
- Cost of shipping
- Cost of assembly and erection

### 2.1.2 Depreciation

Depreciation represents the decline in market value of a piece of equipment due to age, wear, deterioration, and obsolescence. Depreciation can result from:

- Physical deterioration occurring from wear and tear of the machine;
- Economic decline or obsolescence occurring over the passage of time.

In the appraisal of depreciation, some factors are explicit while other factors must be estimated. Generally, the asset costs are known, which include:

- *Initial cost*: The amount needed to acquire the equipment.
- *Useful life*: The number of years it is expected to last.
- *Salvage value*: The expected amount the asset will be sold at the end of its useful life.

However, there is always some uncertainty about the exact length of the useful life of the asset, and about the precise amount of salvage value, which will be realized when the asset is disposed. There are many depreciation methods. Among them, the straight-line method, double-declining-balance method, and sum-of-the-years'-digits method are the most common in the construction equipment industry [2] and will be discussed below. At this point, it is important to state that the term depreciation, as used in this chapter, is meant to represent the change in the assets value from year to year and as a means for establishing an hourly "rental" rate for that asset. It is not meant in the same exact sense as is used in the tax code. The term "rental rate" is the rate the equipment owner charges the project for using the equipment; that is, the project "rents" the equipment from its owner.

In calculating depreciation, the initial cost should include the costs of delivery and start-up, including transportation, sales tax, and initial assembly. The equipment life used in calculating depreciation should correspond to the equipment's expected economic or useful life. The reader is referred to the references at the end of this chapter for a more thorough discussion of the intricacies of depreciation.

#### 2.1.2.1 Straight-Line Depreciation

Straight-line depreciation is the simplest to understand as it makes the basic assumption that the equipment will lose the same amount of value every year of its useful life until it reaches its salvage value. The depreciation in a given year can be expressed by Equation 2.1.

$$D_n = \frac{\text{IC} - S - \text{TC}}{N} \tag{2.1}$$

where $D_n$ = Depreciation in year "$n$"
IC = Initial cost($)
$S$ = Salvage value($)
TC = Tire/track costs($)
$N$ = Useful life (years)
and $D_1 = D_2 = \ldots = D_n$

#### 2.1.2.2 Sum-of-Years'-Digits Depreciation

The sum-of-years'-digits depreciation method tries to model the fact that depreciation is not a straight line. The actual market value of a piece of equipment after it is one year old is less than the amount predicted by the straight-line method. Thus, this is an accelerated depreciation method and models more annual depreciation in the early years of a machine's life and less in its later years. The calculation is straightforward and done using Equation 2.2.

$$D_n = \frac{(\text{Year "}n\text{" Digit})(\text{IC} - S - \text{TC})}{1 + 2 + \ldots + N} \tag{2.2}$$

where $D_n$ = Depreciation in year "$n$"
Year "$n$" Digit = Reverse order: "$n$" if solving for $D_1$ or 1 if solving for $D_n$
IC = Initial cost ($)
$S$ = Salvage value ($)
TC = Tire/track costs ($)
$N$ = Useful life (years)

#### 2.1.2.3 Double-Declining Balance Depreciation

The double-declining balance is another method of calculating an accelerated depreciation rate. It produces more depreciation in the early years of a machine's useful life than the sum-of-years'-digits depreciation method. It does this by depreciating the "book value" of the equipment rather than just its initial cost. The book value in the second year is merely the initial cost minus the depreciation in the first year. Then the book value in the next year is merely the book value of the second year minus the depreciation in the second year, and so on until the book value reaches the salvage value. The estimator has to be careful when using this method to ensure that the book value never drops below the salvage value. Equation 2.3

is used to calculate this type of depreciation and Example 2.1 demonstrates how to apply the calculation to a typical piece of construction equipment.

$$D_n = \frac{2\,(BV_{n-1} - TC)}{N} \tag{2.3}$$

$$\begin{aligned} \text{where } D_n &= \text{Depreciation in year “}n\text{”} \\ TC &= \text{Tire/track costs (\$)} \\ N &= \text{Useful life (years)} \\ BV_{n-1} &= \text{Book value at the end of the previous year} \\ &\quad \text{and } BV_{n-1} > S \end{aligned}$$

**Example 2.1:** Compare the depreciation in each year of the equipment's useful life for each of the above depreciation methods for the following wheeled front-end bucket loader:

- *Initial cost*: $148,000 includes delivery and other costs.
- *Tire cost*: $16,000
- *Useful life*: 7 years
- *Salvage value*: $18,000

A sample calculation for each method will be demonstrated and the results are shown in Table 2.1.

*Straight-line method*: From Equation 2.1, the depreciation in the first year "$D_1$" which is equal to the depreciation in all the years of the loader's useful life.

$$D_1 = \frac{\$148,000 - \$18,000 - \$16,000}{7 \text{ years}} = \$16,286/\text{year}$$

*Sum-of years'-digits method*: From Equation 2.2, the depreciation in the first year "$D_1$" and the second year "$D_2$" are:

$$D_1 = \frac{7}{1+2+3+4+5+6+7}(\$148,000 - \$18,000 - \$16,000) = \$28,500$$

$$D_2 = \frac{6}{1+2+3+4+5+6+7}(\$148,000 - \$18,000 - \$16,000) = \$24,429$$

*Double declining balance method*: From Equation 2.2, the depreciation in the first year "$D_1$" is:

**TABLE 2.1**
**Depreciation Method Comparison for Wheeled Front-End Loader**

| Year<br>Method | 1 | 2 | 3 | 4 | 5 | 6 | 7 |
|---|---|---|---|---|---|---|---|
| SL $D_n$ | $16,286 | $16,286 | $16,286 | $16,286 | $16,286 | $16,286 | $16,286 |
| SOYD $D_n$ | $28,500 | $24,429 | $20,357 | $16,286 | $12,214 | $8,143 | $4,071 |
| DDB $D_n$ | $37,714 | $26,939 | $19,242 | $13,744 | $9,817 | $6,543 | $0 |
| DDB BV | $94,286 | $67,347 | $48,105 | $34,361 | $24,543 | $18,000 | $18,000 |

$$D_n = \frac{2}{7}\ (\$148,000 - \$16,000) = \$37,714$$

and the "book value" at the end of year 1 = \$148,000 – \$16,000 – \$37,714
= \$94,286

However, in year 6, the equation would give an annual depreciation of \$7,012, which, when subtracted for the book value at the end of year 5, gives a book value for year 6 of \$17,531. This is less than the \$18,000 salvage value; therefore, the depreciation in year 6 is reduced to the amount that would bring the book value to be equal to the salvage value of \$6,543 and the depreciation in year 7 is taken as zero, which means the machine was fully depreciated by the end of year 6.

Selecting a depreciation method to use for computing ownership cost is a business policy decision. Thus, this book will not advocate one method over another. The US Internal Revenue Service publishes a guide that details the allowable depreciation for tax purposes, and many companies choose to follow this in computing ownership costs. As stated before, the purpose of developing this figure is to arrive at an hourly rental rate that the estimator can use to figure the cost of equipment-intensive project features of work, not to develop an accounting system that serves to alter a given organization's tax liabilities. While this obviously impacts a company's ultimate profitability, this book separates tax costs from tax consequences, leaving the tax consequences of business policy decisions for the accountants rather than the estimators.

### 2.1.3 Investment Cost

Investment (or interest) cost represents the annual cost (converted to an hourly cost) of capital invested in a machine [2]. If borrowed funds are utilized for purchasing a piece of equipment, the equipment cost is simply the interest charged on these funds. However, if the equipment is purchased with company assets, an interest rate that is equal to the rate of return on company investment should be charged. Therefore, investment cost is computed as the product of an interest rate multiplied by the value of the equipment, then converted to cost per hour of operation.

The average annual cost of interest should be based on the average value of the equipment during its useful life. The average value of equipment may be determined from the following equation:

$$P = \frac{\mathrm{IC}(n+1)}{2} \tag{2.4}$$

where IC = Total initial cost

$P$ = average value

$n$ = useful life, years.

This equation assumes that a unit of equipment will have no salvage value at the end of its useful life. If a unit of equipment will have salvage value when it is disposed of, the average value during its life can be obtained from the following equation and is demonstrated in Example 2.2:

$$P = \frac{\mathrm{IC}(n+1)\ +\ S(n-1)}{2n} \tag{2.5}$$

where IC = Total initial cost
$P$ = average value
$S$ = salvage value
$n$ = useful life, years.

**Example 2.2:** Consider a unit of equipment costing $50,000 with an estimated salvage value of $15,000 after 5 years. Using Equation 2.2, the average value is:

$$P = \frac{50,000(5+1) + 15,000(5-1)}{2(5)}$$
$$= \frac{300,000 + 60,000}{10}$$
$$= \underline{\underline{\$36,000}}$$

### 2.1.4 Insurance Tax and Storage Costs

Insurance cost represents the cost of fire, theft, accident, and liability insurance for the equipment. Tax cost represents the cost of property tax and licenses for the equipment. Storage cost includes the cost of rent and maintenance for equipment storage yards, the wages of guards and employees involved in moving equipment in and out of storage, and associated direct overhead.

The cost of insurance and tax for each item of equipment may be known on an annual basis. In this case, this cost is simply divided by the hours of operation during the year, to yield the cost per hour for these items. Storage costs are usually obtained on an annual basis for the entire equipment fleet. Insurance and tax costs may also be known on a fleet basis. It is then necessary to prorate these costs for each item. This is usually done by converting the total annual cost into a percentage rate, then by dividing these costs by the total value of the equipment fleet. By doing so, the rate for insurance, tax, and storage, may simply be added to the investment cost rate for calculating the total annual cost of investment, insurance, tax, and storage [2].

The average rates for interest, insurance, tax, and storage found in the literature are listed in Table 2.2 [2–5]. These rates will vary according to related factors, such as the type of equipment and location of the job site.

**TABLE 2.2**
**Average Rates for Investment Costs**

| Item | Percentage Average Value |
|---|---|
| Interest | 3–9% |
| Tax | 2–5% |
| Insurance | 1–3% |
| Storage | 0.5–1.5% |

## 2.2 TOTAL OWNERSHIP COST

Total equipment ownership cost is calculated as the sum of depreciation, investment cost, insurance cost, tax, and storage cost. As mentioned earlier, the elements of ownership cost are often known on an annual cost basis. However, while the individual elements of ownership cost are calculated on an annual-cost basis or an hourly basis, total ownership cost should be expressed as an hourly cost.

After all elements of ownership costs have been calculated, they can be added to yield total ownership cost per hour of operation. Although this cost may be used for estimating and charging equipment cost to projects, it does not include job overhead or profit. Therefore, if the equipment is to be rented to others, overhead and profit should be included to obtain an hourly rental rate. Example 2.3 shows how this is done.

**Example 2.3:** Calculate the hourly ownership cost for the second year of operation of the 465 horsepower twin-engine scraper. This equipment will be operated 8 hr/day and 250 days/year in average conditions. Use the sum-of-years digits method of depreciation and as the following information:

- *Initial cost*: \$186,000
- *Tire cost*: \$14,000
- *Estimated life*: 5 years
- *Salvage value*: \$22,000
- *Interest on the investment*: 8%
- *Insurance*: 1.5%
- *Taxes*: 3%
- *Storage*: 0.5%
- *Fuel price*: \$2.00/gallon
- *Operator's wages*: \$24.60/hr

$$\text{Depreciation in the second year} = \frac{4}{15}(186{,}000 - 22{,}000 - 14{,}000) = \$40{,}000$$

$$= \frac{40{,}000}{8(250)} = \$20.00/\text{hr}$$

Investment cost, tax, insurance, and storage cost:

$$\text{Cost rate} = \text{Investment} + \text{tax, insurance, and storage} = 8 + 3 + 1.5 + 0.5 = 13\%$$

$$\text{Average investment} = \frac{186{,}000 + 22{,}000}{2(5)} = \$20{,}800$$

$$\text{Investment, tax, insurance, and storage} = \frac{84{,}000(0.18)}{2000} = \$7.56/\text{hr}$$

$$\text{Total ownership cost} = 16.53 + 7.56 = \underline{\underline{\$24.09/\text{hr}}}$$

## 2.3 COST OF OPERATING CONSTRUCTION EQUIPMENT

Operating costs of the construction equipment, which represents a significant cost category and should not be overlooked, are those costs associated with the operation of a piece of equipment. They are incurred only when the equipment is actually being used. Operating costs of the equipment are also called "variable" costs because they depend on several factors such as the number of operating hours, the types of equipment used, and the location and working condition of operation.

The operating costs vary with the amount of equipment used and job-operating conditions. The best basis for estimating the cost of operating construction equipment is the use of historical data from the experience of similar equipment under similar conditions. If such data are not available, recommendations from the equipment manufacturer could be used.

### 2.3.1 Maintenance and Repair Cost

The cost of maintenance and repairs usually constitutes the largest amount of operating expense for construction equipment. Construction operations can subject equipment to considerable wear and tear, but the amount of wear varies enormously between the different items of equipment used and between different job conditions. Generally, maintenance and repair costs get higher as the equipment gets older. Equipment owners will agree that good maintenance, including periodic wear measurement, timely attention to recommended service, and daily cleaning when conditions warrant it, can extend the life of equipment and actually reduce the operating costs by minimizing the effects of adverse conditions. All items of plant and equipment used by construction contractors will require maintenance and probably also require some repairs during the course of their useful life. The contractor who owns the equipment usually sets up facilities with workers qualified to perform the necessary maintenance operations on the equipment.

The annual cost of maintenance and repairs may be expressed as a percentage of the annual cost of depreciation, or it may be expressed independently of depreciation. The hourly cost of maintenance and repair can be obtained by dividing the annual cost by its operating hours per year. The hourly repair cost during a particular year can be estimated by using the following formula [2]:

$$\text{Hourly repair cost} = \frac{\text{Year digit}}{\text{Sum of years' digit}} \times \frac{\text{Life time repair cost}}{\text{Hours operated}} \tag{2.6}$$

The lifetime repair cost is usually estimated as a percentage of the equipment's initial cost without the cost of tires. It is adjusted by the operating condition factor obtained from Table 2.3. Example 2.4 shows how to use these values in a sample calculation.

**Example 2.4:** Estimate the hourly repair cost of the scraper in Example 2.3 for the second year of operation. The initial cost of the scraper is \$186,000, tires cost \$14,000, and its useful life is 5 years. Assume average operating condition and 2,000 hr of operation per year.
Lifetime repair cost factor = 0.90

$$\begin{aligned}
&\text{Lifetime repair cost factor} &&= 0.90\\
&\text{Lifetime repair cost} &&= 0.90(186{,}000 - 14{,}000) = \$154{,}800\\
&\text{Hourly repair cost} &&= \frac{2}{15}\left(\frac{154{,}800}{2000}\right) = \underline{\underline{\$10.32/\text{hour}}}
\end{aligned}$$

**TABLE 2.3**
**Range Typical Lifetime Repair Costs from the Literature [2–4]**

| Equipment | % Initial Cost without Tires | | |
|---|---|---|---|
| | Operating Conditions | | |
| Type | Favorable | Average | Unfavorable |
| Crane | 40–45 | 50–55 | 60–70 |
| Excavator Crawler | 50–60 | 70–80 | 90–95 |
| Excavator wheel | 75 | 80 | 85 |
| Loader track | 80–85 | 90 | 100–105 |
| Loader wheel | 50–55 | 60–65 | 75 |
| Motor grader | 45–50 | 50–55 | 55–60 |
| Scraper | 85 | 90–95 | 105 |
| Tractor crawler | 85 | 90 | 95 |
| Tractor wheel | 50–55 | 60–65 | 75 |
| Truck, off-highway | 70–75 | 80–85 | 90–95 |

### 2.3.2 Tire Cost

The tire cost represents the cost of tire repair and replacement. Because the life expectancy of rubber tires is generally far less than the life of the equipment they are used on, the depreciation rate of tires will be quite different from the depreciation rate on the rest of the vehicle. The repair and maintenance cost of tires as a percentage of their depreciation will also be different from the percentage associated with the repair and maintenance of the vehicle. The best source of information in estimating tire life is the historical data obtained under similar operating conditions. Table 2.4 below lists the typical ranges of tire life found in the most recent literature on the subject for various types of equipment.

Tire repair costs can add about 15% to tire replacement costs. So, the following equation may be used to estimate tire repair and replacement cost:

$$\text{Tire repair and replacement cost} = 1.15 \times \frac{\text{Cost of a set of tires(\$)}}{\text{Expected tire life(h)}} \tag{2.7}$$

### 2.3.3 Consumable Costs

Consumables are those items that are required for the operation of a piece of equipment that literally gets consumed in the course of its operation. These include but are not limited to fuel, lubricants, and other petroleum products. They also include such things as filters, hoses, strainers, and other small parts and items that are used as the equipment is run.

#### 2.3.3.1 Fuel Cost

Fuel consumption is incurred when the equipment is operated. When operating under standard conditions, a gasoline engine will consume approximately 0.06 gal of fuel per flywheel horsepower hour (fwhp-hr), while a diesel engine will consume approximately 0.04 gal per fwhp-hr. A horsepower hour is a measure of the work performed by an engine.

**TABLE 2.4**
**Range Typical Tire Life from the Literature [2,5]**

| Equipment | Average Tire Life in Hours Operating Conditions | | |
|---|---|---|---|
| **Type** | **Favorable** | **Average** | **Unfavorable** |
| Loader wheel | 3,200–4,000 | 2,100–3,500 | 1,300–2,500 |
| Motor grader | 5,000 | 3,200 | 1,900 |
| Scraper single engine | 4,000–4,600 | 3,000–3,300 | 2,500 |
| Scraper twin engine | 3,600–4,000 | 3,000 | 2,300–2,500 |
| Scraper elevating | 3,600 | 2,700 | 2,100–2,250 |
| Tractor wheel | 3,200–4,000 | 2,100–3,000 | 1,300–2,500 |
| Truck, off-highway | 3,500–4,000 | 2,100–3,500 | 1,100–2,500 |

The hourly cost of fuel is estimated by multiplying the hourly fuel consumption by the unit cost of fuel. The amount of fuel consumed by the equipment can be obtained from the historical data. When the historical data are not available, Table 2.5 gives approximate fuel consumption (gallons/hour) for major types of equipment, and Example 2.5 provides a sample calculation.

**Example 2.5:** Calculate the average hourly fuel consumption and hourly fuel cost for the twin-engine scraper in Example 2.3. It has a diesel engine rated at 465 hp and fuel cost \$2.00/gal. During a cycle of 20 seconds, the engine will operate at full power while filling the bowl in tough ground, requires 5 seconds. During the balance of the cycle, the engine will use no more than 50% of its rated power. Also, the scraper will operate about 45 minutes/hr on average. For this condition, the approximate amount of fuel consummated during 1 hr is determined as follows:

$$\begin{aligned}
&\text{Rated power} : 465\text{ hp}\\
&\text{Engine factor} :\\
&\text{Filling the bowl, 5 seconds/20 second cycle} = 0.250\\
&\text{Rest of cycle, } 15/20 \text{ x } 0.5 = 0.375\\
&\text{Total cycle} = 0.625\\
&\text{Time factor, 45 minutes/60 minutes} = 0.75\\
&\text{Operating Factor, } 0.625 \text{ x } 0.75 = 0.47
\end{aligned}$$

$$\begin{aligned}
&\text{From Table 2.5} : \text{Use "unfavorable" fuel consumption factor} = 0.040\\
&\text{Fuel consumed per hour} : 0.47(465)(0.040) = 8.74\text{ gal}\\
&\text{Hourly Fuel Cost} : 8.74\text{ gal/hr}(\$2.00/\text{gal}) = \underline{\underline{\$17.48/\text{hr}}}
\end{aligned}$$

### 2.3.3.2 Lubricating Oil Cost

The quantity of oil required by an engine per change will include the amount added during the change plus the make-up oil between changes. It will vary with the engine size, the capacity of the crankcase, the condition of the piston rings, and the number of hours between oil changes. It is a common practice to change oil every 100–200 hr [6].

**TABLE 2.5**
**Average Fuel Consumption Factors (gal/hr/hp) [2,5]**

| Equipment | Gallons/hour/horsepower Working Conditions | | |
|---|---|---|---|
| **Type** | **Favorable** | **Average** | **Unfavorable** |
| Loader track | 0.030–0.034 | 0.040–0.042 | 0.046–0.051 |
| Loader wheel | 0.020–0.024 | 0.027–0.036 | 0.031–0.047 |
| Motor grader | 0.022–0.025 | 0.029–0.035 | 0.036–0.047 |
| Scraper single engine | 0.023–0.026 | 0.029–0.035 | 0.034–0.044 |
| Scraper twin engine | 0.026–0.027 | 0.031–0.035 | 0.037–0.044 |
| Tractor crawler | 0.028–0.342 | 0.037–0.399 | 0.046–0.456 |
| Tractor wheel | 0.020–0.028 | 0.026–0.038 | 0.031–0.052 |
| Truck, off-highway | 0.017–0.029 | 0.023–0.037 | 0.029–0.046 |
| Truck, on-highway | 0.014–0.029 | 0.020–0.037 | 0.026–0.046 |

The quantity of oil required can be estimated by using Equation 2.8 [6]:

$$q = \frac{0.006(\text{hp})(\text{f})}{7.4} + \frac{c}{t} \tag{2.8}$$

$$\begin{aligned} \text{where } q &= \text{quantity consumed, gal/hr} \\ \text{hp} &= \text{rated horsepower of engine} \\ c &= \text{capacity of crankcase, gal} \\ f &= \text{operating factor} \\ t &= \text{number of hours between changes} \\ \text{Consumption rate} &= 0.006 \text{ pounds per horsepower} - \text{hour} \\ \text{Conversion factor} &= 7.4 \text{ pounds per gallon} \end{aligned}$$

The consumption data or average cost factors for oil, lubricants, and filters for their equipment under average conditions are available from the equipment manufacturers.

### 2.3.4 Mobilization and Demobilization Cost

This is the cost of moving the equipment from one job site to another. The costs of equipment mobilization and demobilization can be large and are always important items in any job where substantial amounts of equipment are used. These costs include freight charges (other than the initial purchase), unloading cost, assembly or erection cost (if required), highway permits, duties, and special freight costs (remote or emergency). In fact, on a - $3 million earthmoving job, it is not unusual to have a budget from $100,000 to $150,000 for move-in and move-out expenses. The hourly cost can be obtained from the total cost divided by the operating hours. Some public agencies cap the maximum amount of mobilization that will be paid before the project is finished. In these instances, the estimator must check the actual costs of mobilization against the cap. If the cap is exceeded, the unrecovered amount must be allocated to other pay items to ensure that the entire cost of mobilization is recovered.

### 2.3.5 Equipment Operator Cost

Operator's wages are usually added as a separate item after other operating costs have been calculated. They should include overtime or premium charges, workmen's compensation insurance, social security taxes, bonus, and fringe benefits in the hourly wage figure. Care must be taken by companies that operate in more than one state or work for federal as well as state and private owners. The federal government requires that prevailing scale (union scale) wages be paid to workers on its project regardless if the project is located in a union or right-to-work state. This is a requirement of the Davis Bacon Act [5], and most federal contracts will contain a section in the general conditions that details the wage rates that are applicable to each trade on the project.

### 2.3.6 Special Items Cost

The cost of replacing high wear items, such as dozer, grader, and scraper blade cutting and end bits, as well as ripper tips, shanks, and shank protectors, should be calculated as a separate item of operating expense. As usual, unit cost is divided by expected life to yield cost per hour.

## 2.4 METHODS OF CALCULATING OWNERSHIP AND OPERATING COST

The most common methods available are the Caterpillar method, Association of General Contractors of America (AGC) method, the U.S. Army Corps of Engineers (USACE) method, and the Peurifoy method. Each method is described below.

### 2.4.1 Caterpillar Method

The Caterpillar method is based on the following principles [6]:

1. No prices for any items are provided. For reliable estimates, these must always be obtained locally.
2. Calculations are based on the complete machine. Separate estimates are not necessary for the basic machine, dozer, control, and so on.
3. The multiplier factors provided will work equally well in any currency expressed in decimals.
4. Because of different standards of comparison, what may seem a severe application to one machine owner may appear only average to another. Therefore, in order to better describe machine use, the operating conditions and applications are defined in zones.

#### 2.4.1.1 Ownership Costs

These costs are calculated for depreciation, interest, insurance, and taxes. Usually, depreciation is done to zero value with the straight-line method, which is not based on tax consideration, but resale or residual value at replacement may be included for depreciation or tax incentive purposes. Service life of several types of equipment is given in the *Caterpillar Performance Handbook* [6]. Acquisition or delivered costs should include costs due to freight, sales tax, delivery, and installation. On rubber-tired machines, tires are considered a wear item and covered as an operating expense. Tire cost is subtracted from the delivered price. The delivered price less the estimated residual value results in the value to be recovered through work, divided by the total usage hours, giving the hourly cost to project the asset's value. The interest on capital used to purchase a machine must be considered, whether the machine is purchased outright or financed. Insurance cost and property taxes can be calculated in one of the two ways.

#### 2.4.1.2 Operating Costs

Operating costs are based on charts and tables in the handbook. They are broken down as follows:

1. Fuel;
2. Filter, oil, and grease (FOG) costs;
3. Tires;
4. Repairs;
5. Special items; and
6. Operator's wages.

The factors for fuel, FOG, tires, and repairs costs can be obtained for each model from tables and charts given in the *Caterpillar Performance Handbook* [6]. Tire costs can be estimated from previous records or from local prices. Repairs are estimated based on a repair factor that depends on the type, employment, and capital cost of the machine. The operator's wages are the local wages plus the fringe benefits. Table 2.6 is an example of this method application for a truck-mounted crane.

### 2.4.2 USACE Method

This method is often considered as the most sophisticated method for calculating equipment ownership costs because it not only covers economic items, but it also includes geographic conditions. This method generally provides hourly use rates for construction equipment based on a standard 40-hr workweek. The total hourly use rates include all costs of owning and operating equipment except operator wages and overhead expenses. The ownership portion of the rate consists of allowances for depreciation and costs of facilities capital cost of money (FCCM). Operating costs include allowances for fuel, filter, oil, grease, servicing the equipment, repair and maintenance, and tire wear and tire repair [7].

The standby hourly rate is computed from the average condition by allowing the full FCCM hourly cost plus one-half of the hourly depreciation.

#### 2.4.2.1 Ownership Costs

The USACE method operates on the following principles:

1. Depreciation: It is calculated by using the straight-line method. The equipment cost used for depreciation calculation is subtracted by tire cost at the time the equipment was manufactured. Another cost that must be subtracted is salvage value. It is determined from the tables in Appendix D of USACE manual (found in Appendix A of this book) or from local advertisements of used equipment for sale. The expected life span of the equipment is designated from the manufacturers' or equipment associations' recommendations.
2. FCCM: The Department of the Treasury adjusts the cost-of-money rate on in January and July each year. This cost is computed by multiplying the cost-of-money rate, determined by the Secretary of the Treasury, by the average value of equipment and prorating the result over the annual operating hours. It is normally presented in terms of FCCM per hour.

**TABLE 2.6**
**Caterpillar Method: Backhoe Ownership and Operating Cost Example [6]**

| | | | | |
|---|---|---|---|---|
| | | **Caterpillar Method** | | |
| **Backhoe-loader, wheeled** | | | **Average conditions of use** | *** "Factor" from reference** |
| 0.75 CY hoe/1.0CY loader bucket | | | Tires = $2,459 | Freight = $17.43/cwt |
| Equipment horsepower: 100; | | | Fuel cost = $3.00/gal | Sales tax = 5.80% |
| Total expected use = 10,000 hr | | | Estimated annual = 1,360 hr | Discount = 7.5% |
| Weight = 164.5 cwt | | | Total tire life = 2,500 hr | |
| | | **Compute Input Data** | | |
| List price | Includes loader and hoe buckets | | | $159,000.00 |
| Useful life (years) | 10,000 total hr/1,360 hr/yr | | | 7.35 |
| Discount | $159,000(7.5%) | | | $11,925.00 |
| Total sales tax | $159,000(5.8%) | | | $8,530.35 |
| Freight: | 164.5 cwt ($17.43/cwt) | | | $2,867.24 |
| Delivered price | $159,000 − $11,925 + $8,530.35 − $2,459 | | | $153,146.35 |
| Salvage value 10% | $159,000 (10.0%) | | | $15,900.00 |
| | | **Compute Hourly Ownership and Operating Costs** | | |
| | | | | Hourly Ownership Costs ($/hr) |
| Interest | | | 6.8% | $4.32 |
| Insurance | | | 3.0% | $1.92 |
| Taxes | | | 2.0% | $1.28 |
| Depreciation | | | $153,146.35/10,000 hr | $15.31 |
| | | | *Total Hourly Ownership Costs* | ***$22.83*** |
| *Factor | | | | Hourly Operating Costs ($/hr) |
| Equipment | | 0.038 | 0.038(100 hp)$3.00/gal | $11.40 |
| Tires | | - | $2,459/2,500 | $0.98 |
| Repairs | | 0.07 | 0.07($153,146.35)/1,360 hr | $7.88 |
| Operator | | - | | $19.56 |
| | | | *Total Hourly Operating Costs* | **$39.83** |
| | | **Compute Annual Ownership and Operating Costs** | | |
| Annual ownership cost | | | 1,360 hr ($22.29/hr) | $31,048.89 |
| Annual operating cost | | | 1,360 hr ($39.19/hr) | $54,163.54 |
| | | | *Total Annual Cost* | **$85,212.43** |

* Caterpillar Inc. (2017). *Caterpillar Performance Handbook*, 47th ed. Peoria, IL: Caterpillar Inc.

It should be noted that licenses, taxes, storage, and insurance cost are not included in this computation. Instead, they are considered as indirect costs.

### 2.4.2.2 Operating Costs

USACE includes the following operating costs:

1. Fuel costs: Fuel costs are calculated from records of equipment consumption, which is done in cost-per-gallon per hour. Fuel consumption varies depending on the machine's requirements. The fuel can be either gasoline or diesel.

2. FOG: FOG costs are usually computed as percentage of the hourly fuel costs.
3. Maintenance and repair costs: These are the expenses charged for parts, labor, sale taxes, and so on. Primarily, maintenance and repair cost per hour are computed by multiplying the repair factor to the new equipment cost, which is subtracted by tire cost, and divided by the number of operating hours.
4. Hourly tire cost: This is the current cost of new tires plus cost of one recapping and then divided by the expected life of new tires plus the life of recapped tires. It has been determined that the recapping cost is approximately 50% of the new tire cost, and that the life of a new tire plus recapping will equal approximately 1.8 times the "useful life" of a new tire.
5. Tire repair cost: This cost is assumed to be 15% of the hourly tire wear cost.

Table 2.7 is an example of how this method is applied to the same piece of equipment as in Table 2.6.

**TABLE 2.7**
**Corps of Engineers Method: Backhoe Ownership and Operating Cost Example [7]**

| | | USACE Method | |
|---|---|---|---|
| **Backhoe-loader, wheeled** | | **Average conditions of use** | *** "Factor" from reference** |
| 0.75 CY hoe/1.0CY loader bucket | | Tires = $2,459 | Freight = $17.43/cwt |
| Equipment horsepower: 100; | | Fuel cost = $3.00/gal | Sales tax = 5.80% |
| Total expected use = 10,000 hr | | Estimated annual = 1,360 hr | Discount = 7.5% |
| Weight = 164.5 cwt | | Total tire life = 2,500 hr | |
| | | **Compute Input Data** | |
| List price | Includes loader and hoe buckets | | $159,000.00 |
| Useful life (years) | 10,000 total hr/1,360 hr/yr | | 7.35 |
| Discount | $159,000(7.5%) | | $11,925.00 |
| Total sales tax | $159,000(5.8%) | | $8,530.35 |
| Freight: | 164.5 cwt ($17.43/cwt) | | $2,867.24 |
| Delivered price | $159,000 − $11,925 + $8,530.35 − $2,459 | | $153,146.35 |
| Salvage value 10% | $159,000(10.0%) | | $15,900.00 |
| | | **Assemble Input Data Factors** | |
| | *Factor | | |
| Economic key | 45 | | |
| Discount code B | 0.075 | | |
| Tire cost index (TCI) | 0.988 | *TCI yr man/TCI yr use = 2,371/2,400 = 0.988 | |
| Repair factor | 0.944 | Econ Adj Factor = EAF (Economic Adjustment Factor)<br>yr man/EAF yr use = 7,904/7,763 = 1.018<br>Repair factor = RCF(EAF)(LAF) = 0.80(1.018)(1.16) = 0.944 | |
| Fuel factor (equipment) | 0.050 | | |
| Filter-oil-grease (FOG) factor | 0.441 | | |
| Tire wear factor (front) | 0.83 | | |
| | 0.54 | | |

*(Continued)*

**TABLE 2.7 (Cont.)**

| Backhoe-loader, wheeled | | USACE Method Average conditions of use | * "Factor" from reference |
|---|---|---|---|
| Tire wear factor (drive) | | | |
| Repair cost factor (RCF) | 0.80 | | |
| Labor Adj factor (LAF) | 1.16 | | |
| | | **Compute Hourly Ownership and Operating Costs** | |
| | | | *Hourly Ownership Costs ($/hr)* |
| Depreciation | - | [(153,146.35) (1–10%) – (1.16* $2,459)]/10,000 hr | $13.54 |
| Fac Capital Cost Money | 0.617 | (153,146.35)(0.617)(0.034)/1,360 | $2.34 |
| | | *Total Hourly Ownership Costs* | **$15.88** |
| | | | *Hourly Operating Costs ($/hr)* |
| Equipment Fuel | 0.05 | 0.05(100 hp)($3.00/gal) | $15.00 |
| Tires | 1.5; 1.8 | 1.5($245,900)/(1.8)(2,500) | $1.20 |
| Repairs | | [$153,146.35 – (0.988*$2,459)] (0.944)/1,360 hr | $14.23 |
| Filter-Oil-Grease (FOG) | 0.441 | 0.441 ($15.00/hr)1.16 | $7.67 |
| Tire repair cost | 1.5 | 1.5(1.20)(1.16) | $2.08 |
| Operator | - | | $19.56 |
| *Total Hourly Operating Costs* | | | **$59.74** |
| | | **Compute Annual Ownership and Operating Costs** | |
| Annual ownership cost | | 1,360 hr ($15.89/hr) | $21,597.21 |
| Annual operating cost | | 1,360 hr ($59.75/hr) | $81,246.04 |
| | | *Total Annual Cost* | $102,843.25 |

* US Army Corps of Engineers. (2016) *Construction Equipment Ownership and Operating Expense Schedule*, Region I. Document EP 1110-1-8 (Vol. 1), Washington, DC.

### 2.4.3 The AGC Method

This method enables the owner to calculate the owning and operating costs to determine capital recovery [1]. Rather than dealing with the specific makes and models of the machines, the equipment is classified according to capacity or size. For example, this method computes the average annual ownership expense and the average hourly repair and maintenance expense as a percentage of the acquisition costs.

#### 2.4.3.1 Ownership Cost

The ownership costs considered in this method are the same as described in the Caterpillar method; however, replacement cost escalation is also considered. Depreciation is calculated by the straight-line method, and includes purchase price, sales tax, freight, and erection cost, with an assumed salvage value of 10%. Average economic life in hours and average annual operating hours are shown for each size range. Replacement cost escalation of 7% is designed to augment the capital recovery and to offset inflation and machine price increase.

Interest on the investment is assumed to be 7%, whereas taxes, insurance, and storage are taken as 4.5%.

#### 2.4.3.2 Operating Costs

Maintenance and repair costs are calculated based on an hourly percentage rate times the acquisition cost. It is a level rate regardless of the age of the machine. This expense includes field and shop repairs, overhaul, and replacement of tires and tracks, and so on. The FOG costs and operator's wages are not considered in this method. Table 2.8 shows how the AGC method is applied to the crane example.

### 2.4.4 Peurifoy Method

R.L. Peurifoy is considered by many to be the father of modern construction engineering. His seminal work on the subject, now in its ninth edition [4] set the standard for using rigorous engineering principles to develop rational means for developing cost estimates based on equipment fleet production rates. These methods will be discussed in detail in Chapter 5 of this book. Therefore, it is important that Peurifoy's approach to determining equipment ownership costs be included in any discussion of the subject.

#### 2.4.4.1 Ownership Cost

This method assumes the straight-line method for depreciation. The value of the equipment is depreciated to zero at the end of the useful life of the equipment. The ownership costs are based on an average investment cost that is taken as 60% of the initial cost of the equipment. Usually, equipment owners charge an annual fixed rate of interest against the full purchase cost of the equipment. This gives an annual interest cost, which is higher than it should be. Since the cost of depreciation has already been claimed, it is more realistic to base the annual cost of investment on the average value of equipment during its useful life. This value can be obtained by taking an average of values at the beginning of each year that the equipment will be used, and this is the major difference between the Peurifoy method and the other methods. The cost of investment is taken as 15% of the average investment.

#### 2.4.4.2 Operating Costs

Since the tire life is different from that of the equipment, its costs are treated differently. The maintenance cost is taken as 50% of the annual depreciation, the fuel and the FOG costs are included, whereas the operator wages are not included. Table 2.9 finishes by showing how this method is applied to the crane example.

### 2.4.5 Comparison of Costs Calculated by Different Methods

It is interesting to note that each method arrives at a different hourly rental rate for the same piece of equipment. This illustrates the statement made earlier in this chapter that the method used to arrive at a number is largely a business policy decision rather than a technical decision. Table 2.10 is a summary of the four previous examples and furnishes an interesting comparison of the business decisions made by each group.

The first thing that is notable is that the AGC method yields the highest rental rate. Perhaps this is because the AGC is a trade organization for construction contractors and as a result, there is a bias to be conservative in the published method for calculating an equipment rental rate. Pursuing that line of reasoning, the USACE rate is the lowest. As a result, USACE is a large public owner who may have a bias to keep the cost of equipment on its projects as low as possible. The remaining two falls somewhere in the middle

**TABLE 2.8**
**AGC Method: Backhoe Ownership and Operating Cost Example [1]**

| | | AGC Method | | |
|---|---|---|---|---|
| **Backhoe-loader, wheeled** | | **Average conditions of use** | | ***"Factor" from reference** |
| 0.75 CY hoe/1.0CY loader bucket | | Tires = $2,459 | | Freight = $17.43/cwt |
| Equipment horsepower: 100; | | Fuel cost = $3.00/gal | | Sales tax = 5.80% |
| Total expected use = 10,000 hr | | Estimated annual = 1,360 hr | | Discount = 7.5% |
| Weight = 164.5 cwt | | Total tire life = 2,500 hr | | |
| | | **Compute Input Data** | | |
| List price | Includes loader and hoe buckets | | | $159,000.00 |
| Useful life (years) | 10,000 total hr/1,360 hr/yr | | | 7.35 |
| Discount | $159,000(7.5%) | | | $11,925.00 |
| Total sales tax | $159,000(5.8%) | | | $8,530.35 |
| Freight: | 164.5 cwt ($17.43/cwt) | | | $2,867.24 |
| Delivered price | $159,000 – $11,925 + $8,530.35 – $2,459 | | | $153,146.35 |
| Salvage value 10% | $159,000(10.0%) | | | $15,900.00 |
| Acquisition cost | $153,146.35(1–10%) | | | $137,831.70 |
| | | **Compute Hourly Ownership and Operating Costs** | | |
| | | | | *Hourly Ownership Costs ($/hr)* |
| Replacement cost escalation | | | 7.0% | |
| Interest | | | 7.0% | |
| Taxes, insurance, storage | | | 4.5% | |
| Depreciation | | | 15.0% | |
| Total ownership expense | | | 33.5% | |
| *Total Hourly Ownership Costs* | | | [(33.5%/1,360)($137,831.70)]/100 | **$33.95** |
| *Factor | | | | *Hourly Operating Costs ($/hr)* |
| Repair & maintenance expense | | 19.4% | [$137,831.70 (19.4%/1,360)]/100 | $19.66 |
| Operator | | | | $19.56 |
| | | | *Total Hourly Operating Costs* | **$39.22** |
| | | **Compute Annual Ownership and Operating Costs** | | |
| Annual ownership cost | | | 1,360 hr ($33.95/hr) | $46,173.62 |
| Annual operating cost | | | 1,360 hr ($39.22/hr) | $53,340.95 |
| | | | *Total Annual Cost* | **$99,514.58** |

* Douglas, J. (1978) Equipment costs by current methods. *J Con Div* ASCE. 104(02), pp. 191–225

as each has no constituency to protect. Each equipment-owning organization will have its own internal method for arriving at these rates that will satisfy the financial accounting needs of that company. These published methods are primarily used in negotiations between an owner and a contractor to determine if the contractor's internal equipment rates are fair and reasonable.

**TABLE 2.9**
**Peurifoy Method: Backhoe Ownership and Operating Cost Example [6]**

| | **Peurifoy Method** | |
|---|---|---|
| **Backhoe-loader, wheeled** | **Average conditions of use** | * **"Factor" from reference** |
| 0.75 CY hoe/1.0CY loader bucket | Tires = $2,459 | Freight = $17.43/cwt |
| Equipment horsepower: 100; | Fuel cost = $3.00/gal | Sales tax = 5.8% |
| Total expected use = 10,000 hr | Estimated annual = 1,360 hr | Discount = 7.5% |
| Weight = 164.5 cwt | Total tire life = 2,500 hr | Interest = 5.0% |

**Compute Input Data**

| | | |
|---|---|---|
| List price | Includes loader and hoe buckets | $159,000.00 |
| Useful life (years) | 10,000 total hr/1,360 hr/yr | 7.35 |
| Discount | $159,000(7.5%) | $11,925.00 |
| Total sales tax | $159,000(5.8%) | $8,530.35 |
| Freight: | 164.5 cwt ($17.43/cwt) | $2,867.24 |
| Delivered price | $159,000 – $11,925 + $8,530.35 – $2,459 | $153,146.35 |
| Salvage value 10% | $159,000(10.0%) | $15,900.00 |

**Compute Hourly Ownership and Operating Costs**

| | | | |
|---|---|---|---|
| *Factor | | | *Hourly Ownership Costs ($/hr)* |
| Equivalent uniform annual cost | | [($159,000 – $2,459)*5%]<br>(1+0.05$^{-7.35}$ | $20.45 |
| Taxes, insurance, storage | 3.75% | 3.75%($159,000–$2,459) | $1.28 |
| | | *Total Hourly Ownership Costs* | **$24.76** |
| *Factor | | | *Hourly Operating Costs ($/hr)* |
| Fuel | 3% | 0.69(3%)(100 hp)$3.00/gal | $6.21 |
| Fuel combined factor | 0.69 | | |
| Fuel-oil-grease factor | | Use Corps FOG cost | $7.67 |
| Repair & maintenance | 37% | 0.994 [($153,146.35) – (0.998)<br>($2,459)]<br>10,000 hr | $14.23 |
| Tire repair cost | 16% | 16% ($2,459/2,500 hr) | $0.16 |
| Operator | — | | $19.56 |
| | | *Total Hourly Operating Costs* | **$34.58** |

**Compute Annual Ownership and Operating Costs**

| | | |
|---|---|---|
| Annual ownership cost | 1,360 hr ($24.76/hr) | $33,677.03 |
| Annual operating cost | 1,360 hr ($34.58/hr) | $47,034.75 |
| | *Total Annual Cost* | **$80,711.78** |

* Peurifoy, R.L., Schexnayder, C.J., Schmitt, R., and Shapira, A. (2018), *Construction Planning, Equipment and Methods*, 9th ed., New York: McGraw Hill.

**TABLE 2.10**
**Summary of Different Methods for Calculating Equipment Ownership and Operating Costs**

| Item | Caterpillar | USACE | AGC | Peurifoy |
|---|---|---|---|---|
| Ownership cost per hour | $22.83 | $15.88 | $33.95 | $24.76 |
| Operating cost per hour | $39.83 | $59.74 | $39.22 | $34.58 |
| **Total cost per hour** | **$62.66** | **$75.62** | **$73.17** | **$59.35** |
| Annual ownership cost | $31,048.89 | $21,597.21 | $46,173.62 | $33,677.03 |
| Annual operating cost | $54,163.54 | $81,246.04 | $53,340.95 | $47,034.75 |
| **Total annual cost** | **$85,212.43** | **$102,843.25** | **$99,514.58** | **$80,711.78** |

## 2.5 SUMMARY

This chapter has provided information and data to allow the estimator who does not already have an internal method to calculate the cost of owning and operating a piece of construction equipment to do so. The information can be used in several ways. First, it could be used as a reference for setting an internal standardized method for calculating equipment rental rates. Next, it could be used to perform an independent estimate of rates that are being proposed for a given project to determine if they appear to be fair and reasonable. Lastly, it can be used as a mutually agreed standard for calculating these types of rates during contract or change order negotiations. In any event, the estimator must strive to use the best numbers available at the time and to ensure that all the costs of both owning and operating the equipment are included in the final rate.

## REFERENCES


[1] Douglas, J. (1978). Equipment Costs by Current Methods. *Journal of the Construction Division*, 104 (2), pp. 191–225.

[2] Nunnally, S.W. (2007). *Construction Methods and Management*. 7th ed. Patterson,: Pearson Prentice Hall.

[3] Atcheson, D. (1993). *Earthmoving Equipment Production Rates and Costs*. Venice: Norseman Publishing Co.

[4] Peurifoy, R.L., Schexnayder, C.J., Schmitt, R. and Shapira, A. (2018). *Construction Planning, Equipment and Methods*. 9th ed. New York: McGraw Hill.

[5] Clough, R.H. and Sears, G.A. (1994). *Construction Contracting*. 6th ed. New York: John Wiley and Sons, Inc., pp. 384–385.

[6] Caterpillar Inc. (2017). *Caterpillar Performance Handbook*. 47th ed. Peoria, IL: Caterpillar Inc.

[7] US Army Corps of Engineers (USACE). (2016). *Construction Equipment Ownership and Operating Expense Schedule*, Region I. Document EP 1110- 1–8 (Vol. 1), Washington, DC.


## CHAPTER PROBLEMS

1. Compute the total depreciation using the straight-line depreciation method for the following wheeled backhoe:

   - Initial cost: $88,000 includes delivery and other costs.
   - Tire Cost: $2,500
   - Useful life: 5 years
   - Salvage value: $21,000

2. Compute the total depreciation using the double decline balance depreciation method for wheeled backhoe in the above problem.
3. Calculate the hourly ownership cost for the third, fourth, and fifth years of operation of the 465-hp twin-engine scraper describe in Example 2.3.
4. Estimate the hourly repair cost of the backhoe in Problem 1 for the first year of operation. Assume average operating condition and 2,000 hr of operation per year.
5. Calculate the average hourly fuel consumption and hourly fuel cost for the backhoe in Problem 1. It has a diesel engine rated at 100 hp and fuel cost $3.00/gal. During a cycle of 20 seconds, the engine will operate at full power while filling the hoe bucket, which takes 4 seconds. During the balance of the cycle, the engine will use no more than 50% of its rated power. Also, the backhoe will produce for 50 minutes per each hour.
6. Calculate the total hourly costs for the crane described below using:

   a. The Caterpillar method
   b. The Peurifoy method

| | |
|---|---|
| Truck-mounted Crane 150-ton w/260′ Lattice Boom | Equipment Load factors: |
| Equipment horsepower: 207 | Average conditions of use |
| Carrier horsepower: 430 | Crane |
| Estimated annual use in hours: 1,590 hr | Lifting: 0.30 |
| Total expected use in hours: 20,000 hr | Return: 0.53 |
| Tires front: $3,520 total | Carrier |
| Tires drive: $7,040 total | Running: 0.10 |
| Fuel cost: $2.00/gal | Idle: 0.45 |
| Sales tax: 8.7% | |

# 3 Equipment Life and Replacement Procedures

## 3.0 INTRODUCTION

Once purchased and placed into operation, a piece of equipment will suffer wear and tear, resulting in mechanical issues that will reduce its efficiency and increase the cost to keep it in the fleet. Eventually, it will need to be replaced, making the equipment replacement decision an important factor in maintaining a profitable equipment fleet. The point where this decision must be made is called the equipment's life and essentially involves determining when it is no longer economically feasible to repair. This chapter describes three equipment management economic decision-making factors:

- *Equipment life*: Calculating an equipment's economic useful life.
- *Replacement analysis*: Analytic tools to compare replacement alternatives at the end of its useful life.
- *Replacement equipment selection*: Considerations necessary for choosing the best replacement alternative.

Equipment life can be mathematically defined in three different ways: physical life, profit life, and economic life. All three should be considered an equipment repair or replace decision as they provide three different approaches to make the final decision. Depreciation, inflation, investment, maintenance and repairs, downtime, and obsolescence are all included in equipment replacement analysis, allowing the fleet manager to conduct a logical replacement analysis to make the necessary decisions.

Equipment replacement analysis involves computing economic life, alternative selection, and replacement timing for the machine in question. There are five theoretical replacement methods available to the analyst, which include:

- Intuitive method for small equipment fleets.
- Minimum cost method for public agencies.
- Maximum profit method for construction contractors and other fleet owners.
- Payback period method, using engineering economics and generally applicable to all types and sizes of equipment fleets.
- Mathematical modeling method used in computer simulations for optimizing equipment fleet size and composition.

The above methods furnish the theoretical foundation for understanding the empirical methods used in both the public and private sectors. Examples of the practical replacement methods used by Texas, Montana, and Louisiana Departments of Transportation are detailed later in this chapter.

## 3.1 EQUIPMENT LIFE

Construction equipment life can be defined as physical life, profit life, or economic life. The different definitions are illustrated in Figure 3.1. [1] The graph shows that the machine is not

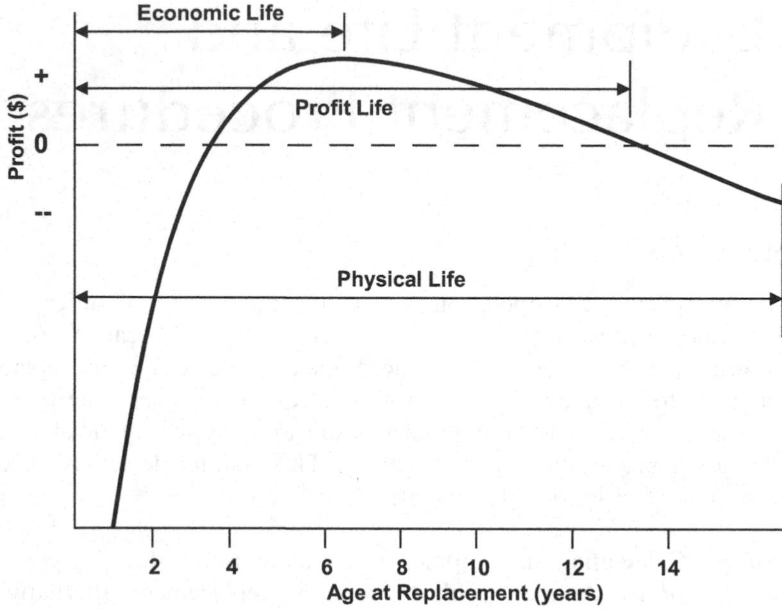


**FIGURE 3.1** Equipment Life Definitions after Douglas [1].

profitable (i.e., earns more than the capital cost of its procurement) early in its physical life. This is followed by a phase where it earns more than it costs to own, operate, and maintain it and ends in a condition where the costs of retaining it exceed the amount it earns.

### 3.1.1 Physical Life

Physical life is reached when the machine can no longer reliably produce at the required level. At this point, it will usually be abandoned or scrapped. Operating and maintenance (O&M) costs increase as construction machinery ages. A machine's physical life and the pace at which its O&M costs increase are a function of the physical conditions of the jobsite on which it is employed, the skill of its operators, and the quality of the maintenance it receives. [1] Experience has shown that regular expenditures for preventive maintenance ameliorate large expenses to replace major operating system components.

### 3.1.2 Profit Life

Profit life is the period over which a machine's earnings are greater than its costs, resulting in an operating loss after that point [1]. The end of an equipment's profit life is characterized by increasingly costly repairs as major components fail and are replaced. Thus, the successful fleet manager tracks each machine's condition to identify when it is nearing the point where it must be replaced before costly major systems fail.

### 3.1.3 Economic Life

The theory of equipment economic life seeks to maximize profits over the equipment's life. Thus, the economic life span attempts to optimize production with respect to profit. Figure 3.1 shows that the economic life is less than physical life, ending at a point in time when the profit generated by a given machine is at its maximum. Hence, replacing the equipment prevents the erosion of its profitability by increasing O&M and repair costs. Fleet managers need precise records of O&M and repair costs to determine the appropriate replacement time at the end of the machine's economic life. The economic life calculation includes ownership costs, operating costs, depreciation, inflation, investment, maintenance, repair, downtime, and obsolescence costs [1,2].

#### 3.1.3.1 Depreciation Costs and Replacement

The dictionary defines depreciation as "a decrease in the value of property through wear, deterioration, or market conditions [3]." The depreciation is the machine's loss in value over the time it is purchased to the time it is replaced. Table 3.1 presents an analysis of the life for a hypothetical piece of equipment, illustrating the process for calculating the hourly cost depreciation and replacement cost. In the table, the book value is the actual amount to be realized on a trade-in, assuming a 5% annual average cost increase. It should be noted that the average hourly depreciation cost is not linear and it decreases as the equipment hours over which it is applied increases.

#### 3.1.3.2 Inflation

Equipment replacement costs are affected by both general economic and industry inflation. Economic inflation is the overall loss in the buying power of the national currency, whereas industry inflation is the change in construction costs due to long- and short-term fluctuations in commodity pricing. For example, the Consumer Price Index is a widely reported inflation index that seeks to model the purchasing power of the United States consumer dollar and constitutes a metric for economic inflation. Industry inflation is associated with specific market sectors and is typically independent of economic inflation. For example, the surprising increase in steel prices during 2004–2005 was specific to the construction industry and hence would need to be considered separately from more general measures of economic activity.

One issue that is often not addressed is the presence of deflation. Prices do fall as well as rise, and certain commodities, such as diesel fuel and liquid asphalt, are more volatile than other more stable commodities. Figure 3.2 shows the change in diesel prices from 2015 to 2019. It clearly shows periods of increasing and decreasing prices. It is impossible to predict the future of volatile commodities like diesel, and as such, the assumptions made for volatile commodities should definitely be tested using sensitivity analysis, allowing the price to

**TABLE 3.1**
**Depreciation and Replacement Costs**

| End of Year | Replacement Cost ($) | Book Value ($) | Loss on Replacement ($) | Cumulative Hours of Use ($/hr) | Cumulative Cost Per Hour ($/hr) |
|---|---|---|---|---|---|
| 0 | 30,000 | 30,000 | 0 | 0 | 0 |
| 1 | 31,500 | 22,500 | 9,000 | 2,000 | 4.50 |
| 2 | 33,000 | 18,000 | 15,000 | 4,000 | 3.75 |
| 3 | 34,500 | 15,100 | 19,400 | 6,000 | 3.23 |
| 4 | 36,000 | 12,800 | 23,200 | 8,000 | 2.90 |
| 5 | 37,500 | 10,600 | 26,900 | 10,000 | 2.69 |
| 6 | 39,000 | 9,100 | 29,900 | 12,000 | 2.49 |
| 7 | 40,500 | 7,900 | 32,600 | 14,000 | 2.33 |
| 8 | 42,000 | 6,800 | 35,200 | 16,000 | 2.20 |

**FIGURE 3.2** Change in US Diesel Fuel Prices.

go both up and down (see section 3.2.3 for details). While the inflation and/or deflation should always be considered in equipment replacement decision-making, its effects can be ignored if the equipment manager is using a comparative analytical method because it can be assumed to affect all alternatives equally [4].

### 3.1.3.3 Investment Costs

Investment costs include interest, insurance, taxes, and license fees that are typically added to equipment's initial purchase cost. Table 3.2 shows that investment costs are often expressed as a percentage of initial equipment cost. Table 3.2 continues the Table 3.1 hypothetical example, demonstrating the calculation of the hourly investment cost. Based on the values shown in Table 3.2, the hypothetical annual investment cost in this example is assumed to be 15%.

### 3.1.3.4 Maintenance and Repair Costs

Maintenance and repair costs are the crux of the equipment replacement decision, flowing from the cost of labor and parts consumed maintaining and repairing the given piece of equipment. This is a dynamic system and impacted by the following parameters:

- Equipment type,
- Equipment age,
- Operating conditions,
- Operator skill,
- Daily care by the operator and maintenance personnel, and
- Frequency and quality of preventive maintenance.

As a result, accurate cost records are essential to be able to estimate maintenance and repair costs in equipment replacement analyses. Table 3.3 provides an example of the hourly maintenance and repair cost calculation [4].

**TABLE 3.2**
**Investment Costs**

| Year | Investment Start of Year ($) | Depreciation ($) | Investment End of Year ($) | Investment Cost ($) | Cumulative Investment Cost ($) | Cumulative Use (hr) | Cumulative Cost Per Hour ($/hr) |
|---|---|---|---|---|---|---|---|
| 1 | 30,000 | 7,500 | 22,500 | 4,500 | 4,500 | 2,000 | 2.25 |
| 2 | 22,500 | 4,500 | 18,000 | 3,375 | 7,875 | 4,000 | 1.97 |
| 3 | 18,000 | 2,900 | 15,100 | 2,700 | 10,575 | 6,000 | 1.76 |
| 4 | 15,100 | 2,300 | 12,800 | 2,265 | 12,840 | 8,000 | 1.61 |
| 5 | 12,800 | 2,200 | 10,600 | 1,920 | 14,760 | 10,000 | 1.48 |
| 6 | 10,600 | 1,500 | 9,100 | 1,590 | 16,350 | 12,000 | 1.36 |
| 7 | 9,100 | 1,200 | 7,900 | 1,365 | 17,715 | 14,000 | 1.27 |
| 8 | 7,900 | 1,100 | 6,800 | 1,185 | 18,900 | 16,000 | 1.18 |

**TABLE 3.3**
**Maintenance and Repair Costs**

| Year | Annual Maintenance and Repair Cost ($) | Cumulative Cost ($) | Cumulative Use (hr) | Hourly Cumulative Cost ($/hr) |
|---|---|---|---|---|
| 1 | 970 | 970 | 2,000 | 0.49 |
| 2 | 2,430 | 3,400 | 4,000 | 0.85 |
| 3 | 2,940 | 6,340 | 6,000 | 1.06 |
| 4 | 3,280 | 9,620 | 8,000 | 1.20 |
| 5 | 4,040 | 13,660 | 10,000 | 1.37 |
| 6 | 4,430 | 18,090 | 12,000 | 1.51 |
| 7 | 5,700 | 23,790 | 14,000 | 1.70 |
| 8 | 6,290 | 30,080 | 16,000 | 1.88 |

#### 3.1.3.5 Downtime

Downtime is the time when a machine cannot be used because of repairs and typically increases as equipment usage increases [1]. Availability is the opposite of downtime. For example, if the equipment's downtime is 10%, its availability is 90%.

The downtime costs are the sum of the ownership cost, operating cost, operator cost, and productivity loss when the equipment is not available for use. Table 3.4 shows how to compute the hourly downtime cost. The productivity loss cost is not included because it is not easily quantified as a dollar value. However, it is included as a weight factor with maximum availability equal to 1.0 and proportionate loss in availability. Productivity measures the equipment's capacity to produce at the planned rate. A reduction in productivity due to downtime generally incurs additional costs to recover the lost production by extending the operating time of equipment after repair or by assigning additional equipment to achieve the required production rate. Table 3.4 shows that if the cumulative costs per hour are calculated, and the productivity factors are known, then the productivity-adjusted, cumulative cost per hour can be found by dividing the cumulative cost per hour by the productivity factor.

**TABLE 3.4**
**Downtime Costs Example**

| Year | Downtime (%) | Operating Cost ($) | Hourly Down-time Cost ($/hr) | Annual Down-time Cost ($/yr) | Cumulative Downtime Cost ($) | Cumulative Hours (hr) | Hourly Cumula-tive Cost ($/hr) | Productivity Factor | Hourly Cumula-tive Cost ($/hr) | Productivity Adjusted Hourly Cumulative Cost ($/hr) |
|---|---|---|---|---|---|---|---|---|---|---|
| 1 | 3% | 7.00 | 0.21 | 420 | 420 | 2,000 | 0.21 | 1.00 | 0.21 | 0.21 |
| 2 | 6% | 7.00 | 0.42 | 840 | 1,260 | 4,000 | 0.32 | 0.99 | 0.32 | 0.32 |
| 3 | 9% | 7.00 | 0.63 | 1,260 | 2,520 | 6,000 | 0.42 | 0.98 | 0.43 | 0.44 |
| 4 | 11% | 7.00 | 0.77 | 1,540 | 4,060 | 8,000 | 0.51 | 0.96 | 0.53 | 0.55 |
| 5 | 13% | 7.00 | 0.91 | 1,820 | 5,880 | 10,000 | 0.59 | 0.95 | 0.62 | 0.65 |
| 6 | 15% | 7.00 | 1.05 | 2,100 | 7,980 | 12,000 | 0.67 | 0.94 | 0.71 | 0.76 |
| 7 | 17% | 7.00 | 1.19 | 2,380 | 10,360 | 14,000 | 0.74 | 0.93 | 0.80 | 0.86 |
| 8 | 20% | 7.00 | 1.40 | 2,800 | 13,160 | 16,000 | 0.82 | 0.92 | 0.89 | 0.97 |

#### 3.1.3.6 Obsolescence

Obsolescence is the reduction in value due to the availability of newer, more productive models [4]. There are two forms of obsolescence: technological and market preference. Technological obsolescence can be measured in terms of productivity. Over the short term, technological obsolescence occurs at a fairly constant rate. On the other hand, market preference obsolescence is a function of customers' preferences and as such, is less predictable and impossible to quantify in a standard calculation. Table 3.5 does not include the costs of market preference obsolescence due to the issues with quantifying its value.

Obsolescence is an important factor in the highly competitive construction industry. The latest technology provides a contractor a competitive edge because enhanced technology usually is associated with enhanced production. Thus, retaining older pieces of equipment that are functioning perfectly well can reduce the contractor's ability to submit competitive bid prices due to lower production rates. Table 3.5 shows the cost increase caused by retaining old equipment that could be replaced with newer ones that produce at a higher rate with lower unit costs.

#### 3.1.3.7 Summary of Costs

The fleet manager can assemble the various types of costs detailed above and determine a machine's economic life by determining the year in which the minimum hourly cost occurs. This process is illustrated in Table 3.6. Table 3.7 takes this notion to the next level and calculates the loss sustained at each year in the equipment's life if it were replaced in that specific year. These analyses lead to the conclusion that the example's minimum hourly cost is $6.82 per hour, and the equipment's economic life occurs at the end of the fourth year. Therefore, the new equipment acquisition process should commence in the fourth year.

Equipment economic life determination is a key factor for applying the various optimum replacement timing methods explained in the following sections.

## 3.2 REPLACEMENT ANALYSIS

Replacement analysis is a tool that permits equipment fleet managers to schedule the equipment replacement decision. This analysis compares the cost of owning the equipment under analysis with the costs of potential replacement alternatives. Both theoretical and practical methods to accomplish this task are covered in this section.

**TABLE 3.5**
**Obsolescence Costs Per Hour for the Life of the Equipment**

| Year | Obsolescence Factor | Hourly Equipment Cost ($/hr) | Hourly Obsolescence Cost ($/hr) | Annual Obsolescence Cost ($/yr) | Cumulative Cost($) | Cumulative Use (hr) | Hourly Cumulative Cost ($/hr) |
|---|---|---|---|---|---|---|---|
| 1 | 0.00 | 7.00 | 0.00 | 0 | 0 | 2,000 | 0.00 |
| 2 | 0.06 | 7.00 | 0.42 | 840 | 840 | 4,000 | 0.21 |
| 3 | 0.11 | 7.00 | 0.77 | 1,540 | 2,380 | 6,000 | 0.40 |
| 4 | 0.15 | 7.00 | 1.05 | 2,100 | 4,480 | 8,000 | 0.56 |
| 5 | 0.20 | 7.00 | 1.40 | 2,800 | 7,280 | 10,000 | 0.73 |
| 6 | 0.26 | 7.00 | 1.82 | 3,640 | 10,920 | 12,000 | 0.91 |
| 7 | 0.32 | 7.00 | 2.24 | 4,480 | 15,400 | 14,000 | 1.10 |
| 8 | 0.37 | 7.00 | 2.59 | 5,180 | 20,580 | 16,000 | 1.29 |

**TABLE 3.6**
**Summary of Cumulative Costs Per Hour**

| | Year | | | | | | | |
|---|---|---|---|---|---|---|---|---|
| **Item** | **1** | **2** | **3** | **4** | **5** | **6** | **7** | **8** |
| Depreciation and replacement ($/hr) | 4.5 | 3.75 | 3.23 | 2.9 | 2.69 | 2.49 | 2.33 | 2.2 |
| Investment ($/hr) | 2.25 | 1.97 | 1.76 | 1.61 | 1.48 | 1.36 | 1.27 | 1.18 |
| Maintenance and repairs ($/hr) | 0.49 | 0.85 | 1.06 | 1.2 | 1.37 | 1.51 | 1.7 | 1.88 |
| Downtime (productivity adjusted) ($/hr) | 0.21 | 0.32 | 0.44 | 0.55 | 0.65 | 0.76 | 0.86 | 0.97 |
| Obsolescence ($/hr) | 0 | 0.21 | 0.4 | 0.56 | 0.73 | 0.91 | 1.1 | 1.29 |
| Total ($/hr) | 7.45 | 7.10 | 6.89 | 6.82 | 6.92 | 7.03 | 7.26 | 7.52 |

**TABLE 3.7**
**Losses Resulting from Improper Equipment Replacement**

| Replaced at End of Year | Cumulative Hours | Hourly Cumulative Cost ($/hr) | Minimum Hourly Cost ($/hr) | Extra Hourly Cost ($/hr) | Total Loss ($) |
|---|---|---|---|---|---|
| 1 | 2,000 | 7.45 | 6.82 | 0.63 | 1,256 |
| 2 | 4,000 | 7.10 | 6.82 | 0.28 | 1,125 |
| 3 | 6,000 | 6.89 | 6.82 | 0.07 | 400 |
| 4 | 8,000 | 6.82 | 6.82 | 0.00 | 0 |
| 5 | 10,000 | 6.92 | 6.82 | 0.10 | 1,005 |
| 6 | 12,000 | 7.03 | 6.82 | 0.20 | 2,439 |
| 7 | 14,000 | 7.26 | 6.82 | 0.44 | 6,134 |
| 8 | 16,000 | 7.52 | 6.82 | 0.70 | 11,12 |

### 3.2.1 Theoretical Methods

Professor James Douglas wrote the seminal work on this topic in his 1975 book *Construction Equipment Policy* [1]. In that work, he posited four different theoretical approaches to establishing an equipment replacement policy based on a rigorous and rational analysis of cost, equipment life, and production. Douglas laid out following four theoretical methods for performing replacement analysis:

- Intuitive method
- Minimum cost method
- Maximum profit method, and
- Mathematical modeling method.

The value of each of the above approaches lies in the fact that each method can be applied based on the business objectives of the given equipment fleet owner. The intuitive method, which relies on common sense to decision-making, provides a baseline against which other methods can be compared. The minimum cost method matches a public construction agency's equipment management policy by focusing on replacing equipment at a point in time where the overall cost of operating and maintaining a given piece of equipment is minimized. The maximum profit method is designed for construction companies and other entities that employ construction equipment in a profit-making enterprise and accordingly must make equipment replacement decisions based on the impact to the bottom line. Finally, the mathematical modeling method fulfills the need for a rigorous analytical approach to inform the replacement decision. This method involves developing stochastic computer simulations to optimize fleet size and composition for large equipment-intensive projects. These will be discussed first combined with a discussion of the payback period method [5], which finds its roots in engineering economic theory. The following example, with current equipment pricing drawn from the USACE *Equipment Ownership Manual EP 1110-1-8* [6], will be used to demonstrate the mechanics of each of the different methods for equipment replacement analysis.

**Example 3.1:** An aggregate producing company presently owns a fleet of 7.5 cubic yard on-highway dump trucks that cost $65,000 each. These trucks are currently one year old, and the annual O&M cost is $30,000 per truck for the first year, increasing by $2,000 each year. The revenue generated by each truck is $70,000 for the first year decreasing by about $1,750 per year thereafter. The owner of the company asks the equipment fleet manager to consider replacing the current dump trucks with a new model that employs a new technology that results in reduced maintenance expenditure. The new proposed replacement trucks are the same size as the current trucks and cost $70,000 each. The annual O&M cost is $30,000 per truck for the first year but only increases by $1,500 per year thereafter. The revenue generated by each truck is the same as for the current model truck. This company uses the double declining balance method for calculating depreciation. The trucks currently in use will be called the "Current Trucks," and the new model will be called the "Proposed Truck" in the tabular examples that follow.

#### 3.2.1.1 Intuitive Method

The intuitive method is a common approach to making replacement decisions due to its simplicity and reliance on individual judgment. It primarily depends on professional judgment. The decision to replace a piece of equipment is often made when it requires a major overhaul or when developing the resource plan for an upcoming equipment-intensive project. Availability of capital is often the decisive factor if the replacement is unplanned.

While the example can be solved with the intuitive method, there is no rational answer for the economic life of the two types of trucks. Retaining the current trucks seems to be better because they are only one year old, and earning revenues at the same rate as the new trucks. As the potential reduction in maintenance costs does not seem to be particularly dramatic, the owner will probably choose to retain the current trucks, which cost $5,000 less than proposed trucks. Because no rational calculation of economic life is made, the potential savings in long-term maintenance and operating cost are not included in the decision [1].

#### 3.2.1.2 Minimum Cost Method

Minimizing equipment cost is a goal for all types of equipment owners. However, it is the key objective for public agencies that maintain fleets of construction equipment, since no revenue is produced to offset the costs. The minimum cost method focuses on minimizing total equipment costs, including both O&M costs and the decline in book value due to depreciation. It furnishes a rational method with which to conduct the objective comparison of alternatives rather than the intuitive method's reliance on professional judgment. For simplicity's sake, the example shown below of minimum cost method does not include many of the ancillary costs discussed in Chapter 2. The reader should determine which of the following factors should also be included: penalty costs for downtime, obsolescence cost, labor cost, tax expenses (consideration of depreciation methods available), and inflation. Tables 3.8 and 3.9 illustrate the process for determining economic life using each alternative. Note that the boxed values are the minimum values for each column.

**TABLE 3.8**
**Average Annual Cumulative Costs of the Current Trucks**

| End of Year (1) | Annual O&M Cost (2) | Book Value | Annual Depr. Expense (3) | Annual Cost (4) = (2) + (3) | Cumulative Cost (5) | Average Annual Cumulative Cost (6) = (5)/(1) |
|---|---|---|---|---|---|---|
| 1 | $30,000 | $39,000 | $26,000 | $56,000 | $56,000 | $56,000 |
| 2 | $32,000 | $23,400 | $15,600 | $47,600 | $103,600 | $51,800 |
| 3 | $34,000 | $14,040 | $9,360 | $43,360 | $146,960 | $48,987 |
| 4 | $36,000 | $8,424 | $5,616 | $41,616 | $188,576 | $47,144 |
| 5 | $38,000 | $5,054 | $3,370 | $41,370 | $229,946 | $45,989 |
| 6 | $40,000 | $3,033 | $2,022 | $42,022 | $271,967 | $45,328 |
| 7 | $42,000 | $1,820 | $1,213 | $43,213 | $315,180 | $45,026 |
| 8 | $44,000 | $1,092 | $728 | $44,728 | $359,908 | $44,989 |
| 9 | $46,000 | $655 | $437 | $46,437 | $406,345 | $45,149 |
| 10 | $48,000 | $393 | $262 | $48,262 | $454,607 | $45,461 |
| 11 | $50,000 | $236 | $157 | $50,157 | $504,764 | $45,888 |
| 12 | $52,000 | $141 | $94 | $52,094 | $556,859 | $46,405 |

**TABLE 3.9**
**Average Annual Cumulative Costs of the Proposed Trucks**

| End of Year (1) | Annual O&M Cost (2) | Book Value | Annual Depr. Expense (3) | Annual Cost (4) = (2) + (3) | Cumulative Cost (5) | Average Annual Cumulative Cost (6) = (5)/(1) |
|---|---|---|---|---|---|---|
| 1 | $30,000 | $42,000 | $28,000 | $58,000 | $58,000 | $58,000 |
| 2 | $31,500 | $25,200 | $16,800 | $48,300 | $106,300 | $53,150 |
| 3 | $33,000 | $15,120 | $10,080 | $43,080 | $149,380 | $49,793 |
| 4 | $34,500 | $9,072 | $6,048 | $40,548 | $189,928 | $47,482 |
| 5 | $36,000 | $5,443 | $3,629 | $39,629 | $229,557 | $45,911 |
| 6 | $37,500 | $3,266 | $2,177 | $39,677 | $269,234 | $44,872 |
| 7 | $39,000 | $1,960 | $1,306 | $40,306 | $309,540 | $44,220 |
| 8 | $40,500 | $1,176 | $784 | $41,284 | $350,824 | $43,853 |
| 9 | $42,000 | $705 | $470 | $42,470 | $393,295 | $43,699 |
| 10 | $43,500 | $423 | $282 | $43,782 | $437,077 | $43,708 |
| 11 | $45,000 | $254 | $169 | $45,169 | $482,246 | $43,841 |
| 12 | $46,500 | $152 | $102 | $46,602 | $528,848 | $44,071 |

The economic life of a machine is the year in which the average annual cumulative cost is minimized. This will result in the lowest cost over the desired period. As shown in Table 3.8, economic life is reached at the end of the eighth year. Table 3.9 shows it to be the ninth year for the proposed truck. The minimum average annual costs for the current truck and proposed truck are $44,989 and $43,699, respectively. Table 3.10 provides a side by side comparison of cumulative average annual costs for both trucks.

In the minimum cost method, the decision to replace equipment is made when the estimated annual cost of the current machine for the next year *exceeds* the minimum average annual cumulative cost of the replacement. In this example, the current truck's estimated

**TABLE 3.10**
**Comparison of Average Annual Cumulative Costs**

| | Current Trucks | | Proposed Trucks |
|---|---|---|---|
| End of Year | Annual Cost ($) | Average Annual Cumulative Cost ($/yr) | |
| 1 | 56,000 | 56,000 | 58,000 |
| 2 | 47,600 | 51,800 | 53,150 |
| 3 | 43,360 | 48,987 | 49,793 |
| 4 | 41,616 | 47,144 | 47,482 |
| 5 | 41,370 | 45,989 | 45,911 |
| 6 | 42,022 | 45,328 | 44,872 |
| 7 | 43,213 | 45,026 | 44,220 |
| 8 | 44,728 | 44,989 | 43,853 |
| 9 | 46,437 | 45,149 | 43,699 |
| 10 | 48,262 | 45,461 | 43,708 |
| 11 | 50,157 | 45,888 | 43,841 |
| 12 | 52,094 | 46,405 | 44,071 |

annual cost for next year (i.e. end of year 2) is $47,600, and the minimum average annual cumulative cost of the proposed truck is $43,853. Thus, if the objective is to minimize costs, this analysis supports replacing the current year-old trucks with the newer model.

### 3.2.1.3 Maximum Profit Method

This method seeks to make the replacement decision by selecting the alternative that maximizes profit. Hence, it is appropriate for organizations that generate revenue and reap profits from their equipment. It works best if the profits associated with a given piece of equipment can be isolated and clearly defined. However, it is not often easy to isolate the equipment-generated profit from the profit associated with the entire project. If reasonable estimates cannot be made, the minimize cost method should be used to make the replacement decision. The previous example is used to demonstrate the maximum profit method. Tables 3.11 and 3.12 illustrate the process to determine the economic life of the two alternatives using profit as the replacement decision metric.

**TABLE 3.11**
**Average Annual Cumulative Profits of the Current Trucks**

| End of Year (1) | Annual Revenue (2) | Annual Cost (3) | Annual Profit (4) = (2) − (3) | Cumulative Profit (5) | Average Annual Cumulative Profit (6) = (5)/(1) |
|---|---|---|---|---|---|
| 1 | $70,000 | $56,000 | $14,000 | $14,000 | $14,000 |
| 2 | $68,250 | $47,600 | $20,650 | $34,650 | $17,325 |
| 3 | $66,500 | $43,360 | $23,140 | $57,790 | $19,263 |
| 4 | $64,750 | $41,616 | $23,134 | $80,924 | $20,231 |
| 5 | $63,000 | $41,370 | $21,630 | $102,554 | $20,511 |
| 6 | $61,250 | $42,022 | $19,228 | $121,783 | $20,297 |
| 7 | $59,500 | $43,213 | $16,287 | $138,070 | $19,724 |

**TABLE 3.12**
**Average Annual Cumulative Profits of Proposed Trucks**

| End of Year (1) | Annual Revenue (2) | Annual Cost (3) | Annual Profit (4) = (2) – (3) | Cumulative Profit (5) | Average Annual Cumulative Profit (6) = (5)/(1) |
|---|---|---|---|---|---|
| 1 | $70,000 | $48,300 | $21,700 | $21,700 | $21,700 |
| 2 | $68,250 | $43,080 | $25,170 | $46,870 | $23,435 |
| 3 | $66,500 | $40,548 | $25,952 | $72,822 | $24,274 |
| 4 | $64,750 | $39,629 | $25,121 | $97,943 | $24,486 |
| 5 | $63,000 | $39,677 | $23,323 | $121,266 | $24,253 |
| 6 | $61,250 | $40,306 | $20,944 | $142,210 | $23,702 |
| 7 | $59,500 | $41,284 | $18,216 | $160,426 | $22,918 |
| 8 | $57,750 | $42,470 | $15,280 | $175,705 | $21,963 |

Tables 3.11 and 3.12 show the necessary data required to calculate the economic lives of the alternatives in the example using the maximum profit method. The economic life of equipment is the year in which the average annual cumulative profit is maximized. This results in higher profits over the period. In Table 3.11, the economic life of the current trucks occurs at the end of the fifth year because the average annual cumulative profit is maximized in that year at $20,511. The maximum average annual cumulative profit of $24,486 is in the fourth year for the proposed trucks in Table 3.12. The proposed trucks should replace the current trucks because the maximum average annual cumulative profit of the proposed trucks, $24,486, is more than that associated with the current trucks, that is, $20,511.

The next issue in this method is to identify the proper timing of the replacement. This occurs when estimated annual profits of the current equipment for the next year of *falls below* the average annual cumulative profit of the proposed replacement. In this example, the current trucks' estimated annual profits never exceed the average annual profit, $24,486, of the proposed model; so, they should be replaced immediately.

#### 3.2.1.4 Payback Period Method

The time required for a piece of equipment to return its original investment by generating profit is the payback period [5]. The capital recovery is calculated using the total after-tax, net savings, and the depreciation tax benefit disregarding financing costs. This method furnishes a metric that is based on time rather than money and allows the comparison of alternatives based on how long it takes for each alternative to recover its investment. The payback period method is useful when it is hard to forecast equipment cash flow due to market instability, inherent uncertainty, and technological changes. This method is founded in engineering economic theory and does not use equipment economic life or other factors beyond the payback period. Therefore, it is recommended that the payback method be used in conjunction with other analysis methods to further inform the final equipment replacement decision. Again, the previous example is used to demonstrate the mechanics of this method.

For the current trucks in Example 3.1, the payback method is calculated as follows:
Initial cost of the current truck = $65,000
Cumulative profits for the first three years = $57,790
Difference=$65,000 − $57,790 = $7,210
Profit of the fourth year = $23,134
Proportional fraction of the third year = $7,210/$23,134 = 0.31
Payback period for the current trucks = 3.31 years

For the proposed trucks, the payback method is calculated as follows:
Initial cost of the proposed truck = \$70,000
Cumulative profits for the first two years = \$46,870
Difference = \$70,000 − \$46,870 = \$23,130
Profit of the third year = \$25,952
Proportional fraction of the third year = \$23,130/\$25,952 = 0.89
Payback period for the proposed trucks = 2.89 years

The above calculation finds that the proposed replacement trucks have a 2.89-year payback compared to the 3.31-year payback period of the current trucks. Thus, the replacement equipment will return its investment to the owner five months faster than the current fleet, indicating replacement is the better option. Combining this knowledge with the previous analysis involving cost and profit makes a clear case for replacing the current fleet with the new model equipped with the latest technology. These three methods combine to provide a powerful set of analytical tools for making this critical decision.

#### 3.2.1.5 Mathematical Modeling Method

The advent of computer applications for construction management problems has furnished a robust tool for addressing the optimization of complex interrelated systems containing a large number of input variables. Modeling construction equipment systems allows the analyst to control the level of complexity of the input and tailor the output to meet the needs of the project. Computer modeling of equipment replacement timing and selection decisions allows increased technical accuracy, as well as enhanced continuity in implementing company equipment management policy since it can be transferred from one manager to the next without a loss in institutional knowledge. A conceptually simple equipment replacement simulation model was developed at Stanford University's Construction Institute in the 1970s using a discounted-cash-flow model [1]. Revenues and costs are assumed to vary as exponential functions. The costs are subtracted from the revenues and then discounted to present value to yield the present worth of profits after taxes.

A mathematical model is a function or group of functions comprising a system. Douglas specifies that the model must include the following factors:

- Time value of money
- Technological advances in equipment (obsolescence)
- Effect of taxes (depreciation techniques, etc.)
- Influence of inflation, investment credit, gain on sale
- Increased cost of borrowing money
- Continuing replacements in the future
- Increased cost of future machines
- Effect of periodic overhaul costs and reduced availability [1].

Other factors important to revenue are increased productivity (productivity obsolescence), availability of machines (maintenance policy), and deterioration of the machine with age. Additionally, in this model, revenues and costs may be classified as follows:

- Revenues from the service of the machines
- Maintenance and operating costs, including annual fixed costs, penalties, and overhead
- Capital costs, including interest on investment, depreciation charges, and interest on borrowed funds
- Discrete costs such as engine, track, and final drive overhauls

- Income and corporation taxes, considering depreciation method, recapture of income on sale, and investment credit [1]

The goal of this method is to maximize the difference between revenue and the expected value of the cost. At this point, the reader is referred to the references at the end of the chapter for the complex mathematical details of Douglas' model itself.

### 3.2.2 Practical Methods

Public and private equipment owners have developed their own policies for making equipment management decisions. They are typically based on empirical data and adjusted based on previous project experience. These methods represent a wealth of knowledge built from decades of equipment management experience. By understanding these methods and combining that knowledge with the analytical methods discussed in the previous section, the fleet manager effectively increases the number of tools available to address the day-to-day issues of managing a fleet of construction equipment.

#### 3.2.2.1 Public Agency Methods

Public agencies do not have a profit motive to consider when setting equipment replacement policy. Thus, their decision criterion relates to minimizing the costs of owning, operating, and maintaining the fleets of equipment that they own and operate. Additionally, public agencies' decision-making abilities are constrained by public law and create a need for increased long-term planning to make timely decisions. The major source of funding for public equipment fleet expenses comes from tax revenues that feed capital budgets [7]. Public purchases of capital equipment must often gain approval from the appropriate authority and be paid for from tax revenues that were collected for this purpose. This creates a constraint on expenditures that is often referred to as the "color of money," where it is possible to have surplus funds that were designated for one purpose in the public coffers while at the same time have insufficient funds to make purchases for another specific purpose [8]. The most common situation is a strict separation of capital expenditures for the purchase of new pieces of equipment from operations and maintenance expenses, which are designated to pay for routine expenses such as fuel and repair parts [8].

Often major capital expenses must pass through an appropriations process where the governing authority reviews and approves a specific sum of money to purchase a specific item. This process may require the agency to identify the need to replace a given piece of equipment a year or more in advance of the need. For example, The City of Minneapolis maintains a five-year capital improvement program (CIP) that provides funding for capital projects. Equipment fleet is not "the [appropriate] asset nature to fund through the City's CIP process" [9]. Thus, Minneapolis maintains a separate five-year funding plan to address major equipment purchases [9]. Theoretically, to get the purchase of a piece of equipment into this budget, the equipment fleet manager is required to make replacement decisions at least five years in advance of the need to provide the time for the City to appropriate the necessary funding. While private contractors often have long-term equipment replacement plans of their own, they are not constrained to executing deviations from that plan because they are in full control of what and when available financial resources are expended. Minneapolis' five-year plan for equipment forces its equipment fleet manager to make decisions in conditions of greater uncertainty than that faced by its private-sector counterpart.

Because a public agency does not have a great deal of financial flexibility, the constraints on the use of available funding can affect the replacement and repair cycles for its equipment fleet. For example, the City of Macomb, Michigan deferred all vehicle and equipment purchases for one year in 2010 due to budget deficits. As a result, in 2011, they were faced with substantially higher maintenance and repair costs [7]. While withholding funds for equipment replacement was an unavoidable fiscal reality, the consequence was that the decision effectively extended the service life of the equipment scheduled to be replaced in 2010 beyond its economic life. The consequences are the potential of owning equipment that is unable to be productively employed because of unacceptably high repair costs that will eventually lead to being disposed of at a salvage value far below the unit's possible market value if it had been repaired the previous year [7].

Experience has also shown that idle equipment deteriorates if it is not operated. Gaskets and seals dry out causing fluid to leak or a gasket to blow when the machine is operated for the first time after a long idle period [10]. Thus, the public sectors' financial constraints have the potential to have an unintended negative impact on an agency's fleet if needed repairs cannot be made, and old equipment cannot be replaced when it reaches the end of its economic life. One can infer from this discussion that from the public fleet manager's perspective, there may be a strong tendency to keep a piece of equipment for as long as possible before replacing it merely because of the administrative burden required to get purchase authority.

Thus, public agencies have evolved an equipment management strategy that is based largely on empirical terms that flow from the experiences of public equipment managers. This is often translated to a specified fixed amount of usage in terms of mileage or engine hours that defines the equipment's economic life regardless of the actual O&M costs that are being experienced on a given piece of equipment. Some agencies also select cost points for equipment O&M costs that are defined in terms of a percentage of book value of the machine at which replacement is directed. Most agencies employ schedules or benchmarks for classes of equipment based on the criteria of age and usage and included lifetime repair costs, as well as the equipment's condition. The following cross-section of the public agency methods used by the Texas, Montana, and Louisiana Departments of Transportation (DOT) illustrate a decent range of differing practical equipment replacement methods and are reviewed in the following sections.

*3.2.2.1.1 Texas Department of Transportation*

The Texas Department of Transportation (TxDOT) has equipment replacement criteria that are based on age, usage (miles or hours), and estimated repair costs. It is the most complex of the three DOTs reviewed in this section and thus, is presented first. TxDOT's equipment fleet is quite large, comprising approximately 17,000 units. This fleet is used to furnish in-house road maintenance and small construction on the state's 301,081 total miles of roads and highways. With a fleet this large, the annual disposal program involves the replacement of approximately 10% of the total fleet [11]. There are twenty-five subordinate districts in TxDOT, and each manages its own portion of the TxDOT fleet. A subjective evaluation of existing equipment for replacement is done at the district level using input from equipment, maintenance, and field personnel. This input is then combined with objective equipment performance data that include age, miles (or hours) of operation, downtime, as well as operating and maintenance costs, to arrive at the final decision on which units to keep and which ones need to be replaced. The replacement decision is made one year before a given piece of equipment hits its target age/usage/repair cost level to allow sufficient time for the procurement of the replacement model.

In 1991, the department fielded the TxDOT Equipment Replacement Model (TERM) to identify fleet candidates for equipment replacement. The model was based on research of other DOT policies and an analysis of actual equipment costs incurred by TxDOT prior to that date. The logic of the model is expressed in the following terms:

> … each equipment item reaches a point when there are significant increases in repair costs. Replacement should occur prior to this point. Ad hoc reports were developed and are monitored annually to display historical cost information on usage and repairs to identify vehicles for replacement consideration. From this historical information, standards/benchmarks for each criteria [sic] are established for each class of equipment.
>
> [11]

Input data for the TERM model comes from TxDOT's Equipment Operations System (EOS), which has historical equipment usage and cost data dating back to 1984. EOS captures an extensive amount of information on all aspects of equipment operation and maintenance. Using the model's logic is relatively simple. First, the EOS historical cost data are processed against three benchmarks for each identified equipment class on an annual basis. There are three criteria that are checked:

1. Equipment age,
2. Life usage expressed in miles (or hours), and
3. Inflation adjusted life repair costs expressed as a percentage of original purchase cost, which has been adjusted to its capital value.

Next, when a given piece of equipment exceeds all three criteria, it is identified as a candidate for replacement. Finally, the owning district makes the subjective evaluation of the given item of equipment, including downtime, condition of existing equipment, new equipment needs, identified projects, and other factors. A final decision on whether to replace is then made. TERM is not meant to replace the knowledge of the equipment manager, but rather to furnish a tool to inform the decision-making process.

*3.2.2.1.2 Montana Department of Transportation*

Like TxDOT, the Montana Department of Transportation (MDT) evaluates its equipment fleet annually to determine which pieces of its equipment fleet should be replaced [12]. It uses the expected annual costs of new equipment as the metric against which current equipment is measured. In calculating this cost, the following factors are considered:

- The expected annual costs of the existing equipment,
- The purchase price of the new equipment,
- Its depreciation, and
- Its expected life.

To be classified as a potential replacement alternative, the new equipment must meet the following criteria: the total costs of owning the equipment for its useful life is equal to the total loss in value for its useful life plus the total costs of operating the equipment over a specified number of years. Time value of money is accounted for using engineering economic theory for the single present worth (SPW) and uniform capital recovery (UCR) mathematical equations. The replacement analysis of MDT uses three equations:

- Equivalent annual costs of new equipment.
- Salvage value.
- Annual cost of an existing unit.

The decision criterion for equipment replacement is that the equivalent annual ownership cost of the new equipment must be less than the annual cost of the current equipment. Thus, this method can first identify economic candidates to serve as alternatives against which the current equipment can be assessed and an objective criterion on which the replacement decision can be made [1].

*3.2.2.1.3 Louisiana Department of Transportation and Development*

Louisiana Department of Transportation and Development (LaDOTD) sponsored a research project conducted at Louisiana State University as a means of determining optimal equipment replacement policy [13]. The project returned with the following decision criteria:

> Disallow the application of maintenance funds for major repairs to equipment that has reached 80% of its economic life or if the repair cost will exceed 50% of the book value of the equipment. [13]

The report uses the same definitions for economic life as used by Douglas [1]. It was anticipated that net savings would be obtained after a four-year period by increasing capital investment to decrease the cost of equipment operations, assuming the use of economic predictions. Accumulated costs for each unit were compared with the limits of the repair costs in order to identify "uneconomical" equipment to need critical repairs. This critical repair method was very effective in verifying the optimum time for changing each unit. The method successfully calculated the optimum replacement point with 96% of certainty and allowed the LADOT to set up the priority ranking of replacement needs. As a result, available funds can be allocated and used effectively [2,9].

### 3.2.3 Sensitivity Analysis on Theoretical Methods

Construction equipment fleet managers must make assumptions to predict future costs. In doing so, variables are introduced into the computations that can influence the outcome of the equipment replacement decision. Therefore, it is important to understand the dynamics of the equipment replacement decision method through sensitivity analysis. Sensitivity analysis permits the equipment manager to test the validity of the assumptions made during the replacement analysis process and determine those input variables that drive the bottom line. Its purpose is to highlight those input variable assumptions that would appreciably change the decision if the assumption is in error. By methodically evaluating the sensitivity of each input variable, the analyst gains insight as to the confidence with which the final decision can be made. In other words, if the outcome is found to be highly sensitive to a given variable, and its assumed value does not have strong historical back-up data, the confidence in the output's correctness drops dramatically. Conversely, if the outcome of the method is found to be insensitive to variations in the input values, then confidence in the answer's correctness is high.

For example, the actual value of fuel costs and operator costs strongly affect the predicted value of future operating costs. Due to inherent fluctuations in the oil market and labor market, these are difficult to predict for the long term. Equipment replacement methods require that these estimates be made for the long-term economic life of the piece of equipment under analysis. Therefore, to increase the confidence in the results, a sensitivity analysis is performed. This involves the following steps:

- Listing the parameters most likely to affect the future cost figures being estimated.
- Determining a probable range over which these parameters may vary.

- Determining the effect on the future cost figures being estimated of the parameters ranging over their probable range.

When a future cost is significantly affected by the ranging variable, the cost estimate is said to be very sensitive to that variable [5]. The sensitivity analyses are preformed on the equipment replacement analysis methods using the information supplied in Example 3.1.

#### 3.2.3.1 Sensitivity Analysis on Minimum Cost Method

Table 3.8 shows the average annual cumulative costs of the current trucks, using a depreciation rate of 40% to calculate the annual depreciation expenses. Also, the annual O&M cost is $30,000 per truck for the first year, increasing by $2,000 each year thereafter. Two input parameters, the annual depreciation rate, and the annual O&M cost are selected for the sensitivity analysis on equipment replacement decision analysis for the current trucks. Sensitivity to the depreciation assumption is tested first. The annual depreciation rate is changed to 20% in the first calculation shown in Table 3.13. Next, the calculation is run using a 60% rate and shown in Table 3.14.

When the depreciation rate is decreased to 20%, the average annual cumulative cost is the minimum of $42,573 in the third year as compared to the eighth year with the original 40% depreciation assumption. With depreciation assumed at 60%, the economic life is at the end of the eighth year compared to the ninth year for the original assumption. Thus, this method is found to be sensitive to the depreciation assumption.

Next, sensitivity to the O&M cost assumption is tested. The annual O&M cost is assumed to increase by $1,000 instead of $2,000. As a result, the minimum average annual cumulative cost changed from $44,989 at the end of the eighth year as shown in Table 3.8 to $40,888 at the end of the eleventh year as shown in Table 3.15.

If the increase in annual O&M cost is assumed to be $3,000, the lowest average annual cumulative cost is $47,828 in the sixth year, as shown in Table 3.16. Thus, the method is found to be sensitive to this parameter as well.

Considering the outcome of the sensitivity analysis on the minimum cost method should drive the equipment manager to make sure that the values used for both the depreciation rate and O&M costs based on historical records. The lesson here is that arbitrary assumptions lacking historical back-up data can yield vastly different answers, which greatly reduces the confidence of the final decision.

**TABLE 3.13**
**Average Annual Cumulative Costs of the Current Trucks (20% Depreciation)**

| End of Year (1) | Annual O&M Cost (2) | Book Value | Annual Depr. Expense (3) | Annual Cost (4) = (2) + (3) | Cumulative Cost (5) | Average Annual Cumulative Cost (6) = (5)/(1) |
|---|---|---|---|---|---|---|
| 1 | $30,000 | $52,000 | $13,000 | $43,000 | $43,000 | $43,000 |
| 2 | $32,000 | $41,600 | $10,400 | $42,400 | $85,400 | $42,700 |
| 3 | $34,000 | $33,280 | $8,320 | $42,320 | $127,720 | $42,573 |
| 4 | $36,000 | $26,624 | $6,656 | $42,656 | $170,376 | $42,594 |
| 5 | $38,000 | $21,299 | $5,325 | $43,325 | $213,701 | $42,740 |
| 6 | $40,000 | $17,039 | $4,260 | $44,260 | $257,961 | $42,993 |
| 7 | $42,000 | $13,631 | $3,408 | $45,408 | $303,369 | $43,338 |
| 8 | $44,000 | $10,905 | $2,726 | $46,726 | $350,095 | $43,762 |
| 9 | $46,000 | $8,724 | $2,181 | $48,181 | $398,276 | $44,253 |

**TABLE 3.14**
**Average Annual Cumulative Costs of the Current Trucks (60% Depreciation)**

| End of Year (1) | Annual O&M Cost (2) | Book Value | Annual Depr. Expense (3) | Annual Cost (4) = (2) + (3) | Cumulative Cost (5) | Average Annual Cumulative Cost (6) = (5) / (1) |
|---|---|---|---|---|---|---|
| 1 | $30,000 | $26,000 | $39,000 | $69,000 | $69,000 | $69,000 |
| 2 | $32,000 | $10,400 | $15,600 | $47,600 | $116,600 | $58,300 |
| 3 | $34,000 | $4,160 | $6,240 | $40,240 | $156,840 | $52,280 |
| 4 | $36,000 | $1,664 | $2,496 | $38,496 | $195,336 | $48,834 |
| 5 | $38,000 | $666 | $998 | $38,998 | $234,334 | $46,867 |
| 6 | $40,000 | $266 | $399 | $40,399 | $274,734 | $45,789 |
| 7 | $42,000 | $106 | $160 | $42,160 | $316,894 | $45,271 |
| 8 | $44,000 | $43 | $64 | $44,064 | $360,957 | $45,120 |
| 9 | $46,000 | $17 | $26 | $46,026 | $406,983 | $45,220 |

**TABLE 3.15**
**Average Annual Cumulative Costs of the Current Trucks (O&M at $1,000)**

| End of Year (1) | Annual O&M Cost (2) | Book Value | Annual Depr. Expense (3) | Annual Cost (4) = (2) + (3) | Cumulative Cost (5) | Average Annual Cumulative Cost (6) = (5)/(1) |
|---|---|---|---|---|---|---|
| 1 | $30,000 | $39,000 | $26,000 | $56,000 | $56,000 | $56,000 |
| 2 | $31,000 | $23,400 | $15,600 | $46,600 | $102,600 | $51,300 |
| 3 | $32,000 | $14,040 | $9,360 | $41,360 | $143,960 | $47,987 |
| 4 | $33,000 | $8,424 | $5,616 | $38,616 | $182,576 | $45,644 |
| 5 | $34,000 | $5,054 | $3,370 | $37,370 | $219,946 | $43,989 |
| 6 | $35,000 | $3,033 | $2,022 | $37,022 | $256,967 | $42,828 |
| 7 | $36,000 | $1,820 | $1,213 | $37,213 | $294,180 | $42,026 |
| 8 | $37,000 | $1,092 | $728 | $37,728 | $331,908 | $41,489 |
| 9 | $38,000 | $655 | $437 | $38,437 | $370,345 | $41,149 |
| 10 | $39,000 | $393 | $262 | $39,262 | $409,607 | $40,961 |
| 11 | $40,000 | $236 | $157 | $40,157 | $449,764 | $40,888 |
| 12 | $41,000 | $141 | $94 | $41,094 | $490,859 | $40,905 |

#### 3.2.3.2 Sensitivity Analysis on Maximum Profit Method

In the maximum profit method, the average annual cumulative profits of the two alternative trucks are driven by the decrease rate of the annual revenue and the change in annual cost, as shown in Table 3.10. In this table, the annual cost is related to the annual depreciation rate, and the maintenance and operating cost. Therefore, the sensitivity analysis on the current trucks for this method will be done using three parameters: the annual depreciation rate, the maintenance and operating cost, and annual revenue. First, as in the previous section, the annual depreciation rate is varied at 20% and then 60% while the annual O&M cost increase rate and the annual revenue decrease rate are fixed to determine the sensitivity of the output to the change in the depreciation assumption. Next, O&M cost increase rate is varied at $1,000 per year and $3,000 per year to check its sensitivity. Finally, with the

**TABLE 3.16**
**Average Annual Cumulative Costs of the Current Trucks (O&M at $3,000)**

| End of Year (1) | Annual O&M Cost (2) | Book Value | Annual Depr. Expense (3) | Annual Cost (4) = (2) + (3) | Cumulative Cost (5) | Average Annual Cumulative Cost (6) = (5)/(1) |
|---|---|---|---|---|---|---|
| 1 | $30,000 | $39,000 | $26,000 | $56,000 | $56,000 | $56,000 |
| 2 | $33,000 | $23,400 | $15,600 | $48,600 | $104,600 | $52,300 |
| 3 | $36,000 | $14,040 | $9,360 | $45,360 | $149,960 | $49,987 |
| 4 | $39,000 | $8,424 | $5,616 | $44,616 | $194,576 | $48,644 |
| 5 | $42,000 | $5,054 | $3,370 | $45,370 | $239,946 | $47,989 |
| 6 | $45,000 | $3,033 | $2,022 | $47,022 | $286,967 | $47,828 |
| 7 | $48,000 | $1,820 | $1,213 | $49,213 | $336,180 | $48,026 |
| 8 | $51,000 | $1,092 | $728 | $51,728 | $387,908 | $48,489 |
| 9 | $54,000 | $655 | $437 | $54,437 | $442,345 | $49,149 |
| 10 | $57,000 | $393 | $262 | $57,262 | $499,607 | $49,961 |
| 11 | $60,000 | $236 | $157 | $60,157 | $559,764 | $50,888 |
| 12 | $63,000 | $141 | $94 | $63,094 | $622,859 | $51,905 |

depreciation rate and O&M cost rate fixed, the decreased annual revenue rates varied between $875 per year and $2,625 per year. Table 3.17 reports the results of the three sensitivity analyses.

Looking first at the sensitivity to the depreciation rate assumption, one can see that varying this rate has a huge impact on the economic life of the truck when defined by maximizing the average annual cumulative profits. The greatest effect is found on the low end of the spectrum. The O&M and revenue rate assumptions also have an impact but are not as great as the depreciation assumption as they only change the economic life by one year.

### 3.2.4 Comparison and Discussion of Sensitivity Analysis Results

It is interesting to note the change in sensitivities as one moves from the minimum cost method to the maximum profit method. Intuitively, there should be some difference as the

**TABLE 3.17**
**Average Annual Cumulative Profits of the Current Trucks**

| End of Year | AACP (20% Depr) | AACP (40% Depr) | AACP (60% Depr) | AACP (O&M Increase by $1,000) | AACP (O&M Increase by $2,000) | AACP (O&M Increase by $3,000) | AACP (Revenue Decrease by $875) | AACP (Revenue Decrease by $1,750) | AACP (Revenue Decrease by $2,625) |
|---|---|---|---|---|---|---|---|---|---|
| 1 | $27,000 | $14,000 | $1,000 | $14,000 | $14,000 | $14,000 | $14,000 | $14,000 | $14,000 |
| 2 | $26,425 | $17,325 | $10,825 | $17,825 | $17,325 | $16,825 | $17,763 | $17,325 | $16,888 |
| 3 | $25,677 | $19,263 | $15,970 | $20,263 | $19,263 | $18,263 | $20,138 | $19,263 | $18,388 |
| 4 | $24,781 | $20,231 | $18,541 | $21,731 | $20,231 | $18,731 | $21,544 | $20,231 | $18,919 |
| 5 | $23,760 | $20,511 | $19,633 | $22,511 | $20,511 | $18,511 | $22,261 | $20,511 | $18,761 |
| 6 | $22,632 | $20,297 | $19,836 | $22,797 | $20,297 | $17,797 | $22,485 | $20,297 | $18,110 |
| 7 | $21,412 | $19,724 | $19,479 | $22,724 | $19,724 | $16,724 | $22,349 | $19,724 | $17,099 |

maximum profit method has one additional parameter, and the introduction of the additional parameter would be expected to change the mathematical dynamic of the analysis. The sensitivity analysis on the minimum cost method leads to the conclusion that the increase in the rate of the annual O&M cost is more sensitive than the depreciation rate. In other words, the replacement analysis based on the minimum cost method can be more affected by the change of the annual increase rate of the O&M cost than that of the annual depreciation rate. Sensitivity analysis output can be described visually using a tornado diagram. Figure 3.3 is the tornado diagram for this analysis. The amount of change in parameter value shifts the output value from its centroid, which is based on the expected values of the parameters being varied, which implies the level of sensitivity. Thus, the length of the output range that is produced by the change in input variable is roughly proportional to the level of sensitivity. So, as the range bar for O&M costs is longer than the one for the depreciation rate in Figure 3.3, the average minimum annual cost is most sensitive to this parameter.

Figure 3.4 is the tornado diagram for the maximum profit method, and it clearly shows that the annual depreciation rate is the most sensitive of the input parameters. Thus, if the equipment owner wants to maximize the average annual cumulative profit, they need to find the ways of controlling the annual depreciation rate, which will be more effective than to try to prevent annual revenue from decreasing. Sensitivity analysis gives the equipment owner a "feeling" for how accurate the estimates that are being made in this important step can be. It adds objective analytical information to the process and, in doing so, decreases uncertainty while increasing confidence in the final solution.

Hence, a wide range of equipment replacement decision-making method choices are available to the analyst, and equipment owners should carefully decide which methods they will use in this process, focusing on those parameters that can be controlled to either minimize cost or maximize profits. They can then back this up by considering the results of a sensitivity analysis done on the assumptions that were made in the chosen method, and thereby feel more confident that they have indeed made the correct decision based on the available facts.

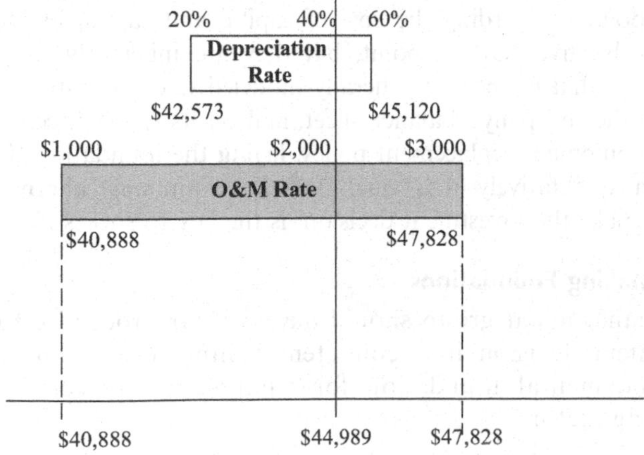


**FIGURE 3.3** Tornado Diagram for Minimum Cost Sensitivity Analysis on the Current Trucks.

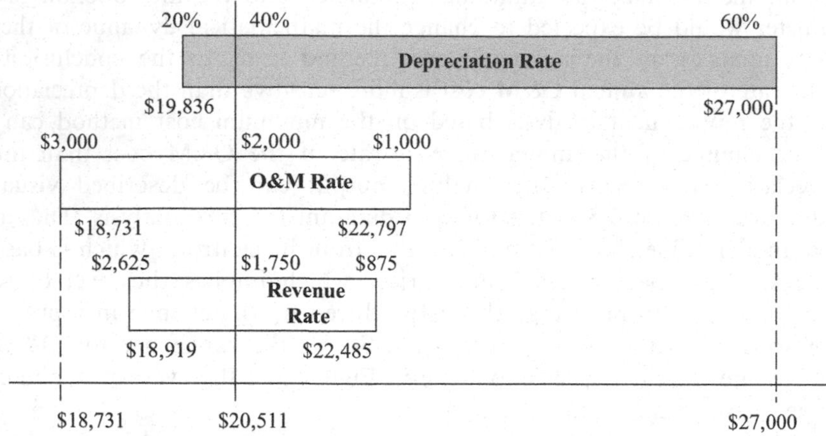


**FIGURE 3.4** Tornado Diagram for Maximum Profit Sensitivity Analysis on the Current Trucks.

## 3.3 REPLACEMENT EQUIPMENT SELECTION

Picking the right piece of equipment to replace an existing one is a complicated decision that involves more than running the numbers to see if the new model will add value to the bottom line. Given the explosive growth in machine technology and information technology, selecting the wrong replacement alternative can be a costly mistake not only in terms of higher than expected ownership costs due to lower than expected production but also in the loss of market share that occurs when a company's operating costs exceed the industry norms. Therefore, the remainder of this chapter will be devoted to discussing the issues that should also be considered after the mathematical models are complete, and the economic answers are on the table.

### 3.3.1 Replacement Decision-Making

Timing the replacement is a difficult question that requires a thorough examination of company strategies and policies regarding the cost of capital and capital budgeting. The previous methods furnish an objective starting point, but they are inherently simplified, disregarding many important factors that cannot be generally modeled like tax status, the effect of owning capital equipment on the company's balance sheet, and on its stock price. Thus, when developing a process for an equipment replacement policy, laying the foundations for decision-making, which involves both qualitatively and quantitatively examining alternatives and selecting a means in which to make the investment decision, is the key to success.

#### 3.3.1.1 Decision-Making Foundations

Every equipment management group should have a clear procedure to help it make the equipment replacement decision in a consistent manner every time the topic must be addressed. The fundamental foundation for equipment replacement decision-making includes the following factors:

- Identify the decision-maker
- Define the defender (the current equipment) and the challengers (potential replacements)
- List the qualitative and quantitative decision factors

Industrial engineers use the term "Defender-Challenger Analysis" when methodically comparing alternatives using engineering economic theory [5]. This term works well in equipment replacement decision-making, and for the following discussion the existing piece of equipment will be called the "Defender", and the potential replacement candidates will be called the "Challengers." It is important to define exactly what these alternatives are and what each consists of in terms of technology, capacity, productivity, and safety before starting the analysis. It may be expedient to take a given base model of equipment and develop several challengers that have different components and qualities. In this way, a logical analysis of the different "bells and whistles" can be accomplished, and each can be compared to the defender to determine if adding a given optional component adds value that is commensurate with the added cost of the option.

Finally, a means for evaluating qualitative factors should be developed and used after the quantitative analysis is complete. The qualitative factors can be used in several ways. First, they can be used as a "tiebreaker." In other words, if two alternatives were very close together quantitatively, the alternative that furnishes the greatest number of qualitative advantages would be selected. The second way would be to assign some form of numerical weighting to each qualitative factor and incorporate an evaluation of those factors into the quantitative analysis using utility theory [5] or some other analytic method to quantify the inherently qualitative feature of an alternative. Finally, the qualitative factors can be separated into two groups: factors that are required and factors that are merely desired. A required factor on a new dump truck might be a factory-installed global positioning system (GPS) unit to allow the company to track the location of its vehicles using a previously purchased GPS system that is currently in operation. A desired factor might be a preference for a given manufacturer's vehicle based on that company's good reputation for service. In this case, a challenger that did not have all the required factors would be eliminated at the outset of the analysis as unacceptable. Then, the desired factors would be used as the tiebreaker in the same fashion as the first method. Examples of qualitative factors include the availability of a given replacement, its strategic value for potential growth and expansion in the company, and the ability to take advantage of market opportunities for preferred financing and other perquisites.

#### 3.3.1.2 Examination of Alternatives

When a piece of equipment is determined as needing replacement, there are always five alternatives that can be considered.

- Overhaul the existing equipment
- Rent a new piece of equipment
- Lease a new piece of equipment
- Purchase a new piece of equipment
- Purchase a used piece of equipment.

The benefits and costs of each alternative should be considered throughout the decision process. Each alternative should be weighed on a common scale for both quantitative and qualitative factors.

#### 3.3.1.3 Decision to Invest

The final decision to invest (or not invest) in a replacement should be made within the framework of capital budgeting decisions and include a quantitative analysis of cost and the time value of money. Equally as important in the decision process are qualitative factors and their impact on the firm. As a final check, the decision-maker should ensure that the decision passes the common sense test by including all important decision factors.

### 3.3.2 General Factors

Once the decision to buy new equipment is made, the equipment manager should consider following four factors [4]:

- Machine productivity
- Product features and attachments
- Dealer support
- Price

#### 3.3.2.1 Machine Productivity

Every equipment owner wants to buy the optimum size and the best quality at the lowest cost. It is important to select the size of machine that will deliver the best productivity for a given job. Chapter 5 in this book furnishes several analytical methods to help make this decision. Additionally, the owner's past experience is very good factor for supporting the mathematical output. The equipment dealer should have the latest data on machine capability under various operating conditions, which can then be used in the models shown in Chapter 5. Additionally, before purchasing, the equipment manager should differentiate the primary usage of the machine from its secondary usage. For example, a tracked excavator is primarily used to dig trenches and other excavations. However, when used on a pipe installation crew, it can also serve secondarily as the means for picking the pipe off a truck and placing them in the trench. Focusing on the major required function of the machine makes it easier to determine the proper size or capacity and as well as any required machine attachments.

#### 3.3.2.2 Product Features and Attachments

Selecting the right equipment with adequate attachments not only increases productivity but also decreases downtime. For example, wheel loader production can be increased by adding automatic bucket controls, special-purpose buckets, and optional counterweights [4]. The equipment manager should be careful not to add special attachments that do not enhance the economics of the overall system. Qualitative factors such as safety must also be considered when considering attachments and special product features. Factors such as mechanical compatibility with other types of equipment that enhance the ability of the maintenance crew to perform its duties often pay-off in reduced downtime and reduced spare parts costs.

#### 3.3.2.3 Dealer Support

Dealer support is a vital piece of the ability of a piece of equipment to achieve its rated production rates. The ability to get spare parts in a timely manner, the availability of service facilities and qualified technicians, and the transparency of the dealer's website all play an important part in ensuring maximum equipment availability. From the day the equipment is purchased until the day it is traded-in on a new piece, the performance of the dealer determines whether that machine will perform as anticipated. The dealer's reputation for user-friendly support and customer-oriented action is a qualitative factor that can ultimately make or break a fleet of heavy construction equipment's profitability. Thus, this factor should be given special priority in the final equipment purchase decision.

#### 3.3.2.4 Price

The equipment replacement decision-making methods detailed in earlier sections of this chapter require a purchase price and a salvage value as input. While this might be the final factor considered in machine selection, it becomes the fundamental factor that will drive the final decision. Resale price, maintenance and repair costs, and the cost of special features and attachments should be factored into the decision as well. A life cycle cost mentality should be used when looking at prices. A machine with a lower initial price might also be more expensive to operate and maintain. A purchase price should be coupled with satisfactory performance as well as dealer parts and service support to ensure that actual equipment availability meets the assumptions made in the analysis. When all the factors have been weighed, then the equipment manager is ready to arrive at the best decision. Frein summed up the price issue in the following quotation:

> The total cost of owning and operating a machine, and not the machine price, should be the decision maker in equipment selection.
>
> [4]

## 3.4 SUMMARY

This chapter defined and discussed three types of equipment life: physical life, profit life, and economic life. It touched on the concepts of depreciation and replacement, inflation, investment, maintenance and repairs, downtime, and obsolescence that impacted the equipment replacement decision. Replacement analysis was introduced by demonstrating theoretical replacement methods by a continuing example, and practical replacement methods were also described. The concept of sensitivity analysis was applied to two of the theoretical methods to demonstrate gain accuracy and confidence in the output of the analyses. Finally, the decision-making process for replacement equipment selection was introduced in a step-by-step fashion and the four general factors that should be considered after the replacement decision is made were explained.

## REFERENCES


[1] Douglas, J. (1975). *Construction Equipment Policy.* New York: McGraw-Hill, Inc.

[2] Popescu, C. (1992). *Managing Construction Equipment*. Austin: C&C Consultants, Inc. pp. 4.1–4.49.

[3] Harcourt, H.M. (2016). *The American Heritage® Dictionary of the English Language*, Fifth Edition by the Editors of the American Heritage Dictionaries. Boston: Houghton Mifflin Harcourt.

[4] Frein, J.P. (2012). *Handbook of Construction Management and Organization*. Berlin: Springer Science + Business Media. pp. 137–298.

[5] Gransberg, D.D. and O'Connor, E.P. (2015). *Major Equipment Life-Cycle Cost Analysis* (No. MN/RC 2015-16). Minnesota Department of Transportation, Research Services & Library. St. Paul, Minnesota.

[6] US Army Corps of Engineers (USACE). (2016). *Construction Equipment Ownership and Operating Expense Schedule*, Region I. Document EP 1110- 1-8(Vol. 1). Washington, DC.

[7] Antich, M. (2010). Top 10 Challenges Facing Public Sector Fleets in CY-2011. *Government-Fleet*. www.government-fleet.com/channel/fuel-management/article/story/2010/11/top-10-challenges-facing-public-sector-fleets-in-cy-201/page/2.aspx. (Accessed November 23, 2019).

[8] Lang, J. (2008). Financial Management. American Public Works Association. www.apwa.net/c/documents/education/institutes/ppt/finance.ppt. (Accessed November 29, 2014).

[9] City of Minneapolis (COM). (2014). Financial Management Policies. www.minneapolismn.gov/policies/policies_financial-management-cover. (Accessed November 23, 2019).

[10] Motors, M. (2014). Bringing a Car Out of Storage. Tech Articles, Moss Motors, Ltd. www.mossmotors.com/SiteGraphics/Pages/out_of_storage.html. (Accessed November 23, 2019).

[11] Texas Department of Transportation (TxDOT). (2011). *TxDOT Equipment Replacement Model (TERM) Software Manual.* Austin: TxDOT. https://library.ctr.utexas.edu/Presto/content/Detail.aspx?ctID=UHVibGljYXRpb25fMTE2MTA=&rID=ODAxMg==&ssid=c2NyZWVuSURfMTQ2MDk=. (Accessed November 22, 2019).

[12] Cooney, R.C. and Paxton, M.C. (2016). *Strategic Enterprise Architecture Design and Implementation Plan for the Montana Department of Transportation*, Final Research Report FHWA/MT-16-007/8238-001. Helena, MT: Montana DOT, pp. 41–46.
[13] Ray, T.G. (1999). *Development of an Approach to Facilitate Optimal Equipment Replacement* (No. FHWA/LA. 88/329). Baton Rouge: Louisiana State University.

## CHAPTER PROBLEMS

1. For the equipment life graph shown in Figure 3.5, answer the following questions:

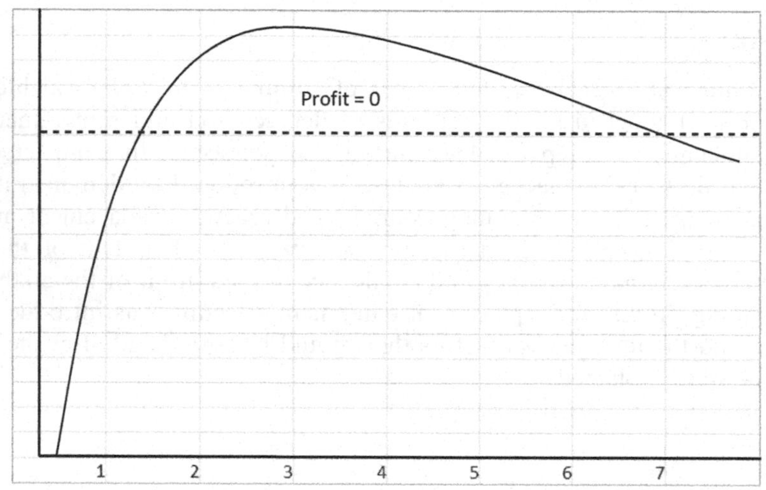


**FIGURE 3.5** Equipment Life Problem Given Information.

   a. What is the physical life?
   b. What is the profit life?
   c. What is the economic life?

2. A construction company owns four 3-year-old self-propelled double drum smooth rollers that cost $130,000 when new. The annual O&M cost for each roller is $20,000 for the first year, increasing by $1,000 each year. The revenue allocated to the roller is $25,000 for the first year, decreasing by about $1,000 per year thereafter. The company is considering replacing the current rollers with a new model that employs intelligent compaction technology and cost $155,000 each. The annual O&M cost is $22,000 per roller for the first year, increasing by $1,100 per year thereafter. Due to the intelligent compaction technology, the revenue generated by the new roller is expected to be $40,000 the first year, decreasing by about $1,000 per year thereafter. This company uses the double declining balance method for calculating depreciation. Determine whether or not to replace the current roller with a new model using:

   a. Minimum cost method
   b. Maximum profit method

# 4 Earthmoving, Excavating, and Lifting Equipment Selection

## 4.0 INTRODUCTION

Construction equipment selection is the process of picking the right tool for the given job. It seeks to ensure that the given piece of equipment is configured in a manner that maximizes its production potential and minimizes downtime. Selecting the right piece of equipment for any given task involves considering a great number of factors that affect production, efficiency, and cost. This chapter will provide the reader a detailed discussion of each of those considerations along with practical examples of how they are applied in the equipment selection decision.

## 4.1 BASIC CONSIDERATIONS FOR EQUIPMENT SELECTION

Equipment costs rank second to labor costs in terms of uncertainty and impact on the anticipated profit of a construction project. Selecting the right piece of equipment, like the right person for the job, directly affects field productivity, which directly influences profitability. Using a machine with insufficient capacity will reduce productivity. Increasing machine capacity can increase productivity but may negatively affect profitability due to the increased operating cost of the oversized machine. Not only do proper individual machine capacities need to be determined, but the capacities of machines that support one another in the production process must also be matched. Pairing machines with mismatched capacities is not efficient and will decrease production, increasing unit costs for the work.

### 4.1.1 Matching Equipment Characteristics

The first equipment selection step involves matching the right machine to the work activity. The work activity includes all factors associated with the specific physical task. Each piece of construction equipment is specifically designed by the manufacturer to perform certain mechanical operations that accomplish the work activity. Mechanical operations are typical for each classification of earthmoving, excavation, and lifting equipment. For instance, all front-end loaders work the same way. They are built to scoop at ground level, carry the load, hoist the load, and dump the bucket forward. Regardless of the manufacturer, loaders' mechanical operations are similar. Using a front-end loader to excavate a deep hole would not be a proper use of the machine. Failure to match the machine to the work task usually results in operating inefficiency and placing the machine at risk due to improper use. The same can be said for various makes and models of mobile cranes. They all basically work the same and are designed to lift and swing loads.

Two types of failure can occur for all equipment. First, structural or mechanical failure occurs when the machine is overloaded or stressed beyond its components' physical capabilities. A crane's boom might buckle if overloaded, or a bucket loader's hydraulics might blow if asked to function in an excavation role rather than the materials loading role it was designed for. Secondly, stability failure occurs when the machine is overloaded or placed in a situation where it cannot remain balanced and upright. A crane hoisting an unbalanced load might overturn, though the boom remained intact, or a loader whose bucket is overloaded and traveling on an uneven surface might nose-dive on the uneven surface, though all parts of the machine are intact and still operable. As a result, matching machines to the task will decrease these types of failures and becomes a primary goal of the equipment selection decision.

When selecting a piece of equipment, an important consideration is the availability of the timely service, maintenance, and repair for the given machine. The right machine not only matches mechanical functions, but also required power, capacity, and control. Equipment discussed in this book are standard pieces of construction equipment with parts and service readily available. Dealer or rental agency location proximity and staff competency will influence downtown and turnaround for service.

### 4.1.2 Matching the Machine to the Materials

The physical properties of clay, gravel, organic matter, rock, sand, or silt to be moved or excavated also have a direct influence on the type and capacity of equipment selected for a specific work activity. The ease or difficulty of removing and handling the required material directly influences machine productivity. This will also determine the capacities and types of buckets, blades, and attachments or accessories. The soil's cohesive characteristics will influence how much can be put in a bucket, blade, bowl, or bed. For example, a front-end loader bucket will hold more slightly wet sandy clay soil than dry sand. The composition of the soil and moisture content affects the heaped capacity that the bucket can hold, or the blade can push.

### 4.1.3 Matching the Machine to the Traveled Surface

The type and condition of the working surface and the distance to be traveled affect the choice of tires or tracks. This will be discussed in more detail in an upcoming section in this chapter. Desired productivity is also a major influence on earthmoving, excavating, and lifting equipment selection. Meeting the schedule for the quantity of work to be accomplished is the goal. The required hourly production of a piece of machinery is primarily determined by the amount of work to be done and how fast it must be done. The amount of time the contractor must spend on excavation or earthmoving will greatly influence the size of machinery chosen for the work [1]. If there is a large volume of material that must be moved quickly, a large piece of machinery will probably be more efficient. Lifting production is heavily dependent on ground and on-structure craft support efficiency. Lifting capacity and vertical hoist speed are the primary equipment influences on lifting production.

### 4.1.4 Equipment Selection Relationships

The following basic relationships exist for equipment selection:

- As equipment productivity increases, so does the initial purchase price, operating, and maintenance costs.
- As equipment capacity increases, so does the hourly production.
- As equipment productivity increases, the unit cost ($/cubic yard, $/square foot, $/ton, $/load) for the work decreases.

Equipment selection demands that these considerations be addressed in every decision. As efficiency is achieved, the unit cost decreases, making the contractor more competitive for large quantities of work. The ultimate objective is to optimize a project's equipment fleet with the desired budget and schedule. Whether owning or renting, equipment must earn more money than it costs to be profitable. Equipment fleet profitability is most influenced by being able to keep the equipment busy and properly maintained.

## 4.2 EARTHMOVING AND EXCAVATING CONSIDERATIONS

A major influence on equipment fleet productivity is whether the working machine moves on tracks or tires. Both means offer advantages and disadvantages based on working and surface conditions.

### 4.2.1 Tracks and Tires

The usable force available to perform work is dependent on the coefficient of traction of the work surface and the weight carried by the running gear or wheels. The amount of tractive force necessary to push or pull a load is important for sizing the right machine. Manufacturers provide rimpull or drawbar pull tables for most of their equipment models showing tractive power that can be delivered at specified operating speeds. This information can be used to verify a machine's ability, or capacity to work in certain job conditions (primarily rolling or surface resistance and grade resistance) and achieve the desired production.

Coefficients of traction vary based upon the travel surface. They measure the degree of traction between the wheel or track and travel surface. Slick surfaces have lower coefficients of traction than rougher surfaces (assuming both surfaces are relatively level and flat). Coefficients of traction for rubber-tired vehicles range from 0.90 for a concrete surface, 0.20 for dry sand to 0.12 for ice. Typically, coefficients of traction tables are available in equipment performance handbooks. The better the traction generated by the piece of equipment on the travel surface, the shorter the travel time and less wear and tear on the piece of equipment. Simply stated, maximum tractive effort (drawbar or rimpull in pounds) equals the equipment weight multiplied by the coefficient of traction of the travel or work surface.

This formula calculates the maximum amount of force that can be generated for a load on a surface. Excess tractive effort generated by the equipment will cause the tires or tracks to spin. Overloading will also cause this. The machine's engine provides the power to overcome the resistances and move the machine. The engine must be sized or matched to meet the tractive effort required to the capabilities of the machine. The model selected is appropriate if it can generate enough tractive effort to do the work task without overburdening the machine.

#### 4.2.1.1 Tracked equipment

Tracked equipment is designed for work activities requiring high tractive effort (drawbar) or the ability to move and remain stable on uneven or unstable surfaces. Tasks such as pushing over trees, removing tree stumps, or removing broken concrete flatwork require lots of pushing force. The tracked bulldozer is ideal for this type of work. Tractive effort results from the track cleats or grousers gripping the ground to create the force necessary to push or pull dirt, material, or another piece of equipment. Tracked equipment is most efficient when used for short travel distances less than 500 ft. Figure 4.1 shows a typical piece of heavy construction equipment running on tracks. Most loaders on construction sites run on tires.

Tracks can be metal or rubber. Metal tracks are more durable and can withstand much greater abuse than rubber tracks. Heavy-duty earthmoving equipment will almost always run on metal tracks. Rubber tracks are lighter and best for smaller equipment working in organic matter and surfaces requiring minimal disturbance. Tracks come in varying widths and thicknesses. The width of the track shoe determines the ground pressure. The wider the track the more surface area covered and the wider the load distribution. Wider track shoes have greater flotation on the work surface. The heavier the track, the more power required to make it move. Narrow track shoes are better for harsh irregular hard work surfaces.

**FIGURE 4.1** Tracked Loader.

Shoes are typically designed with single or double grousers. Single grouser shoes are better for developing traction, and double grouser shoes typically are less damaging to travel or work surfaces.

It should be noted that tracked equipment typically marks or gouges the surface on which it is operating. Skid-steer types of equipment (bulldozers and loaders) will gouge the surface with the track cleats when they turn. To avoid damaging a surface, contractors often lay plywood down on which the tracked equipment can maneuver, use rubber or padded tracks or use a tired piece of equipment.

#### 4.2.1.2 Tired equipment

Tired equipment is more mobile and maneuverable than tracked equipment. Machines can achieve greater speed and therefore are better for hauling. However, pulling ability is reduced to reach a higher speed. Tired equipment is more efficient than tracked equipment when the distance is greater than 500 ft. The tire diameter and width, tread design, and inflation pressure influence the ability to roll. The larger the tire, the more power required to make it roll. Tread and track design influence the ability to grip the travel surface. Deeper more pronounced tread grips better. The inflation pressure also influences how much resistance the tire has on the travel surface. The less inflation pressure, the greater the surface area covered by the tire, the harder it is to roll and more buoyant the equipment.

Rolling resistance is the resistance of a level surface to a uniform velocity motion across it. It is the force required to shear through or over a surface. An example is a truck tire developing friction on the road surface as it turns. Rolling resistance has two components. Surface resistance results from the equipment trying to roll over the travel surface material. Penetration resistance results from the equipment tires sinking into the surface. Obviously, this resistance will vary greatly with the type and condition of the surface over which the equipment is moving. Simply put, soft surfaces have a higher resistance than hard surfaces.

On a hard surface, a highly inflated (hard) tire has less rolling resistance than a soft tire with less internal air pressure because the tire surface area that is contacting the road surface is lower. A hard tire has greater rolling resistance in sand than a soft tire because it will sink deeper into the rolling surface. The rolling resistances shown in Table 4.1 are adapted

**TABLE 4.1**
**Rolling Resistances [2]**

| Surface | Rolling Resistance (lbs/ton) |
|---|---|
| Asphalt/concrete | 40 lbs/ton |
| Maintained dirt | 50 lbs/ton |
| Poorly maintained dirt | up to 120 lbs/ton |
| Packed sand/gravel | 60 lbs/ton |
| Loose sand/gravel | up to 200 lbs/ton |

from the book entitled *Construction Equipment Management* [2]. The table shows several surfaces and their rolling resistance. Rolling resistance is expressed in pounds of resistance per ton of vehicle or equipment weight. The rolling resistance is greater for a loaded piece of equipment than when it is unloaded. Use the loaded weight of the equipment (equipment including fuel and lubricants plus load) in tons when calculating resistance.

When there is no appreciable penetration into the travel/operating surface, the rolling resistance is about 40 lbs/ton. The weight of the equipment should include the load. When a tire sinks in the mud until it is stable, the rolling resistance as it tries to climb out of the rut increases about 30 lbs/ton (2,000 lbs) for each inch of penetration. Example 4.1 calculates the amount of tractive effort a scraper must generate to overcome the resistance of the working surface:

**Example 4.1:** Calculate the tractive effort generated by a 92,000 lbs loaded scraper traveling on a maintained dirt haul route where the tires sink about 2″ into the travel surface.

92,000 lbs/2,000 lbs/ton = 46 tons.
Rolling resistance = 46 tons (50 lbs/ton) = 2,300 lbs.
Penetration resistance = 2″ (46 tons) (30 lbs per ton per inch) = 2,760 lbs.
Total tractive effort = 2,300 lbs + 2,760 lbs = 5,060 lbs.

With this number, the equipment fleet manager can refer to the manufacturer's performance specifications to select a piece of equipment that can generate the necessary power (in this case rimpull) to overcome this resistance.

Grade resistance is the force-opposing movement of a vehicle up a frictionless slope and does not include rolling resistance. The effort required to move a vehicle up a sloping surface increases approximately in proportion to the slope of the surface. The effort required to move a vehicle down a sloping surface decreases approximately in proportion to the slope of the surface. For slopes less than 10%, the effect of grade increases for a plus slope and decreases for a minus slope. The required tractive effort increases or decreases 20 pounds/gross ton of weight for each 1% of grade as shown in Example 4.2.

**Example 4.2:** Calculate the total resistance if the scraper in the previous example must haul up a 3% grade on part of the haul route.

Grade resistance = 3% (46 tons) (20lbs/ton/%grade) = 2,760 lbs
Total resistance = rolling resistance + penetration resistance + grade resistance.
Total resistance = 2,300 lbs + 2,760 lbs + 2,760 lbs = 7,820 lbs.

The typical total resistance of surfaces over which tired equipment must work should be considered when choosing equipment. This will influence how big the engine should be to power the equipment to overcome the resistances, the type and size of tires, and other operating decisions. Table 4.2 shows basic work requirements and whether tracks or tires are typically the best solution.

### 4.2.2 Buckets and Blades

Buckets come in many shapes and sizes. Most can be easily replaced or changed quickly in the field. The shape of the bucket and the teeth or penetration edge is greatly influence by the material that is to be excavated or moved. A bucket designed for moving loose gravel should not be used to dig into hard material. As the material to be worked becomes harder, typically buckets become slimmer and more elongated. Loaders, backhoes, and excavators typically have standard buckets that can be used for a wide range of material types and uses. Buckets can have jaws or apparatus for grasping irregular shaped loads such as concrete chunks with rebar protruding or jaws that can be used to cut structural members for demo.

The size of the bucket and ultimate payload must be matched to the power of the equipment. Weight represents the safe operational pounds that the excavating, hauling, or moving unit can accommodate. Placing a large bucket on a piece of equipment with a small capacity engine will not be efficient. This will overburden the equipment and wear the engine out prematurely. Manufacturer's suggestions should be followed for bucket size selection. A broad bucket requires more power to push through material than a narrow bucket. However, broad larger buckets are ideal for loose sand or gravel moving.

Buckets vary in width, depth, and structure depending on the match to the power of the machine and the type of material being excavated or moved. Narrow sleek buckets with teeth are designed for penetration of a hard surface. Material moving buckets are typically wider and may not have teeth. The need for penetration power is dependent upon the density of the digging surface. Most equipment models have a standard bucket or range of types and sizes specified for the machine. The bucket typically is included as part of the purchase price. Most equipment has specially designed bucket and attachment systems so that the bucket can be changed easily and quickly. Figure 4.2 shows basic bucket shapes and teeth designed for the type of digging work to be done.

**TABLE 4.2**
**Track or Tire Choice**

| Requirement | Best Choice |
|---|---|
| High tractive effort required | Tracks |
| Low tractive effort required | Tires |
| Stable work surface | Tires |
| Unstable work surface | Tracks |
| Short push or travel distance | Tracks |
| Long push or travel distance | Tires |
| Muddy work conditions | Tracks |
| Side sloping | Tracks |
| Loading heavy unstable loads (dump truck) | Tracks |
| Maneuverability required | Tires |
| Speed required | Tires |

**FIGURE 4.2** Typical Caterpillar Hoe Buckets.

Bucket 1 is for digging in moderate to hard, abrasive materials. Pieces welded on the side near the teeth help penetration and holding the load. Bucket 2 is for digging fragmented rock, frozen ground, and highly abrasive compacted materials. It is taller and thinner than bucket 1. The extra pieces on the front bucket edges protect the bucket sides. Bucket 3 is for digging hard rock and work areas where material is undisturbed or poorly prepared. The thin streamline curved design and sharp irregular teeth configuration make penetration easier. Bucket 4 is for bank forming, ditch cleaning, and finishing and loose material movement. There are no teeth.

Along with the bucket, the bucket teeth or tips are very indicative of the type of work that the equipment is set-up to do. Teeth might be permanently part of the bucket, attached by bolts, welded, or some other means. The teeth may have added tips. If teeth are not permanently attached, they can be easily replaced as their edges wear out. Like the bucket, teeth selection is greatly influenced by the density of the material to be excavated or moved. The penetration of the bucket into the digging surface is easier using sharper, longer, and narrower teeth. Figure 4.3 shows teeth or tip options that are offered on Caterpillar excavating equipment.

Short teeth are used for regular penetrating and breaking apart material. Long teeth are used for chiseling into and breaking apart a packed surface. Heavy-duty long teeth are wider like a chisel and used for breaking apart a packed surface. Heavy-duty abrasion teeth cover more surface area and are used for fitting under a load or breaking apart a larger area. Penetration teeth are used for heavy duty penetrating and breaking apart dense material.

Bucket payload can be measured by volume or weight. Volume is typically stated as "struck measure" of loose volume. Struck measure is volume contained in the bucket after material excess is scraped off flush with the top of the bucket (excavator) or the bowl (scraper or dump truck) or heaped at a specific angle of repose meaning that the soil will support or cling together when piled and maintain this configuration. Figure 4.4 is from the *Caterpillar Performance Handbook* [3] and shows these two measures.

The heaped capacity of a bucket or bowl can be calculated using the fill factor for the type of soil being moved or excavated. Different soils have different fill factors. Stickier soil has a greater fill factor, therefore more can be heaped into the bucket or bowl. The amount

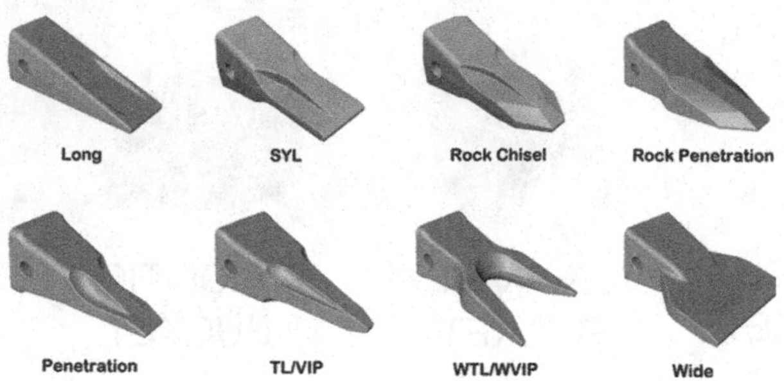


**FIGURE 4.3** Teeth Options.

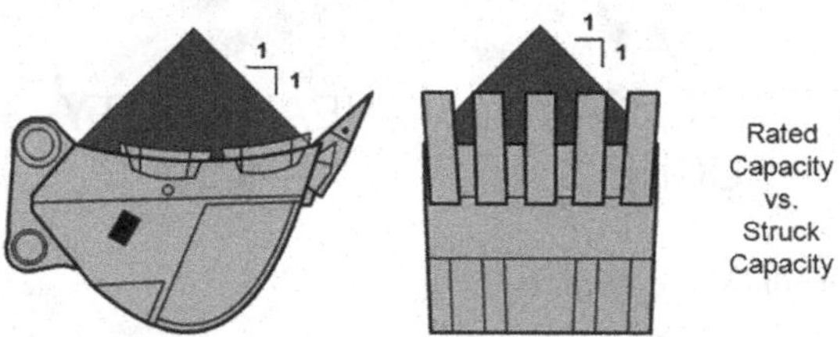


**FIGURE 4.4** Struck and Heaped Bucket Capacity.

of moisture in the soil will influence the fill factor. The average bucket payload equals the heaped bucket capacity times the bucket fill factor. Table 4.3 is adapted from fill factors listed in the *Caterpillar Performance Handbook* [3] and shows different types of soils and respective fill factors.

Weight represents the safe operational pounds that the hauling or moving unit can accommodate. The pounds per loose cubic yard of material times the heaped bucket cubic yard capacity can be used to determine the load weight. Each machine is rated for how many pounds it can structurally lift and remain stable. This information is provided by the manufacturer's equipment specifications.

Like buckets, blades should match the expected work task. A typical blade configuration is like a "C" from top to bottom. As the blade is moved forward and tilted, the bottom of the blade acts as a cutting edge and the top edge rolls the material forward. It is like the material "boils" in front of the blade. Different types of material accumulate in front of the blade differently. Like bucket payload, blade payload is influenced by the type of soil being moved. Basic blade classifications are shown in Table 4.4.

**TABLE 4.3**
**Typical Fill Factors [3]**

| Type of Soil | Fill Factor |
|---|---|
| Moist loam or sandy clay | 100–120% |
| Sand and gravel | 100–120% |
| Hard clay | 85–100% |
| Broken rock | 75–90% |
| Rock | 60–75% |

**TABLE 4.4**
**Common Dozer Blades**

A Blade

C Blade

U Blade

S Blade

Other types of blades are made for specific uses such as tree cutting, land clearing, and deep cutting and penetration [4]

- Straight – "S": used primarily for shallow surface removal, land clearing; designed to push material for short distances, stumps and demo; versatile, light weight and maneuverable, handles a wide range of materials.
- Angle – "A": used primarily for side casting material; excellent for drainage ditch excavation, wider than an "S" blade; used for fine grading and surface removal; not recommended for rock or hard digging surfaces.
- Universal – "U": used for moving big loads over longer distances; curved shape and side and top extensions reduce the spillage of loose material; best suited for lighter materials.
- Cushion – "C": used primarily with scrapers for "on the go" push-loading; can be used for lighter excavation and other general tasks.

Figure 4.5 illustrates the method suggested by the *Caterpillar Performance Handbook* [3] for calculating a bulldozer blade load in the field using the grouser marks in the work surface as a reference for measurement.

To do this measurement, the operator makes a normal pass, pushes a pile of soil, and measures it using this process [5]:

- Measurement Step 1: Measure the average height (*H*) of the pile in feet. Hold the tape vertically at the inside edge of each grouser mark. Sight along the top of the pile at these points and record the observed *H*. Calculate the average *H* from the two observations.
- Measurement Step 2: Measure the average width (*W*) of the pile in feet. Hold the tape 0-end of the tape vertically above one side of the pile at the inside edge of a grouser mark. Sight downward and measure at the corresponding opposite side

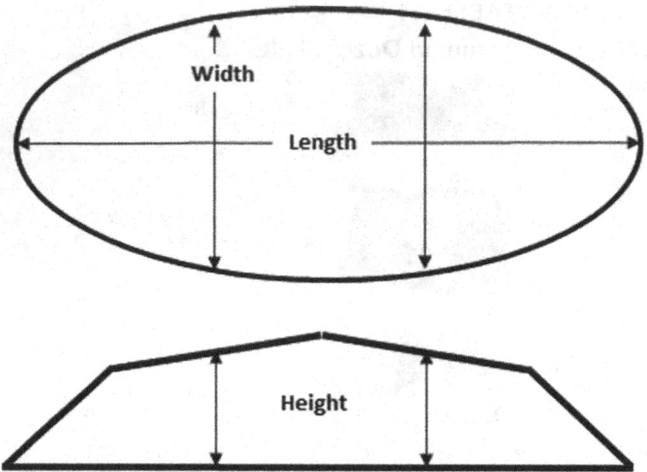


**FIGURE 4.5** Dozer Blade Payload.

of the pile. Do this at the inside edge of the other grouser mark. Calculate the average *W*.

- Measurement Step 3: At the greatest length (*L*) of the pile in feet hold the tape horizontally over the pile with the 0-end of the tape over one end. Sight downward and measure at the corresponding opposite point at the other end of the pile.

$$\text{Blade load (lcy)} = 0.0138\ (H)(W)(L) \tag{4.1}$$

$$\text{Blade load (bcy)} = \text{Blade load lcy (lineal feet of the cut)} \tag{4.2}$$

Another method can be used to field measure and determine a typical blade load. It is based on measuring the pushed pile similarly. It should be noted that *L* in this formula is the length or width of the blade, not the measured length or width of the pile.

$$\text{Volume} = (0.375)(W)(H)(L)/27\text{cf/cy} \tag{4.3}$$

Knowing the actual field measured blade load based on the actual soil type being excavated or pushed can be used to determine a more exact estimated hourly production. If exact production calculations have to be made, measuring and averaging several excavated blade loads in the field is advisable.

There are three basic blade adjustments that the operator can control for operation. The blade can pitch, allowing the operator to vary the angle of attack of the blade's cutting edge with the ground (digging). The blade can angle, allowing the operator to turn the blade so that it is not perpendicular to the direction of travel (side casting). The blade can tilt, allowing the operator to move the blade vertically to permit concentration of tractor driving power on a limited length of blade (sloping).

### 4.2.3 Accessories and Attachments

Today there are accessories and attachments available for every major type of equipment for practically every type of construction activity. Excavators, backhoes, and loaders are designed to use many attachments making them very versatile machines to have on a project. Equipment suppliers should be consulted to see what is available for a specific task. Accessories and attachments must be chosen based on compatibility with the machine size and attachment setup. Performance specifications might be consulted for the attachment's designed working ranges and capabilities. Attachments or accessories are often rented and used on a one-time basis for a short duration.

Accessories utilizing wireless and computer technology include systems to remotely and automatically control blade lift and equipment control, security and tracking systems, and alert or warning systems for maintenance, repair, fuel and lubricant levels, and diagnostics. Development of technology-based accessories will continue as equipment is designed for more specific applications and different working environments and conditions.

### 4.2.4 Earthmoving and Excavating Work

Equipment manufacturers publish equipment performance manuals containing design specifications, performance criteria, and projected costs for using their equipment. The information is typically organized by equipment type and model. Manuals also include guidelines and suggestions for equipment use, equipment accessories, and other related information. For specific equipment information, the performance manual should be consulted for actual specifications and operating capacities.

"Earthmoving" is typically one of the early activities in a project's schedule. Underground utility work or foundation preparation is not started until the rough earthwork or earthmoving is done. Large projects with many mobile pieces of equipment create safety hazards due to noise and dust. Minimizing these factors for the safety of personnel onsite is a major management responsibility, and the contractor should have a plan to control this.

The following "rules of thumb" based on hauling distance should be considered when selecting earthmoving equipment. These are guidelines, and jobsite conditions may influence actual criteria. If the distance that the material must be moved is less than about 500′ then a bulldozer or loader might be used. Bulldozers cut and push the surface material using a blade. Often, a bulldozer is the first piece of equipment on the job. Loaders are not very effective for excavating because they are designed to carry and load excavated material one bucket at a time. If the distance is 350′ to 500′, but less than about 2 miles, then a scraper might be used. The scraper can excavate, haul, and dump. If the material must be moved further than 2 miles, then the best choice is to use front-end loaders to load the excavated soil into dump trucks and haul it to the required location.

Excavation tasks can range from merely removing the top layer of soil and vegetation as the job starts or digging a huge hole in which to build the foundation for a skyscraper. Excavating equipment is typically used to dig, move or load earth. Some machines do excavation as well as earthmoving. The scraper and bulldozer are examples of machines that excavate and move earth. Each type of excavating machine digs differently and is used for specific work activities and site and work conditions. Regardless of the type of project, excavation is necessary for site preparation, underground utility installation, foundation construction, and landscaping in building construction. The same equipment is required for the majority of the earthwork associated with heavy civil projects like highways and water treatment plants. While a single piece of machinery may be able to accomplish some of the task, a combination of different pieces of equipment is normally required.

Hoes or excavators dig below grade or below the tires or tracks of the digging machine. The area covered by the reach of the boom and arm of the excavator is called the digging envelope. How far the tip of the bucket teeth can reach below the machine's tracks to remove a bucketful of material is the digging depth. The deeper the hole, the longer the required reach and the greater the stress placed on the boom and digging arm or stick of the machine. Typically, as the excavation depth increases, so does the size of the required machine. Figure 4.6 shows a typical side view of the designed working envelope for an excavator.

The figure shows the depth and height that the boom and arm can reach. When selecting an excavator, the boom and arm lengths and configurations must be included in the decision. The optimum digging depth for a boom and arm is about 60% to 70% of the machine's reach below grade. To dig a trench 10′ deep, the excavator would have a rated digging depth of about 17′.

#### 4.2.4.1 Earthmoving and Excavating Work Constraints

The maximum depth of cut is the rated depth that the blade or bucket can cut into the soil in one pass. A pass can be considered one time through the cycle to fill the bucket or blade. An efficient operator will set the bucket or blade just deep enough so that when one motion or cycle is complete, the bucket or blade will be filled to its rated capacity. Whether a blade or bucket, the deeper the cut, the more effort required by the equipment to push the blade edge or bucket teeth into and through the material to be excavated. The deeper the penetration, the faster the blade or bucket will fill. This shortens the push distance or reduces the extension of the excavator's boom reducing the cycle time and increasing the production. The trade-off is higher operation costs (more fuel, lubricant, higher maintenance) because the machine must work harder to dig deeper. Optimum motor efficiency (most economical operation) is achieved when the equipment excavates at the optimum depth of cut for that size of motor, blade, or bucket and soil type.

Crowding force is the operational force required to push the edge or the bucket teeth into the material face. It can be done mechanically, like an excavator or front shovel, or by driving into the material face like a loader. Breakout force is the operational force necessary to break material apart once the bucket teeth or edge is set. Breakout force is developed by curling the bucket downward "to the machine," like the excavator, or upward "away from the machine" like the front shovel. Greater force is required to break hard packed material

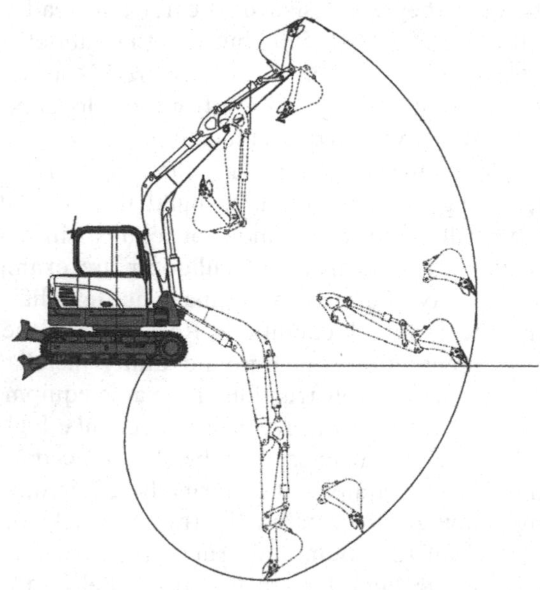

**FIGURE 4.6** Caterpillar Excavator Digging Envelope.

versus loose sandy material. Typically, this means greater power and larger, more durable mechanical components.

The horizontal angle in degrees (plan view) between the position of the bucket when it is digging and its position when it discharges its load is the angle of swing. If the angle of swing is increased, the digging and dumping cycle time is increased, thus reducing production and costing more. Ideal production is achieved when the angle of swing equals 90. For maximum efficiency when setting up an excavator to dig a trench and load dump trucks with the excavated material, the loading path or spot of the trucks should be perpendicular to the direction of the excavator's tracks at 90 or less. Figure 4.7 shows an excavator working on a surface below grade called a "bench."

Note how a flat surface is created next to the hole where a concrete drainage culvert is being built. If the soil is unstable, benching the edges of the hole may be required instead of sloping. It is like stair-stepping the excavation. The bench can also serve another purpose. In this case, the bench also provides a surface for equipment to move and work. By working on this lower surface, the excavator's digging depth can be increased. Note the backhoe and bulldozer in the background. Several excavators could be placed on ascending benches to accommodate the depth of the dig. The bottom excavator is actually doing the excavation. As it digs, it swings and dumps the spoil on the next bench up. Other excavators sitting on the benched areas pass the soil up out of the hole to be stored, spread, or hauled away.

Front shovels excavate above grade or into a material face or pile above the operating surface. Their production cycle is similar to an excavator. Dig, backtrack, dump, reposition, and start over. Shovels digging into dense material typically operate on tracks. Shovels used for material rehandling where digging is not required might operate on tires. Front-end loaders operate similarly to front shovels, but are made for scooping at ground level, not excavating. They are classified similarly by their upward scooping motion. For optimum depth of cut, the bucket should be filled when it reaches the top of the face in one pass. This is dependent on the type of material and the size of the bucket. Optimum digging height for most shovels is between 40% and 50% of the rated maximum digging height. Breakout force is developed by crowding the material away from the shovel by pushing the bucket teeth into the material face and curling the bucket upward and toward the machine.

**FIGURE 4.7** Excavator on a Bench.

#### 4.2.4.2 Earthmoving and Excavating Work Activities

The following discussion focuses on three common earthmoving and excavating work activities that require heavy equipment. The listed equipment packages are groups of heavy equipment that typically work together to perform this work. Each piece of equipment in the package plays a specific role in the series of activities required to perform the work in the most efficient and effective manner. Equipment packages will vary based on the volume of work, desired productivity, equipment availability, and specific work conditions and needs.

Rough site excavation or site leveling is accomplished in the following sequence [6]:

1. A surveyor stakes the area outlining the perimeter of the work and details the depth of the cuts or fills.
2. A motor grader or bulldozer strips the surface of vegetation and debris.
3. A bulldozer with a ripper makes a pass over the area to be cut. It is advisable to rip a couple of inches deeper than the actual cut to be made. The loosened soil provides better traction for the bulldozer than a hard, dense, undisturbed surface.
4. A bulldozer removes the topsoil layer pushing it into an out of the way stockpile to be spread when final grading and landscaping is done.
5. The bulldozer pushes material into piles to be moved to areas of the site needing fill or to be stockpiled at a remote location.
6. Several scrapers assisted by a bulldozer are used for mass surface excavation and to haul the excavated material to another location on site.
7. A motor grader spreads the dumped material at the new location.
8. Excavators dig rough detention areas.
9. Excavated soil is loaded into dump trucks by front-end loaders to be hauled off site.
10. A bulldozer and motor grader are used to finish the rough leveling of the site and detention areas.

Trench excavation for underground utilities is done in the following sequence:

1. A surveyor stakes the route of the trench and details the depth of cut.
2. A bulldozer and motor grader are used to grub, clear, and stabilize the surface.
3. An excavator or backhoe is used to scoop the material from the trench and pile it parallel to the trench. Dense soils might require the use of a trencher.
4. A forklift is used to unload, move, and hold the pipe while it is prepared for installation.
5. The excavator lifts and places the pipe in the trench.
6. A front-end loader is used to backfill the trench when the installation is complete.

Foundation excavation and backfill has the following sequence:

1. A surveyor stakes the foundation perimeter and details the depth of cut once the rough excavation is complete.
2. Excavators dig the hole and place the soil next to the hole.
3. Front-end loaders are used to move this soil to an on-site stockpile.
4. Or excavators dig the hole and directly load the dump trucks.
5. Or front-end loaders are used to load dump trucks hauling the soil to a remote location.
6. Backhoes are used to excavate tighter areas and in the hole.
7. When the foundation is complete front-end loaders move and dump stockpiled soil to backfill against the foundation.

8. Small skid-steer loaders are used to carry material to confined areas and places where less fill is required.
9. The backfill is then compacted against the foundation walls.

Table 4.5 lists common earthmoving and excavation work activities and the types of equipment typically included in the equipment package.

## 4.3 EARTHMOVING EQUIPMENT SELECTION

Earthmoving equipment included in this discussion are

- Bulldozers
- Front-end loaders
- Motor graders
- Scrapers
- Trucks

### 4.3.1 Bulldozers

A bulldozer is a tractor unit with a blade attached to its front. The blade is used to push, shear, cut, and roll material ahead of the tractor. It is an ideal surface earthmover that performs best at about 3 mph. Each model of bulldozer has an operating range for blade size and adjustment. Larger machines have greater operating ranges than smaller machines. A larger machine can pitch and tilt deeper than a smaller machine typically. For heavy civil work, bulldozer blade widths can range from 8′ to 22′, and operating weights can range from about 7 tons to over 120 tons. Maximum digging depth ranges from about 1.5′ to 2.5′.

Figure 4.8 shows a dozer that was working with the mega-terrain leveler shown in Figure 4.9 excavating the foundation hole for a high-rise condominium building in Austin, Texas.

The hard clay-like soil had to be ground up, pushed into piles by the bulldozer, and then loaded by front-end loaders into dump trucks to be hauled away. This was more efficient than using excavators due to the denseness of the soil.

The bar connecting the blade to the body of the bulldozer is parallel to the travel surface and just above it. This positioning can deliver maximum forward force to push a pile of material or a scraper. Note the pistons connected to the blade. These pistons control the tilt of the blade, how deep the penetration and the cut. The frame or cage over the operator is the rollover protection system (ROPS).

The single shank (can be multiple smaller shanks) ripper attachment is used to penetrate tough surface material. It is lowered into the work surface just like the blade. If the work surface is hard and compacted, the dozer will make a pass with the ripper first, breaking up the soil and follow with a pass using the blade. This is often the first activity done on an undisturbed surface prior to cutting and filling. Shorter ripper shanks are used for disturbing dense material, thus minimizing breakage. Longer shanks are used for disturbing less dense more organic material. The depth of ripper penetration in dense material will influence how fast the bulldozer will travel. Ripping production is based on the dozer speed and distance covered in a time unit measure (ft/min).

The elevated sprocket configuration is not as common as the elliptical configuration seen on most bulldozers. The final drives (round gear above the middle bar) are higher above the work area than on an elliptical track. This helps isolate the drive from ground impacts. This is much better suited for rough, rocky, uneven working conditions.

**TABLE 4.5**
**Earthmoving and Excavating Work Activities and Equipment Packages**

| Activity | Dozer | Loader | Grader | Scraper | Dump Truck | Backhoe | Excavator | Front Shovel |
|---|---|---|---|---|---|---|---|---|
| Excavating above grade | | | | | | | | x |
| Excavating below grade | x | | | x | | x | x | |
| Grubbing | x | | | | | | x | |
| Heavy ripping | x | | | | | | | |
| Light ripping | | | x | | | | | |
| Tree stump removal | x | | | | | | x | |
| Topsoil removal/storage | x | | x | x | | | | |
| Rough cutting | x | | | x | | | x | |
| Rough filling | x | x | | x | x | | | |
| Finish grading | | | x | | | | | |
| Foundation excavation | | | | | | x | x | |
| Foundation backfilling | | x | | | | x | x | |
| Footing excavation | | | | | | x | x | |
| Road base construction | x | x | x | | x | | | |
| Temporary road construction | x | x | x | | x | | | |
| Haul road maintenance | | | x | | | | | |
| Culvert placement | x | | x | | x | x | x | |
| Earth berm/dam construction | x | | x | | x | | | |
| Drainage ditch maintenance | | | | | | x | x | |
| Haul less than 500′ | x | x | | | | | | |
| Haul 500′ to 2 miles | | | | x | | | | |
| Haul over 2 miles | | | | | x | | | |
| Soil windrowing | x | | x | | | | | |
| Soil spreading | x | x | x | x | x | | | |
| Excess loose soil removal | | x | | | x | | | |
| Deep trench excavation | | | | | | | x | |
| Shallow trench excavation | | | | | | x | | |
| Trench backfilling | x | x | | | | x | x | |
| Utility pipe placing – small | | | | | | x | x | |
| Utility pipe placing – large | | | | | | | x | |
| Trench box placement/ movement | | | | | | x | x | |
| Debris/trash removal | | x | | | x | | x | |
| Rock removal | x | x | | | x | | x | |
| Asphalt paving removal | x | x | | | x | | x | |
| Concrete removal | x | x | | | x | | x | |
| Structure demo | x | x | | | x | | x | |
| Assisting scrapers | x | | | x | | | | |
| Towing other equipment | x | x | | | | | | |
| Concrete placement – bucket | | | | | | | x | |
| Crane pad construction | x | | x | | x | | | |
| Detention pond excavation | x | | | x | | | x | |
| Benching | x | | x | | | | x | |
| Side sloping | x | | | | | | | |

**FIGURE 4.8** Elevated Sprocket Bulldozer with Ripper.

**FIGURE 4.9** Terrain Leveler.

Along with surface excavation, bulldozers are typically used with scrapers for excavation. A typical production cycle for excavation or pushing a scraper is to position, push, backtrack, and maneuver into position to make contact and push again. Speeds are typically low and influenced by the type and density of material being excavated or moved. Speed empty is usually the maximum that can be achieved in the travel distance.

#### 4.3.1.1 Bulldozer Production

Time components used for most tractor powered equipment production cycles are similar to those of the working bulldozer described in this problem as are typical production cycles. Exact components will vary based on the work setup and operation. The typical production cycle is the following:

1. The bulldozer positions to start excavation.
2. The bulldozer scoops or digs for the length of a pass to fill the blade.
3. The bulldozer hauls the load by rolling, crawling, pushing, or pulling.
4. The load is discharged or dumped at the desired location.
5. The bulldozer repositions to exit the dump site.
6. The bulldozer backtracks to the loading location.

The most efficient and safe production is achieved when the machine is going forward. The time component for each part of the cycle is influenced by many factors, primarily by speed and distance of travel. The production cycles for bulldozers, front-end loaders, motor graders, scrapers, and trucks are all similar to the cycle described above. There are similarities in the example problems for each of these types of earthmoving equipment. Bulldozer production is typically not dependent on other equipment. Production is based on the width of the blade, the depth of cut and the travel, backtrack and return times. Example 4.3 illustrates this principle. Dozer production with scrapers will be discussed in the scraper section of this chapter.

**Example 4.3:** Note: Performance criteria and costs used in all example problems are hypothetical. A Case 750K bulldozer with an 8′ blade is to be used to excavate and push fairly loose material. According to the soils report the material to be moved has a 23% swell factor. When the dozer arrives at the jobsite, a series of test blade loads are excavated to estimate a typical load. The average $H = 4'$, the average load width in front of the blade = 6′ and the load length = 9′. The load time suggested by the manufacturer is 0.08 min. Once the blade goes through the cut, the haul push is 200′ with an average speed of about 2.6 mph. Backtrack distance is 240′, and the dozer will travel at a speed of about 3.2 mph. Once back to the hole, the dozer takes about .06min to reposition. The Case dealer suggests an O&O cost of about $55/hour. Your operator costs about $23/hour with contractor outlay. What is the unit cost for the work if there is about 1,200 bcy of surface material that must be moved by the 750K?

1. How much material (lcy) can be moved in 1 production cycle?

$$V = [(0.0138)(6')(4')(9')] = \underline{\underline{2.98 \text{ lcy/blade load}}}$$

2. How much material (bcy) can be moved in 1 production cycle? The quantity take-off is in bcy so the lcy load must be converted to bcy.

$$3.98 \text{ lcy}/1.23 \text{ lcy/bcy} = 3.67 \text{ bcy/production cycle}$$

3. What is the cycle time for 1 production cycle?

$$\begin{aligned} \text{Haul time} &= 200'/[(3.2\,\text{mph})(88'/\text{min/mph})] = 0.87 \text{ min} \\ \text{Backtrack time} &= 240'/[(3.2\,\text{mph})(88'/\text{min/mph})] = 0.85\,\text{min} \\ \text{Cycle time} &= \text{load} + \text{haul} + \text{backtrack} + \text{reposition} \\ &= 0.08\,\text{min} + 0.87\,\text{min} + 0.85\,\text{min} + 0.06\,\text{min} = 1.86\,\text{min} \end{aligned}$$

4. What is the work hour productivity if the operator works 50 min per 60-min hour?

$$\begin{aligned} \text{Work hour productivity} &= [(\text{load volume (bcy)})(50\,\text{min})]/\text{cycle time} = \\ &= [(3.67\text{ bcy/load})(50\,\text{min})]/1.86\text{ min/cycle} \\ &= 98.66\text{ bcy/work hour} \end{aligned}$$

5. How long will it take to move the 1,200 bcy?

$$1{,}200 \text{ bcy}/98.66 \text{ bcy/hour} = 12.2 \text{ hours. Use 13 work hours.}$$

6. How much will it cost?

$$\$55/\text{hour} + \$23/\text{hour} = \$78/\text{hour} \times 13 \text{ hours} = \$1{,}014$$

7. What is the unit cost to perform the work?

$$\$1{,}014/1{,}200 \text{ bcy} = \$0.845/\text{bcy to move the material with the Case 750K bulldozer}$$

### 4.3.2 Front-End Loaders

Front-end loaders typically are tractor powered and operate on tires. They are typically articulated and very maneuverable, making them ideal for constricted areas. They are used primarily for material moving and re-handling. They are ideal for scooping and hauling material in storage piles to where it is to be permanently placed or loading it into dump trucks. Loaders are ideal for dumping soil back into the hole after the necessary below grade work is done. Tracked loaders may be required for extreme surface conditions demanding greater traction or stability. Every concrete or asphalt batch plant has a tired front-end loader to stock the feed to the batch hopper with aggregate or sand.

Fixed cycle times for loaders (raise, dump, and lower the bucket) range from about 9 seconds to about 20 seconds, depending on the size of the loader. General-purpose bucket capacities range from about 0.75 to 18 lcy. Figure 4.10 shows a small loader that is ideal for confined spaces and smaller loads.

FIGURE 4.10 Skid-Steer Loader.

These small loaders are very maneuverable and are ideal for use in constricted limited working areas. They are often used for moving sand within slab forms or with fork attachments to carry brick, mortar, or sand. They are excellent for surface movement of small amounts of material. These machines are commonly referred to as a "bobcat," the model name for small skid steer loader manufactured by Melrose. Readily available attachments include augers, cold planers for light milling, landscape tillers and rakes, trenchers, vibratory compactors, and brooms.

#### 4.3.2.1 Loader Production

Loaders are used with feed hoppers and dump trucks. The loader is sized by the demand of the feed hopper or the size and number of dump trucks that must be filled [7]. Production cycle components are similar, whether running on tracks or tires. The work surface stability will influence the cycle time. Small skid-steer loaders are very maneuverable, and cycle times might be less. Example 4.4 demonstrates how to calculate a loader production cycle.

**Example 4.4:** A Cat 950G wheel loader with a 4.25 lcy heaped bucket is to be used to move fairly loose stockpiled material onto a conveyor running under the road that is carrying the material to another part of the site. The material will be used to fill that side of the project site. The Cat Performance Manual suggests a cycle time (load, dump, maneuver) of about 55 sec for the way you have the work setup and an O&O cost of about $31/hour. Your operator costs about $23/hour. The conveyor will haul about 280 lcy/hour. Will the production of the loader keep up with the conveyor?

1. How much material (lcy) can be moved in 1 production cycle?

$$V = 4.25 \text{ lcy/cycle}$$

2. What is the cycle time for 1 production cycle?

$$\text{Cycle time} = 55 \text{ sec/cycle}/60 \text{ sec/min} = 0.92 \text{ min}$$

3. What is the work hour productivity if the operator works 50 min per 60-min hour?

$$\begin{aligned}\text{Work hour productivity} &= [(\text{load volume (lcy)})(50\,\text{min})]/\text{cycle time}\\ &= [(4.25\text{ lcy})(50\text{ min})]/0.92\text{ min/cycle}\\ &= 231\text{ lcy/work hour}\end{aligned}$$

Based on this calculation, the 950G will be short about 49 lcy/hour. Assuming one can't change the work layout, options to meet the necessary 280 lcy/hour production include a larger bucket on the 950 (for this situation, probably not feasible), a larger loader or two smaller loaders.

4. What is the daily cost for using this loader if the conveyor runs 11 hours/day?

$$\$31 + \$23 = \$54/\text{hour} \times 11 \text{ hours/day} = \$594/\text{day}$$

5. What is the unit cost/day to use the 950G loader?

$$231 \text{ lcy/hour (11 hours/day)} = 2{,}541 \text{ lcy/day}$$
$$\$594/\text{day}/2{,}541 \text{ lcy/day} = \$0.233/\text{lcy}$$

### 4.3.3 Motor Graders

This type of equipment has been around since the start of road building, powered by a team of oxen, mules, or horses. The need for a smooth stable travel surface has always been an important part of a road system. In some parts of the country, a motor grade is called a "maintainer." This name is appropriate because this equipment is typically used to maintain grade and a smooth surface for rural nonpaved travel roads or haul routes on construction sites. The grader is a long tractor-driven piece of equipment with a blade mounted underneath, as shown in Figure 4.11.

The blade consists of a mold board to which replaceable cutting edges are attached and used to push material straight ahead or to the side of the grader blade at the desired level. While a grader can be used for light surface excavation, it is designed to move material to create a level surface. The blade is attached to a ring underneath the frame. This ring can be swiveled vertically, and the casting angle of the blade is adjusted on it. The blade can be angled to shape road banks. Standard blade widths range from 12′ to 14′, and speed in mid-range gear is approximately 6 mph. The front tires are made to allow them to lean to resist the force created when the blade is cutting and side casting the material.

As the material is pushed ahead of the blade, it fills voids in the surface over which it is moving. Excess material is pushed into other surface voids or to the side. When the material is cast to the side of the grader, this is called a windrow (row of piled material). Usually, a front-end loader will follow behind the grader to scoop up excess material in the windrow if necessary. Laser level readers can be attached to the blade so that the operator can establish a desired elevation using the level signal, not having to rely on feel and experience as much. The grader depth of cut is adjusted based on the signal setting.

**FIGURE 4.11** Motor Grader.

#### 4.3.3.1 Motor Grader Production

On large material moving jobs, the motor grader operator typically controls the movement, spotting, and leveling of delivered fill. This operator is like an equipment group foreman. The delivered fill might be from the other side of the project site or imported from off-site. Since soil is leveled and compacted by lift, where and how much soil should be dumped must be managed for efficient spreading as it is delivered. In road base construction, the grader is typically the last major earthmoving equipment used during compaction operations.

Graders are usually setup to run in linear or rectangular patterns. Production is measured in the area covered in a certain amount of time (square feet/hour, cubic feet/hour). When grading linearly, the operator usually has the blade dropped (cutting) until the end of the pass, turns, drops the blade, and grades the opposite direction. When grading an area for a parking lot, two methods of the rectangular pattern can be used. Using the "back and forth" method, the operator travels with the blade down to the end of the pass, picks up the blade, backtracks, drops the blade, and starts over. Using the "looping" method, the operator drops the blade for the pass, lifts the blade at the end of the pass, turns the grader in an arc the other direction, and drops the blade for this pass. The process is repeated until the area is covered with the grader traveling in a forward loop or oval as many times as necessary to cover the surface. To set the "looping" coverage pattern, the turning radius of the motor grader must be considered. The turning radius is typically listed in the performance spec for the model.

Grader production for road maintenance is a linear process [6]. For mass earthmoving projects, grader production must be matched to production of other equipment (usually scrapers or dump trucks) in the equipment package dumping material to be spread in lifts as illustrated in Example 4.5 Linear grading productivity is estimated using:

$V$ = average grading speed

$W$ = grading width (width of blade if pushing straight ahead perpendicular to travel direction)

OF = the operating factor

$N$ = the number passes required to cover the area to be graded (width of the road base)

**Example 4.5:** A Volvo G740B motor grader with a 14′ blade is to be used to grade the material on a 66′ wide and 9,800′ long road base area. The effective grading width is 12′. The average speed will be around 3mph. The number of passes required is 2 to reach the desired smoothness.

1. What is the work hour productivity if the operator works 50 min per 60-min hour?

$$\begin{aligned}\text{Work hour productivity} &= [(V)(5{,}280'/\text{mile})\ (W)\ (\text{OF})]/[(9\ \text{sf/sy})\ (N)] \\ &= [(3\ \text{mph})\ (5{,}280'/\text{mile})\ (12')\ (0.83)]/[(9\ \text{sf/sy})\ (2)] \\ &= 8{,}765\ \text{sy/hour}\end{aligned}$$

2. How long will it take to grade the road base?

$$(66' \times 9{,}800')/9\ \text{sf/sy} = 71{,}867\ \text{sy}/8{,}765\ \text{sy/hour} = 8.2\ \text{hours}$$

#### 4.3.3.2 Box Blades

Figure 4.12 shows a box blade attachment mounted on the rear of a tractor. The ripper teeth inside the box are raised. This is probably the most universally used piece of equipment for finish grading and contouring. It is typically the last piece of equipment on-site prior to landscaping. Inside the box are ripper teeth for disturbing the soil. As the ripper teeth dig into the soil, the soil boils up into the box where it is broken apart as the tractor

**FIGURE 4.12** Box Blade.

moves forward. The operator can regulate the depth of the ripper teeth, as well as the bottom of the box. As the tractor moves forward, the box is raised, and the soil is evenly spread or the soil fills voids as the tractor drags the box over the surface. The outside rear of the box can be used to push soil around like a bulldozer as well. This equipment is not suitable for hard or rocky soils.

### 4.3.4 Scrapers

Scrapers are designed to excavate, load, haul, and dump loose material. The greatest advantage is their versatility. They can be used for a wide variety of material types and are economical for a range of haul distances and conditions. They are a compromise between a bulldozer, excavator, and dump truck. Scrapers are articulated, tractor powered, and pull a bowl that holds the soil. A blade is mounted on the bottom of the bowl that cuts into the travel surface and the disturbed soil flows into the bowl as the scraper moves forward. Figure 4.13 shows the tractor, bowl, and chain operating. Scrapers can self-load or be assisted by another scraper or a bulldozer.

Scrapers are classified in the following categories:

1. Single engine: A tractor pulling a bowl that can operate under its own power or be push assisted. This is the most common type of scraper on large earthmoving jobs.
2. Tandem or twin engine: This type has a second engine mounted in the rear and can develop greater power. Ideal for steeper hauls at greater speeds. Typically cost about 30% more than a conventional scraper.

**FIGURE 4.13** Elevating Scraper.

3. Push–pull scraper: This type is designed with a push block mounted on the rear and a bail mounted on the front to assist other scrapers or be pushed by other scrapers. They are ideal for dense soil excavating projects when a dozer is not utilized for pushing.
4. Elevating: These are self-contained loading and hauling units. The chain elevator serves as a loading mechanism. The extra weight of the loading mechanism is a disadvantage during the haul cycle, but this type is ideal for short haul situations where the ratio of haul time to load time is low. Elevating scrapers are often used for utility work, dressing-up behind high production spreads, or shifting material during fine-grading operations. The chain breaks the soil as it enters the bowl, making it easier to discharge. These units can also be push assisted.

Wheeled scrapers have the potential for high travel speeds on favorable haul roads and can go up to 30 mph. The demand on the scraper's power is the greatest when digging in hard clay. The operator needs to set the blade just deep enough so that when the pass through the hole is complete, the bowl is full. Typically, wheeled scrapers need bulldozer support to provide the extra tractive effort needed for economical and efficient loading. To reduce the effort that the scraper must exert to load and get out of the hole, a bulldozer is an economical and efficient pusher.

Figure 4.14 shows three techniques typically used for push assisting a scraper through the hole, back track loading, chain loading, and shuttle loading. The technique used should be determined based on the quantity of work and specific site considerations. Avoiding repositioning and keeping the scraper traveling forward will optimize production time. Backtrack loading is the most common type of technique for multiple scrapers and a single pusher.

Heaped scraper capacities range from about 15 to 44 cy. Load ratings range from about 18 to 52 tons. Maximum depth of cut ranges from 13″ to 17″. Maximum depths of spread range from 14″ to over 22″.

#### 4.3.4.1 Scraper Production

To load the scraper, the front end of the bowl (nearest the cab) is lowered until the attached cutting edge penetrates the travel surface. As the scraper moves forward, the

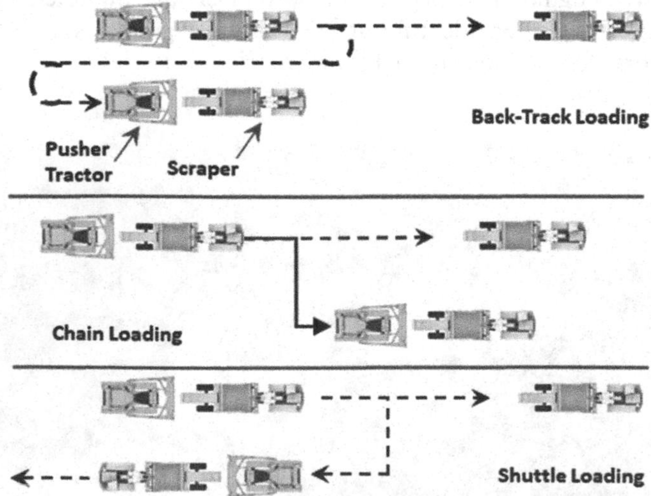


**FIGURE 4.14** Scraper Loading Techniques.

front apron of the bowl is raised so that a strip of excavated earth can flow into the bowl. The amount of excavated material depends on the depth of penetration of the cutting edge. The scraper moves forward until the bowl is full. The blade is lifted, and the apron closes. Ripping (bulldozer with ripper shanks) or tilling (tractor pulling a plow) the lift to be excavated prior to the scraper making a pass can increase scraper production. Sometimes applying water will also loosen the excavated material.

To dump the load, the cutting edge is set above the discharged material, raising the apron. The material is forced out by means of a moveable ejector mounted at the rear of the bowl. The size of the apron opening regulates the amount of material discharged and the material lift depth.

The capacity of the scraper bowl can be measured by capacity or weight. Whichever one is exceeded, will determine the loaded capacity of the scraper. Scraper volume is measured in two ways in loose cubic yards. Struck volume is the loose cubic yards that a scraper would hold if the top of the material were struck off even at the top of the bowl. Heaped volume is the loose cubic yards that a scraper would hold with the material heaped and sloping above the sides of the bowl. The heaped volume accounts for the fill factor.

The cycle time for a scraper is estimated by adding the fixed times to load, dump, turn around, and spot for the next cut, and the variable or travel times to haul full and return empty. Scraper rimpull, speed, and gradability performance can be verified by referring to the rimpull, speed, and gradability curves for the model. The expected performance of the scraper can be compared to these operating requirements of the work. Load times vary based on power, bowl capacity, and site conditions and range from 0.4 to 1.0 min typically. Maneuver and spread or maneuver and dump times range from 0.6 to 0.7 min. Additional maneuvering (spotting) when approaching the cut might be required and should be added if necessary.

Scrapers are often loaded with the assistance of a bulldozer (sometimes called a "pusher dozer"). This is done by having the dozer make contact with the back bale of the scraper as it starts into the hole. The dozer provides most of the pushing power to not only make the cut, but also to transport the bowl through and out (boost) of the cut full of material. This greatly optimizes what a bulldozer is designed to do and greatly reduces the power needed by the scraper to excavate and get started hauling when fully loaded. It is an ideal pairing of equipment to optimize the capabilities of both.

To determine the number of scrapers that can be matched to one pusher dozer, the pusher cycle time must be determined. This cycle time includes match up and make contact with the rear of the scraper, push it through the hole, boost it out of the hole, and maneuver to match up to the next scraper coming through the hole. This is demonstrated in Example 4.6.

**Example 4.6:** A Cat D631E Series II wheel tractor scraper assisted with a D9R bulldozer is to be used to move material about 4,200′ to be used to build a detention pond at the entry a subdivision. The D9 has ripped the soil in the area to be excavated about 18″ deep. The D9R will push the scraper until it is out of the hole. Once full, the scraper's average haul speed is 10 mph. The return route is 4,400′ and the average return speed is 14 mph. The rated heaped capacity of the scraper bowl is 31 lcy. The estimated load time according to the performance manual is 0.6 min. The estimated dump time is about 0.7 min. The Caterpillar estimated hourly O&O costs for moderate conditions for the D9R is $86/hour and for the D631E is $87/hour. The projected O&O cost includes the operator for this calculation.

1. What is the work hour productivity if the operator works 50 min per 60-min hour?

   Work hour production = [(rated capacity) (operational efficiency)]/cycle time

   Haul time = 4,200′/[(10 mph)(88′/min/mph)] = 4.77 min

   Backtrack (return) time = 4,400′/[(14 mph)(88′/min/mph)] = 3.57 min

   Cycle time = load + haul + dump + return

   = 0.6 min + 4.77 min + 0.7 min + 3.57 min = 9.64 min/cycle

   Production = [(31 lcy)(50 min/hour)]/9.64 min/cycle = 160.8 lcy/hour

2. How many scrapers will the D9 support?

   Pusher cycle time = $1.4\,L_s + 0.25$ min

   $L_s$ = the load time of the scraper = 0.6 min

   Boost time = 0.1 min; maneuver time = 0.15 min; boost + maneuver = 0.25 min

   Return time = 40% of the load time

   Pusher cycle time = 1.4(0.6) + 0.25 min = 1.09 min/pusher cycle

   Number of assisted scrapers = (9.64 min/scraper cycle)/(1.09 min/pusher cycle)

   = 8.8 scrapers

   If this is rounded up to nine scrapers working with this D9, a scraper will have to wait a short time before hitting the hole. If it is rounded down to eight scrapers, there should be no delay for any scraper to hit the hole. It is probably advisable to round down and use eight scrapers.

3. How many hours will it take to excavate and haul 20,600 bcy of soil with swell factor of 15% using the D9R and eight D631 scrapers?

   Amount of soil excavated/hour = 8 scrapers (160.8 lcy/hour)

   = 1286.4 lcy/hour [(20,600 bcy) (1.15%swell)]

   = 23,690 lcy/1,286 lcy/hour

   = 18.5 hours

4. How much will it cost to excavate and move this material?

   [(18.5 hours) (1 dozer) ($86/hour)] + [(18.5 hours) (8 scrapers) ($87/hour)]= $14,467

5. What is the unit cost for the work?

   $14,467/20,600 bcy = $0.72/bcy

### 4.3.5 Trucks

Trucks are an extremely important part of the earthmoving and material-moving process. They are basically a tractor and a trailer with sides. Like the rest of the equipment categories, there is a wide range of trucks based upon hauling conditions and need. Typically, trucks are sized by trailer volume. The larger and heavier the load, the larger a tractor you need to pull the trailer. Trucks are typically used with excavators and loaders for excavation and soil haul-off or delivery. Compared to other earthmoving equipment, they can obtain high travel speeds. Rough terrain trucks have frames, suspension systems, and motors designed to traverse rough surfaces and radical travel grades. Trucks designed for hauling on the highway are designed for less rigorous conditions.

Two basic considerations for choosing a truck trailer are the method of dumping, and the class of material hauled. Trucks may dump from the rear (the most common), from the bottom (belly dump), or from the side depending on the type of material and work activity. Common rear dump trucks are typically not articulated, but larger rough terrain trucks are typically articulated for greater maneuverability. Figure 4.15 shows a bottom or belly dump truck. When the gate is opened at the bottom of the bed, the load is spread in an even windrow as the truck moves forward. The windrows spread easier than a pile. A grader will typically follow to spread the material for compaction.

Topsoil, select fill, clay, sand, and aggregate are typical building materials transported by truck. Material considerations include the size and shape of material pieces and cohesiveness of the material. The size to weight relationship will influence the volume of the trailer matched to the tractor. Size and shape will influence how the material will pack in the trailer. If the material is cohesive, the trailer shape should be conducive for the material to be easily discharged when dumped. Rounded edges keep material from compacting in corners. Capacities of construction hauling trucks range from 6 lcy to gigantic trucks used in mass earthmoving or mining. Typical rear dump trucks used on construction sites are 9 or 12 lcy.

#### 4.3.5.1 Truck Production

Dump truck production is similar to the other earthmoving equipment cycles. Trucks however are typically dependent on another piece of equipment to load them. Truckloads are rated by volume and weight. Trucks must be permitted to operate on public highways and streets. Production cycles have fixed and variable times. Typical cycle fixed times include loading, dumping, and spotting to load and spotting to dump times. It should be noted that the loading time is the time required by the piece of equipment loading the truck. The loading time equals the number of cycles required to load the truck times the estimated cycle time. The number of loader cycles to fill a truck equals the volume of the truck divided by the volume of the loader bucket per cycle. Trucks are usually loaded by front-end loaders or excavators.

**FIGURE 4.15** Belly Dump Truck.

Spotting to load or dump and wait or delay times are influenced by job conditions, work setup, and management of the process.

Turn and dump time in moderate conditions for end-dump trucks are about 1.3 min. Spotting time is about 0.3 min. Turn and dump time in moderate conditions for belly or bottom dump trucks are about 0.7 min. Spotting time is about 0.5 min. Note that belly dumping takes less time to turn and dump, but because the placement is more precise than rear dumping, the spotting takes a bit longer. A load from a rear dump truck must be followed by a dozer or loader to knock it down for the grader to spread. A windrow discharged from a belly dump truck can be followed by the grader, making spreading and compacting faster.

The variable times include hauling and return. If the travel route is on the construction site, delays will probably be minimal. If hauling is done on public roads, then delays become much less predictable. Traffic in a metropolitan area can drastically influence production cycle time. Driver time management can also greatly influence production. Example 4.7 provides the sequence of calculations to determine the required number of haulers.

**Example 4.7:** A Cat 950G wheel loader equipped with a 2.3 lcy bucket and a 0.2 min cycle time is to be used to load a Cat D30D articulated truck with a heaped capacity of 21.6 lcy. It takes about 1.3 min to dump the load.

1. How long does it take to load the truck?

$$21.6\,\text{lcy}/2.3\,\text{lcy} = 9.4\ \text{cycles/truck load} = \text{round-off to 9 cycles/load}$$
$$= (9\ \text{cycles})\ (2.3\ \text{lcy/cycle}) = 20.7\ \text{lcy/load}$$
$$(9\ \text{cycles})\ (0.2\ \text{min/cycle}) = 1.8\ \text{min to load the truck}$$

2. How much material can be hauled in 1 work hour by the D30D?

$$\text{Work hour production} = [(\text{rated capacity})\ (\text{operational efficiency})]/\text{cycle time}$$
$$\text{Haul time} = 9{,}200'/[(25\ \text{mph})\ (88'/\text{min/mph})] = 4.18\ \text{min}$$
$$\text{Return time} = 9{,}200'/[(30\ \text{mph})\ (88'/\text{min/mph})] = 3.48\ \text{min}$$
$$\text{Cycle time} = \text{load} + \text{haul} + \text{dump} + \text{return}$$
$$= 1.8\ \text{min} + 4.18\ \text{min} + 1.3\ \text{min} + 3.48\ \text{min} = 10.76\ \text{min/cycle}$$
$$\text{Hourly production} = [(20.7\ \text{lcy})\ (50\ \text{min/hour})]/10.76\ \text{min/cycle} = 96.2\ \text{lcy/hour}$$

3. How many trucks will the 950G support?

Number of trucks or haulers = hauler cycle time/hauler time at the load site (spotting and loading)

Number of haulers = the hauler cycle time/loading and spotting time

Number of D30D trucks supported by the 950G = 10.76 min/cycle/1.8 min to load = 5.97 or 6 haulers

## 4.4 EXCAVATING EQUIPMENT SELECTION

Excavating equipment included in this discussion are:

- Excavators
- Backhoes
- Front shovels

### 4.4.1 Excavators

The excavator combines digging and lifting abilities. Excavators come in a wide range of sizes. Bucket size, boom length, and operating speed are primary considerations for choosing the proper excavator. Typically, the faster the operating speed, the faster the machine can load, swing, dump, return, and dig (the normal excavator production cycle). Excavators are ideal for digging and dumping into a dump truck or a pile.

Excavators are ideal for underground utility construction. For trenching, the operator fills the bucket and dumps it to the side above grade. With the excavator in the same path, the bucket side and bottom can also be used to scrape the material back into the trench and compact it after the work is done. Another reason that the excavator is ideal for underground utility construction is its lifting ability. Most buckets have an "eye" for securing rigging. Pipe can be easily rigged and placed in the trench. If necessary, the load can be picked-up and "walked" to the placement point. Obviously, the excavator should be rated for the load.

Excavators can accommodate numerous attachments such as pinchers for lifting logs or pipes, a jackhammer for busting-up concrete or compacted soil or a magnet for metal material moving. Excavator attachments are similar to backhoe attachments and are run by hydraulics. Along with the many attachments, excavators can be equipped with long reach booms, demolition arrangements, different shoe selections, and different quick coupler systems. Bottom dump buckets permit more accurate loading of narrow trucks and reduce spillage.

Heaped bucket capacities range from very small, 0.1 cy, to extremely large for mass excavation, over 7 cy. Most excavators accommodate a range of bucket sizes. Maximum digging depths range from about 7′ to 34′, depending on the boom and stick lengths and combinations. Lifting capacities over the front of the excavator range from about 1,300 pounds to over 64,000 pounds.

#### 4.4.1.1 Excavator Production

Excavators are ideal for mass rough excavation below grade. Small excavators can be used for small shallow holes and big excavators for big deep holes. The proficiency of the operator will greatly affect production (perhaps more so than with other earthmoving equipment). Excavators are mobile but run at a peak speed of about 3.5 mph. The travel path or work surface must be relatively stable and flat. Small rubber tracked or tired excavators are ideal for work in limited area or height spaces. Most excavators will support different combinations of stick, boom, and attachments. Obviously, they must fit the machine and match the power that can be generated to dig, lift, or run the attachment.

The weight of the soil in the bucket can sometimes make the excavator unstable. The rated load should not exceed 75% of the tipping load. Attention to setup and load weight is essential. If the bucket of soil is too heavy, the bucket pass should not be so deep and less soil excavated per production cycle. Bank weights for common earth materials per the *Caterpillar Performance Handbook* [3] are shown in Table 4.6.

An excavator can be used to dig into a vertical face of material, but because of the downward motion, it is not ideal for scooping. This technique can be used to knock material loose from the face to the travel surface and then be scooped and hauled by a loader. They are ideal for demolition as the reach of the stick and boom keeps the machine a safe distance from falling or shifting debris above grade. Excavators can also be used for material handling too. A typical excavator production cycle is fill the bucket (load), raise the load above grade or to the necessary height, swing the load to the dump point, dump the load, swing the empty bucket back to the excavation point, drop the bucket, and start the cycle over.

Figure 4.16 shows the basic parts of an excavator. The *Caterpillar Performance Handbook* [3] recommends that excavator setup plans for truck loading consider the following:

**TABLE 4.6**
**Common Earth Material Bank Weights [3]**

| Material | Bank Weight (lbs/bcy) |
|---|---|
| Dry clay | 3,100 |
| Wet clay | 3,500 |
| Dry clay and gravel | 2,800 |
| Wet clay and gravel | 3,100 |
| Loam earth | 2,600 |
| Dry gravel | 2,850 |
| Dry, loose sand | 2,700 |
| Wet sand | 3,500 |
| Shale | 2,800 |

**FIGURE 4.16** Excavator Parts.

1. The bench height or distance from bucket insertion to the surface on which the excavator sits should equal about the stick length for stable material. This is the optimal height and allows the excavator to be above the dump truck while loading, minimizes lifting the bucket, reducing cycle time and wear.
2. The truck should be positioned where the truck rail (edge of the bed) is below the boom stick hinge pin (connection of the boom to the stick).
3. The truck should position as close to the centerline of the excavator as possible when aligning for loading.
4. The excavator should position for digging so that the stick is vertical when the bucket is full and curled with the load.
5. It is recommended that the operator boom-up when the bucket is 75% through the curl (digging motion to the machine) cycle.

The loading time is the time required by the piece of equipment loading the truck. Loading time = number of cycles required to load the truck (the estimated cycle time)

$$\text{Number of loader cycles to fill a truck} = \frac{\text{volume of the truck (lcy)}}{\text{volume of the loader (lcy)/cycle}}$$

Trucks are usually loaded by front-end loaders or excavators. Spotting to load or dump and having to wait is a delay time influenced by job conditions, work setup and management of the process as shown in Example 4.8.

**Example 4.8:** A Cat 320C excavator equipped with a 1.96 lcy heaped bucket is used to dig in sandy clay soil. It takes about 0.33 min per bucket load dumped into a fleet of Cat D30D articulated trucks. Each truck carries a heaped capacity of 21.6 lcy. It takes about 5 min to haul and dump the load, return, and position for reloading.

1. How long does it take to load one D30D? (Assume the bucket fill factor for the sandy clay is approximately 1.0.)

   (21.6 lcy/truck load)/(1.96 lcy/bucket load) = 11.02 cycles/truck load
   = round-off to 11 cycles/truck load
   = (11 cycles)(1.96 lcy/cycle)
   = 21.56 lcy/truck load
   (11 cycles) (0.33 min/cycle) = 3.63 min to load the truck

2. How much material can be hauled in 1 work hour by one D30D?

   Work hour production = [rated capacity (operational efficiency)]/cycle time
   Cycle time = load + (haul + dump + return)
   = 3.63 min + 5 min = 8.63 min/cycle
   Estimated hourly production = [(21.6 lcy) (50 min/hour)]/8.63 min/cycle
   = 125.1 lcy/hour/D30D

3. How many D30Ds will the 320C support?

   Number of trucks or haulers = hauler cycle time/hauler time at the load site (spotting and loading)

   $$\text{Number of D30D trucks supported by the 320C} = \frac{\text{(8.63 min/hauling cycle)}}{\text{(3.63 min/loading cycle)}}$$
   = 2.37 haulers

   Three haulers would dictate that one hauler will be waiting to be loaded most of the time. With two haulers, the excavator will be idle for a portion of every cycle. Based on the time available for the excavation, a larger excavator might be considered to better accommodate three D30Ds. When determining whether to round up or down the number of supporting equipment, it is best to keep the most expensive piece(s) of equipment working. It is more cost effective to allow the less expensive piece of equipment to be idle. Because of the small amount of excavation, use 2 D30D trucks to haul the 1,500 bcy of soil that must be excavated.

4. What is the 320Cs hourly production?

$$\text{Hourly production} = [(1.96 \text{ lcy/bucket})\ (50 \text{ min/hour})]/0.33 \text{ min/cycle}$$
$$= 297 \text{ lcy/hour}$$

It should be noted that with this scenario, the number of D30D's determines the hourly production, not the excavator.

Two trucks operating can load and haul about 125 lcy/truck (2 trucks) = 250 lcy/hour

Therefore, the excavator production will be limited to this amount.

5. How long will it take to complete the excavation?

Convert the 3, 500 bcy to lcy: 1, 500 bcy (1.15% swell) = 1, 725 lcy to be moved
1, 725 lcy/250 lcy/hour = 6.9 hours = 7 hours

6. What is the unit cost for excavating and hauling the material to another location?

O&O Hourly cost for a 320C working in moderate conditions = \$22/hour
O&O Hourly cost for a D30D working in moderate conditions = \$42/hour
Operators are paid \$23/hour including contractor outlay
(3 operators) (\$23/hr) = \$69/hour
Hourly O&O cost for this equipment package = \$22 + [2 trucks (\$42/hour)]
= \$106/hour

(\$106/hour + \$69/hour) 7 hours = \$1, 225 cost
\$1, 225/1, 500 bcy = \$0.817/bcy

### 4.4.2 Backhoes

Backhoes are probably the most common piece of construction equipment found on commercial construction projects. They come in many sizes and are ideal for light excavation, trenching, and material moving and loading. Backhoes can be used as a hoe or a loader and can accommodate many different accessories and attachments for different operations. One of the backhoe's greatest strengths is the many attachments that can be used to increase its versatility on a job site. Simple efficient systems are designed for easy connection of most attachments. Many times, if the contractor does not need the attachment full time, it can be rented as needed. Figure 4.17 shows the hoe part is located on the back of the machine (backhoe).

The operator drives and operates the loader bucket from the front seat and operates the hoe from the rear seat. Backhoes are designed to operate using outriggers for stability. Outriggers are spread on the digging end (excavator). The scooping bucket supports the front end. All four wheels are off the ground when digging. The backhoe is ideal for light underground utility construction. The hoe can be used for trenching and lifting like the excavator. The bucket can be used for hauling material and backfill. A large backhoes maximum digging depth is about 16′, loader bucket capacity is about 1.5cy, and the maximum lift capacity is over 4 tons.

Figure 4.18 shows JCB's backhoe demonstration at CONEXPO in Las Vegas in 2001. It is a choreographed show called the "Dancing Diggers." CONEXPO is the largest equipment show in the world. Most related construction heavy equipment is showcased there. Contractors and suppliers come from all over the world. The show involves acres of the latest construction equipment and associated technology. Most manufacturers showcase their latest equipment, from buckets to cranes, to engine parts to attachments of every kind.

FIGURE 4.17 Backhoe.

FIGURE 4.18 Dancing Diggers.

#### 4.4.2.1 Backhoe Production

When using the loader bucket to move material, production is figured in the same manner as a front-end loader. When using the excavator bucket, production is figured like an excavator. The cycle components are the same. Times may be slightly less because of the enhanced maneuverability and size difference of the backhoe compared to a larger loader or excavator. Backhoes are made for lighter work than typical loaders or excavators. They are purchased for their multi-use capacity. They need a level and stable work surface and enough area for proper outrigger placement.

### 4.3.3 Front Shovels

As stated earlier in this chapter, front shovels operate in much the same manner as front-end loaders. They are designed to dig above grade into the face of the excavation, not to scoop at ground level. These shovels typically operate on tracks for better traction when pushing the bucket into the face to be excavated. Front shovels rotate like an excavator. The work typically entails filling the bucket, rotating, and dumping the bucket contents into a pile or a truck. Front shovels are typically not very mobile and do not carry the load like a loader. The typical production cycle is like an excavator, and production is calculated in the same manner. Bucket sizes range from over 6 to 36 cy, and the vertical digging envelope can reach almost 50 ft. Some shovels are equipped with bottom-dump buckets to reduce wear on the machine and provide greater dumping and loading accuracy.

## 4.5 LIFTING CONSIDERATIONS

The Power Crane and Shovel Association (PCSA) is responsible for establishing many of the operating and lifting criteria for this heavy construction equipment. Published technical bulletins and other information are available through this organization [8]. An excellent resource for mobile crane information is the *Mobile Crane Manual* published by the Infrastructure Health and Safety Association of Ontario, Canada [9].

Manufacturers publish model-specific tables to be used for checking lifts. These tables are not interchangeable for different models of equipment. Crane failure can result from stability failure (proper setup) or from structural failure (components of the crane). It is extremely important that tables published for a specific model of equipment are used only for checking lifts by that piece of equipment. Both the stability and structural capacities must be verified. As the size of the equipment increases, typically, the lifting capacity, cost, and need for a competent and experienced operator do too.

### 4.5.1 Placing a Load

Figure 4.19 graphically shows the variables and forces when placing a load with a lattice boom crane. For most lifting problems, the same loading principles apply to all types of cranes regardless of the type, boom configuration, or size. Lifting variables include the load weight and shape, boom length, horizontal distance from the centerline of the crane to the placement point, angle of boom, and the lifting quadrant for picking and placing the load. Forces determined by these variable forces when a lift is made include the tipping moment and stabilizing moment. The physical location of instability related to the crane's body is called the tipping axis. This is also the center of gravity when the crane is loaded and swinging. The tipping axis location varies with the load and counterweight relationship. The crane is stable when the stabilizing moment exceeds the tipping moment.

It is necessary to know the rated lift or load capacity of equipment prior to making a lift. This is basically how much weight a crane can lift with a given length boom, which is setup a certain horizontal distance from the placement point, which in turn, creates a specific angle of the boom to the ground. For a vertical mast forklift, the load capacity is determined by how much the forklift can lift to a certain height. This type of forklift can only tilt the mast a very short distance, so they are assumed to be stable within this range. This information can be found in the manufacturer's lifting capacity tables and should be consulted prior to crane or forklift selection.

The tipping condition is the point of tipping for a particular crane and setup when the overturning moment of the load (the load is too heavy or too dynamic in the air) becomes greater than the stabilizing moment of the machine (machine weight and counterweights). The crane's leverage is greater than the load's leverage. The tipping load is the load that produces this condition. The crane operator must avoid this condition and consult load charts and weight tables to ensure a safe lift if necessary. Stability failure is caused by trying to lift too heavy of a load or extending the boom too far or at too much of an angle

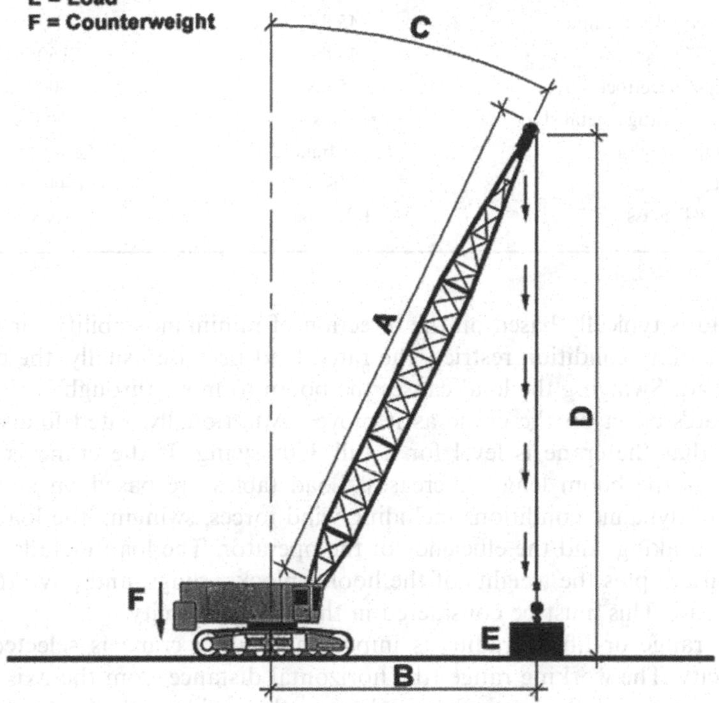


**FIGURE 4.19** Loading Forces.

with the load, causing the machine to tip over in the direction of the load or the direction the load takes it.

Table 4.7 lists common building materials found on a commercial construction project, weights of each piece, and if applicable pallet or bundle weights. The proper amount of counterweight and the balanced transference of the load weight to the ground via tracks, outriggers, or stabilizers provide the machine stability. It should be noted that the boom and crane components (pendants, attachments, and pulleys) have a limited amount of structural strength to accommodate the load. Loading past this amount might cause structural failure before the machine becomes unstable.

The sweep area is the total area that the crane can swing over. This area is divided into parts called quadrants determined with respect to the position of the boom. Load capacity varies depending on the quadrant position of the boom and load with respect to the machine's undercarriage. Typically, one quadrant of the four will sustain the greatest lifting capacity. Crawler crane quadrants are usually defined by the longitudinal centerline of the machine's crawlers – over the sides, over the drive end of the tracks, or over the idler end of the tracks; the ideal lifting position. Wheel-mounted crane quadrants are usually defined by the configuration of the outrigger locations – over the sides, over the rear, or over the front.

**TABLE 4.7**
**Common Material Weights**

| Material | Unit Weight (Each) | Pallet/Bundle Weight |
|---|---|---|
| #5 Rebar | 20 lbs | 2,000 lbs |
| CY 3,500 psi concrete | 3,800 lbs | NA |
| Light 8 × 16 concrete block | 28 lbs | 2,015 lbs |
| Heavy 8 × 16 concrete block | 32 lbs | 2,305 lbs |
| 3° × 7° metal door frame | 45 lbs | NA |
| 4 × 8 × 1/2″ sheetrock | 45 lbs | 3,150 lbs |
| 4 × 8 × 3/4″ sheetrock | 55 lbs | 3,300 lbs |
| 2 × 4 × 8′ 25-gauge metal stud | 2 lbs | 200 lbs |
| Composite shingles | 85 lbs/bundle | 2,040 lbs |
| Roof felt | 90 lbs/roll | 1,440 lbs |
| 8′ × 8′ × 3/4″ glass | 1,250 lbs | NA |

The rated load is typically based on the direction of minimum stability for the mounting. The minimum stability condition restricts the rated load because usually, the crane must lift and swing the load. Swinging the load causes the boom to move through various quadrants, changing the load's effect on the crane as it moves. Additionally, rated loads are based on the assumption that the crane is level for a full 360° swing. If the crane is not level, the effect is greater as the boom length increases. Load tables are based on static conditions. Cranes operate in dynamic conditions including wind forces, swinging the load, the hoisting speed, hoist-line braking, and the efficiency of the operator. The load includes the weight of the item being lifted, plus the weights of the hooks, blocks, slings, and any other items used in hoisting the load. This must be considered in the lifting capacity.

The working range or lifting radius is important once a crane is selected based upon rated load capacity. The working range (the horizontal distance from the axis of rotation of the crane to the center of the vertical hoist line or the tackle with the load applied) considers the boom length necessary to lift the load along with required rigging, the required height at a certain distance from the center of rotation. At a certain boom length, as the distance from center of rotation increases, the angle of the boom decreases. As this angle of the boom decreases, the lifting capacity also decreases (tipping moment increases). Because of this condition, the placement of the equipment prior to the lift is extremely important. The lifting equipment should be placed as close to the lifting and placement points as possible. By getting as close as possible, the lifting radius is reduced, and the lifting capacity increased. If this cannot be done, then perhaps a larger capacity crane may be needed, or if the lifting capacity is adequate, a longer boom may be required.

Lifting capacities should not exceed the following percentages of the tipping loads assuming the crane is properly setup.

- Crawler track – 75%
- Rubber tire mounted – 85%
- Machines on outriggers – 85%

### 4.5.2 The Operator

It is particularly important to have a qualified and competent crane operator when considering equipment selection and operation. Risks incurred during lifting are different from risks incurred during excavating or earthmoving. Lift failure often results in great damage, both to the workers and the built project. If an accident occurs, the operator and condition of the equipment will be the primary considerations of the accident investigation, especially if the lift is unique or the job conditions are difficult.

The operator is ultimately responsible for the coordination and execution of a lift. The operator is the final person to have a chance to determine whether a successful lift can be made. In most cases, the operator must be certified to operate the specific type of lifting equipment. The operator should do the following:

- Verify the equipment capacity
- Verify the rigging capacity
- Inspect the condition of the equipment
- Coordinate the lift
- Execute the lift

### 4.5.3 Mobilization and Setup

Mobilization is the process of transporting and preparing the equipment for use at the jobsite. Typically, forklifts and cranes are transported on a tractor–trailer unit to the project. More than one truck may be needed if the equipment must be disassembled into components to be legally transported. As the size of the lifting equipment increases, the time and cost to dismantle, load, obtain haul route permits, and reassemble the equipment increases. Special permits are often required to transport oversized equipment loads on public highways. Mobilization planning also includes providing any support equipment, such as a forklift or mobile crane necessary for assembly. These should be incorporated into both the schedule and budget. For rental equipment, transport and setup is usually included in the cost.

Before hoisting a load, the equipment should be leveled. If the crane is not level, the lifting capacity is reduced in the downhill direction. Setup should be on firm stable ground or timber mats to resist settling as the load is lifted and maneuvered. If the machine uses outriggers for stabilization, they should be fully extended and leveled. The load of the machine should be transferred to the ground through the outriggers only. Figure 4.20 shows the outrigger foot on a plywood pad. Typically, several layers of plywood or timbers are used to support the outriggers. Equipment should be setup as close to the placement point as possible, and clearance for the boom should be checked from every point. Any potential constraints on the free movement of the boom, like overhead utility lines, should be avoided. The travel surfaces must be stabilized and level if the crane must move with the boom extended. Many times timber pads must be placed in front of the crane by support equipment to provide a stable level travel surface during crane travel.

### 4.5.4 Booms

A lattice boom consists of a framework of structural steel pieces that form a box. It is cable suspended and acts as a compression member. The structure is lightweight to provide extra lifting capacity. This boom is usually transported in sections that are assembled at the site. Crawler and tower cranes typically have lattice booms. Most heavy lifting is done with lattice booms. Figure 4.21 is a typical lattice boom, and Figure 4.22 shows common boom ends available for Manitowoc cranes. Each is designed for a specific purpose.

Hammerhead booms are most typically used for heavy lifting, such as tilt-up concrete construction. Offset or tapered ends are usually found on lighter lifting cranes, such as steel erection or material stocking. Figure 4.23 shows that a telescoping boom works in the same manner as a retractable telescope. As lift height is needed, the boom is telescoped or extended.

FIGURE 4.20 Outrigger Setup.

FIGURE 4.21 Lattice Boom.

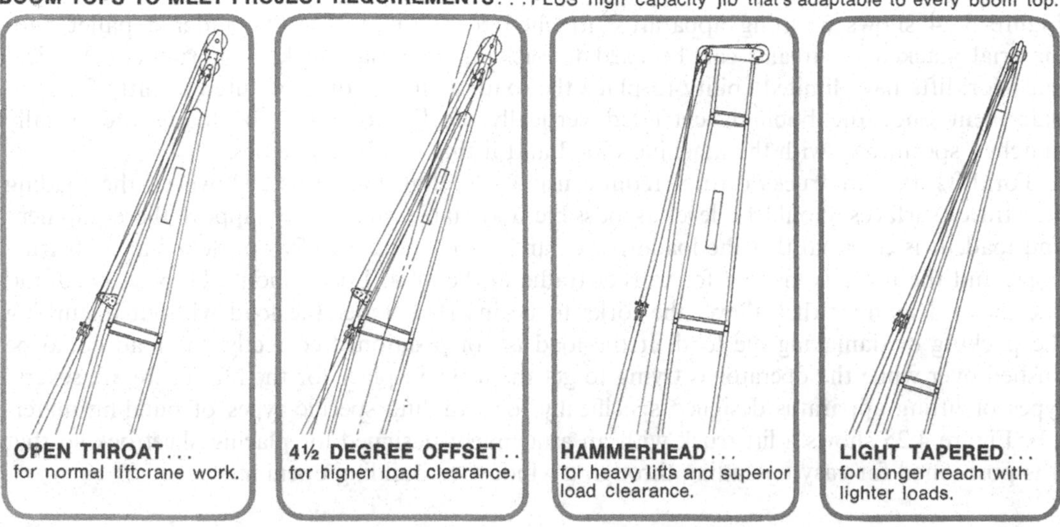


**FIGURE 4.22** Common Lattice Boom Tips.

**FIGURE 4.23** Telescoping Boom.

This boom acts as a bending member when lifting. The boom is ready for lifting when it arrives at the site. Mobile hydraulic cranes, sky track type lifters, and some personnel lifts also use telescoping booms. Moderate to medium lifting can be done with telescoping booms. The equipment manager should keep in mind that a less expensive lattice boom crane often has the same lifting capacity as a larger, more expensive telescoping boom crane.

### 4.5.5 Forks

Figure 4.24 shows a lifting apparatus attached to a boom that fits under a pallet with material stacked on it and can be used to pick up material stacked on runners. Vertical mast forklifts have limited ability to place the load, as it can only be tilted slightly for load placement once the boom is extended vertically. Forks are rated for loads and usually matched specifically with the machine's load and lifting height capacities.

Forklifts and lift trucks do not require usually leveling before use. However, the loading and travel surfaces should be level as possible and stable enough to support the equipment and load. It is essential that the loading area and travel route are clear of debris and obstructions, and the route is marked for worker traffic in the area. Loads should be positioned and packed in a manner that allows the forks to be inserted under the load without disturbing the packing or damaging the load. If the load is not positioned correctly, the load could be pushed over while the operator is trying to get the forks in place for the lift. There are several types of lifting apparatus designed specifically for handling specific types of building materials. Figure 4.25 shows a lift truck with an attachment designed for placing sheetrock so that it is positioned for easy unloading through a window or building opening.

### 4.5.6 Rigging

Rigging is the apparatus that is used to attach the load to the crane cable so that it can be picked up and moved. "Riggers" are a craftsperson with special skills and essential on a project where loads must be lifted above grade and placed. For a successful lift, the proper rigging is as important as the choice of a crane with adequate lifting capacity. The rigging must be rated for the same load as the crane. The weight of the rigging must be included in the total weight of the load. The following factors should be considered in field rigging operations:

- Inspect rigging visually prior to every use.
- Store the rigging in a manner that protects its structural integrity.
- Use qualified personnel to rig loads, especially loads to be transported over other workers or the structure.

**FIGURE 4.24** Forklift.

FIGURE 4.25 Sheetrock Attachment.

- Lift the load slowly a few feet off of the ground to tighten the rigging. Make sure the load is not going to shift once it is in the air.

Prior to a lift the following rules should be observed:

- Verify the total weight of the load.
- Determine the center of gravity of the load. Not rigging to the center of gravity can imbalance the crane when the load is lifted.
- Protect the load from the rigging with padding or other material if necessary.
- Attach tag lines if necessary while the load is on the ground. These are ropes attached to the load so it can be maneuvered from the ground at lift and maneuvered from the placement position while it is still in the air being lowered.

Once the load is rigged, lift the load a few inches and check the crane's stability and the rigging to make sure the load is stable. When the lift starts, swing the boom slowly and steadily. Avoid jerky starts and stops to avoid swinging the suspended load. Hoist the cable slowly and steadily as the lift is made to avoid swinging the load. Four lift factors for rigging that should be evaluated prior to a lift:

1. The size, weight, and center of gravity of the load.
2. The number of legs and the angle the sling makes with the horizontal line when attached to the load.
3. The rated capacity of the rigging.
4. The history of care and previous use of the rigging.

Typical rigging apparatus includes a length of material with connecting devices on the ends. This is called a sling. Types of common sling materials include:

- Alloy steel chains: preferable for lifting hot materials.
- Wire rope: economical, readily available, resistant to abrasion, and flexible. Defined by the lay or way in which the wire strands are woven together. Field lubrication is necessary to prevent rusting.
- Natural and synthetic fiber rope: inexpensive, pliant, grips the load well, typically does not mar the load surface. Not suitable for heavy loads.
- Synthetic webbing: typically does not mar the load surface, straps typically have end pieces designed for easy connection to the load.
- Nylon: used for loads in alkaline or greasy conditions; is resistant to chemicals.
- Dacron: used for loads with high concentrations of acid or high-temperature bleach.
- Polyester: used for loads with bleaching agents or when minimum stretching is required.

How a load is bundled or packaged will influence the type of sling used. Square edges on a load rigged with chain or wire as the sling should be protected to avoid damage as the load is lifted.

Many types of connectors are available. Hooks and shackles are very common and typically connected to wire rope or the crane cable for easy connection of load rigging. These are rated for capacity, just like all other lifting equipment.

Figure 4.26 shows a "spreader bar" attached to the hook at the end of the hoist line. Note the end of the hammerhead boom. The rigging attached to the tilt-up panel is attached to the spreader bar at the pulleys. As the crane lifting cable is retracted, it lifts the spreader bar and the panel. The wire rope rigging loop connected to the panel is run over the pulleys on the bottom of the bar. If necessary, the rigging can reposition over the pulley as the panel is lifted.

This rigging setup is typically used to lift concrete panels in tilt-up construction. Note the cables attached to the bottom of the metal bar. This "spreader" configuration is ideal for attaching the cables on the bottom of the bar to a concrete panel at

**FIGURE 4.26** Crane Using a Spreader Bar.

engineered points. Part of the design consideration for the panel is building enough strength into the panel so that it can be tilted up and moved to its proper location. By distributing or spreading the load symmetrically, there is less stress on the panel when it is hoisted. Not only must the integrity of the rigging be checked, but the integrity of the spreader bar or any apparatus used to attach to the load should be inspected prior to the lift too.

### 4.5.7 Jibs

Many cranes use jibs bolted to the end of the boom to increase the working range. Jibs come in many lengths and configurations and are assembled like booms. Jibs can be attached to both lattice and telescoping booms. The advantage of the jib is that it can be operated independently from the boom without having to increase the angle of the boom (lower the boom). The jib is like an extra boom that is hinged to the main boom. The same lift forces that are placed on the boom are placed on the jib. Instead of being transferred to the crane cab and counterweights, they are transferred to the boom end. The jib weight should be included when evaluating the boom capacity.

As shown in Figure 4.27, most jibs have a gantry, backstay, and forestay pendants like the main boom. These pendants raise or lower the jib as needed (jib offset). A secondary hoist cable runs through the end of the jib and has a ball and hook on the end like the main hoist cable. The same lifting variables and ultimate forces that apply to booms, apply to jibs. If using a jib, prior to the lift, the structural capacity of the jib should be verified using the manufacturer's load tables based on the specific crane to which it is attached and the setup. Jib lengths can reach over 220′. Length and configuration will vary with the capacity of the boom and size of the crane.

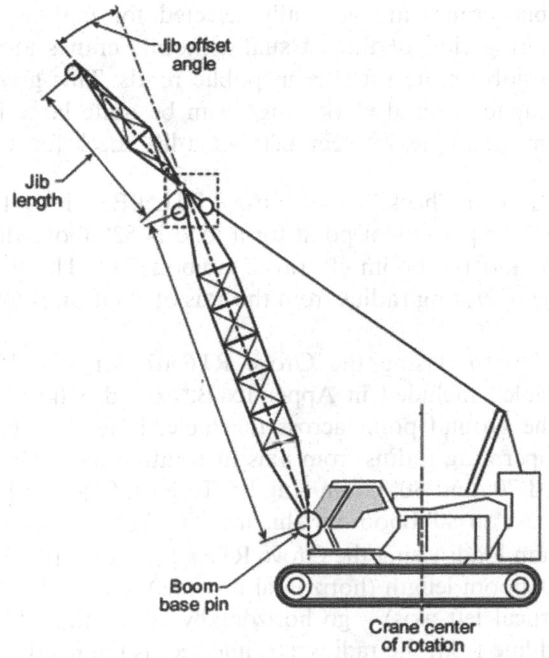


FIGURE 4.27 Common Jib.

### 4.5.8 Hoist Speed

The hoist speed is the velocity with which the hoist cable can be extended or retracted. The capacity of the crane's motor to turn the drum on which the hoist cable is coiled dictates this speed. Raising or lowering the cable with a load creates a dynamic condition that is compounded by the speed with which it is done. Many times the load is raised as the boom is swung or extended adding to the dynamic conditions. The shape and compactness of the load and wind conditions also influence how fast the cable can be reeled in or out.

For high production work activities where the placement height is high, fast hoist speed will decrease the production cycle time. The placement height distance influences the hoist time the same way the travel distance affects the travel time for a groundmoving machine. The manufacturer's specifications for the specific crane being used should be consulted for hoist speed capacities.

## 4.6 LIFTING EQUIPMENT SELECTION

Lifting equipment included in this discussion are:

- Hoisting: mobile, crawler, and tower cranes
- Material moving: forklifts, concrete pumps
- People moving: personnel lifts

### 4.6.1 Cranes

Cranes discussed in this section are considered typical and found on most construction jobs that require hoisting. There are other specialty types of cranes and many different combinations of booms and jibs that are not discussed.

#### 4.6.1.1 Telescoping Boom Mobile Cranes

Telescoping boom mobile cranes are generally selected for making one lift or a limited number of lifts in a short period of time. Usually, mobile cranes are rented. They can be driven assembled to the job or the lift site on public roads. This greatly reduces the setup time and cost. Lifting capacities and work ranges can be quite large if necessary, but most lifting is light to medium. Example 4.9 demonstrates a lift check for a typical crane.

**Example 4.9:** Conduct a "Lift Check" on the Grove RT650E series – 105 ft. Main Boom for the following condition. The placement point for a load is 52′ above the ground. The load is 2′ tall. The rigging is 6′ and the boom clearance is about 10′. The height from the ground should be about 70′. The operating radius from the axis of rotation is 50′.

1. Select the boom length: Using the Grove RT650E series – 105 ft. Main boom "working range table" included in Appendix B, extend a horizontal line from the 70′ height from the ground point across the table. Extend a vertical line from the 50′ point on the operating radius from axis of rotation axis. The two lines intersect between the curved 70′ and 80′ boom lengths. To be safe and provide enough working range use at least an 80′ boom for this lift. The boom angle is about 44°.
2. Determine maximum load: Using the Grove RT650E series – 105 ft. Main boom "load chart," find the 80′ boom length (horizontal top axis) – drop down this column. Find the 50′ radius (vertical left axis) – go horizontally across the table. Where the boom length column and line from the radius axis intersect is the maximum capacity for this setup, for this crane, with this boom. The maximum load = 14,200 lbs or 7.1 tons.

3. Note that this is the maximum capacity. If the weight of the load exceeds this amount, then recheck with a closer radius (move the crane closer so there is less boom angle). If getting closer is impossible, then use a larger capacity crane.

#### 4.6.1.2 Lattice Boom Crawler Cranes

Lattice boom crawler cranes are used on most types of construction projects. They are versatile and have many attachments available to perform a variety of tasks, including draglines and clamshells for excavation, pile drivers, dynamic compactors, demolition wrecking balls, augers for drilling holes, and magnets for moving metal objects. The following are a sampling of common boom configurations:

- A guy derrick crane uses a back boom as a derrick that can be anchored temporarily to other structures to counterweight the load as it is lifted and placed. The lifting cable comes from the back of the cab of the crane, over the derrick boom and then through the lifting boom to the load, thus transferring the compressive force of the load to the derrick. This crane can boost capacity 800% over a basic crawler crane.
- A crawler tower crane is less costly than a fixed tower crane. The main boom is vertical with a luffing boom attachment. The compressive load is transferred to the crane cab and counterweights down this vertical boom. Maximum boom and jib combination are approximately 480′.
- The sky horse configuration is similar to the guy derrick, except the back boom is shorter than the lifting boom. It is not temporarily secured during the lift. This crane can approximately triple the capacity of a standard crane.
- A ringer lift attachment at the base of a crawler crane is used for heavy lifting. The ring helps stabilize the crane to the lifting surface. This crane has a sky horse boom configuration with a luffing jib attachment. Note the great amount of counterweight attached for balance. The counterweight is supported on the structural ring. Ringer lift cranes are used to lift and swing extremely heavy loads.

Figure 4.28 is adapted from an illustration in the *Mobile Crane Manual* [9] published by the Infrastructure Health and Safety Association in Ontario, Canada. It shows the basic parts of a lattice boom crane.

Crawler crane can be unstable due to the moment created at the end of the boom by the load, forcing these cranes to move slowly. They also must travel on a level stable surface. When a stable travel surface is not available, the crane can be moved on wooden pallets placed to reduce the surface imperfections and to distribute the load. The equipment manager will determine how and where the crane travels to its lifting position when planning a lift. Additionally, the details of where it travels with its load must also be planned. Example 4.10 demonstrates a lift check for another typical crane.

**Example 4.10:** Conduct a "Lift Check" for a Manitowoc Model 777 with a No. 78 Main Boom on the following situation. The placement point for a load is 130′ above the ground. The load is 6′ tall. The rigging is 8′ and the boom clearance is about 16′. The height above ground should be about 160′. The distance from centerline of rotation is 100′.

1. Determine a minimum boom length: Using the Manitowoc Model 777 with No. 78 Main Boom "heavy-lift boom diagram" included in Appendix B, extend a horizontal line from the 160′ height above ground point across the table. Extend a vertical line from the 100′ point on the distance from centerline of rotation axis. The two lines

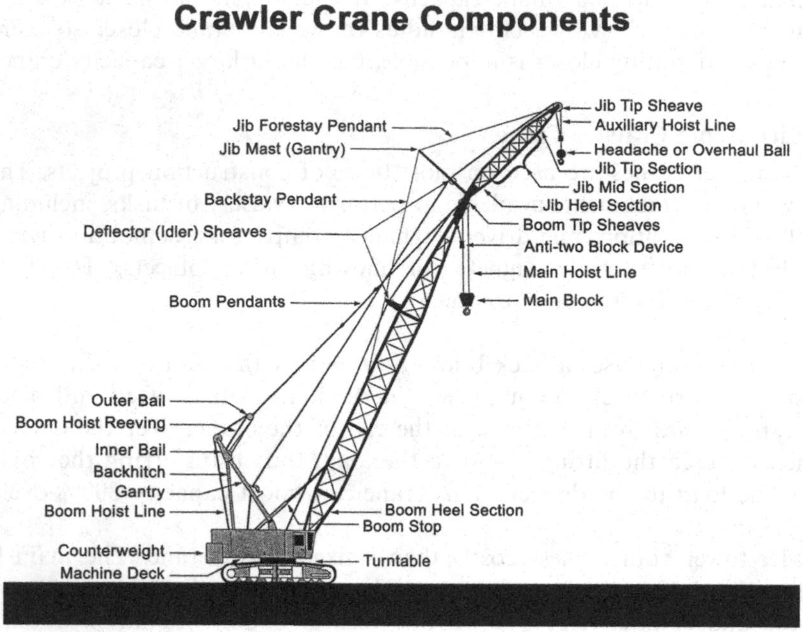

**FIGURE 4.28** Parts of a Lattice Boom Crane.

intersect almost right on the curved 180′ line. To be safe and provide enough working range, use at least a 190′ boom for this lift. The boom angle is about 57°.

2. Determine the lift capacity: Using the Manitowoc Model 777 with No. 78 Main Boom "heavy-lift load charts," find the 190′ boom length (horizontal top axis) – drop down this column. Find the 100′ radius (vertical left axis) – go horizontally across the table. Where the boom length column and line from the radius axis intersect is the maximum capacity for this set-up, for this crane, with this boom. The lift capacity = 27,600 lbs or 13.8 tons.
3. Note that this is the maximum capacity. If the weight of the load exceeds this amount, then recheck with a closer radius (move the crane closer so there is less boom angle). If getting closer is impossible, then use a larger capacity crane.

#### 4.6.1.3 Tower Cranes

Tower cranes are designed to work in congested areas. These cranes are a lifting device on top of a tower or mast. When a pick is made, the same lifting forces occur in the same manner as other cranes. Counterweight must be provided to balance the load. Compression is transferred down to the ground. The major difference, however, is that a tower is used to transfer the load to the ground instead of outriggers, tires or tracks. Lifting from up in the air can be both more demanding and complicated than lifting while sitting on the ground.

Manufacturers classify tower cranes as top slewing, bottom slewing, self-erecting, and special application. Slewing refers to where a crane turns about a fixed point. Cranes can be bottom slewing or top slewing. Only the boom rotates on top slewing cranes. The tower and boom rotate on bottom slewing cranes. The most common type of top slewing tower crane is the horizontal boom. This boom provides optimum coverage. These cranes use a trolley system that positions the hoist line and load by rolling on the bottom of the boom. The longer the boom, the greater the coverage; however, the lifting capacity decreases as the load on the trolley nears the end of the boom.

All tower cranes have the same basic parts:

- A power source
- A base that is fixed (concrete) or movable (rail tracks)
- A tower or mast
- A boom or jib combination
- A hoist cable and motor system
- A pendant cable system
- A gantry system (tower top or intermediate)
- A turntable mounting for the boom and operator's cab
- A counterweight system
- An operator's cab

Horizontal, luffing, and articulated luffing boom configurations are common for tower cranes.

Figure 4.29 shows a horizontal jib tower crane. The crane configuration has a forward jib and a rear jib with a trolley running on the forward jib controlling the hoisting cable.

Figure 4.30 shows an articulated jib tower crane. Note the pivot point between the first boom and the second boom or jib. Note the configuration of the pendant cables. Also, note the lifting cable is on a trolley like a horizontal jib crane; the bottom boom can be moved in and out at an angle – the top boom can be raised or lowered for load clearance as the trolley is manipulated (by repositioning the jib excess hook reach can be converted to

**FIGURE 4.29** Liebherr 390HC.

**FIGURE 4.30** Liebherr 112 HC-K.

added hook height). While this is happening, the crane can also be rotated 360° if necessary. These cranes are primarily used for tower construction or restricted jobsites; this is the "luffed" position. The articulated jib crane can also be operated in the horizontal position for maximum working range (like a horizontal jib crane).

Figure 4.31 shows Potain luffing boom tower crane. The advantage of using the luffing boom cranes is obvious when compared to using horizontal boom cranes. This setup is like a lattice boom crane on the ground, except the crane is on a tower. Luffing cranes are considered as special application tower cranes. These types of cranes are excellent for restricted working areas.

Tower cranes typically are placed in a concrete foundation, but can be mounted on the structure or on rails. A professional engineer must certify the foundation or any bracing or connections to an existing structure supporting the tower crane. If a service is erecting the crane, they will usually provide the base construction details. Since tower cranes typically do not move, the tower location and boom length determine the covered area. Typically, an electrical power source must be run to the crane base and up the tower. Provisions for the security of the electrical service must be made.

Towers have a maximum freestanding height. Counterweight elements must be accurately weighed, and the weight clearly and durably marked on each element and entered in the equipment record system, or on the erector's checklist that must then be available at the workplace. Erection, climbing, or self-erecting and dismantling must be part of the tower crane strategy. The tower crane may be attached temporarily to the structure, decreasing the moment at the top of the tower when a load is hoisted. This gives the crane greater stability and the tower can be built higher.

**FIGURE 4.31** Liebherr 500 HC-L.

An access ladder is fixed in position on the mast of the tower crane. Each tower crane jib must have a continuous catwalk with a handline from the mast to the tip. An anemometer to measure wind speed is mounted on the crown, apex, or operator's cab of each tower crane. A mobile crane is used to assemble the jib, machinery section, and counterweights.

Erecting a tower crane is typically a substantial construction process that requires other equipment such as forklifts, crawler cranes, or, most likely, mobile cranes. Pieces are typically hauled to the jobsite and then for erection. If the crane base is not part of a structure, a separate base must be constructed prior to erection. One of the considerations in the tower crane strategy is final disposition of the crane base after the crane is dismantled.

The maximum unsupported jib height is 265 ft. The crane can have a total height much greater than 265 ft if it is tied into the building as the building rises around the crane. The crane can lift the most weight when the loads are closest to the tower. The weight decreases as the load moves farther from the tower.

The area of coverage based on the lift capacity must be detailed for the whole site by reviewing the site plan. Sometimes multiple towers are necessary to get adequate coverage. Using tower cranes is typically safer and more efficient than having to move crawler cranes multiple times. There is an economy for the combination of crane capacity, coverage, and the number of cranes. Crane requirements for heavy one-time lifts are somewhat different than requirements for everyday lifting performed by most tower cranes. For example, it is usually more economical to have a tower crane that can lift 90% of the loads on hand every day for an extended period and pay a rental fee for a larger capacity crane to make the lifts that exceed the tower crane's capacity than to install a tower crane that can lift 100% of the required loads.

Most project lifting strategies use a combination of crane types. As a result, a site plan must be developed to select locations, capacities, heights, and numbers of tower cranes. Final locations must also accommodate material delivery, staging, and rigging. Example 4.11 illustrates a lift check for a third crane.

**Example 4.11:** Conduct a "Lift Check" a Potain MD 485B-M20 for the following situation. The placement point for a load is 130′ above the ground. The load is 6′ tall. The rigging is 8′ and the boom clearance is about 16′. The height above ground should be about 160′. The hook radius is 100′.

1. Select the jib: Using the Potain MD 485B-M20 – maximum capacity = 44,092 lbs "working range diagram" included in Appendix B to determine the required jib/boom length for the lift. It should be noted that this P485B has a maximum free-standing hook height (HH) of 286′1″. The hook height or tower can be adjusted based on the highest placement point to be serviced. Obviously, there is no need to have more than an adequate amount of tower as the higher the tower the more moment created by the lift. The 286′1″ HH is more than enough to reach the 160′ height above ground. The jib selected is the 246′1″ long L75.
2. Determine the lift capacity: Using the Potain MD 485B-M20 – maximum capacity = 44,092lb "rated load chart" SM-DM Trolley included in Appendix B to in pounds for this crane. Find the 246′1″ jib length (horizontal top axis) – drop down this column. Find the 100′ hook radius (vertical left axis) – go horizontally across the table. Where the jib length column and line from the hook radius axis intersect is the maximum capacity for this set-up, for this crane, with this jib. The lift capacity = 25,836 lbs or 12.9 tons.

### 4.6.2 Forklifts

Forklifts are ideal for loading/unloading delivery trucks and moving material around the job site on the ground. They are used for placing material on the structure or on scaffolds up to three stories. Above that height, a crane is typically required. Instead of securing the load to a hoist line using rigging, the load rests on forks maneuvered securely under the load on the ground. Paths should be made and controlled around the jobsite for material movement by forklifts. There are two basic types of forklifts. One has a vertical mast, and the other has a telescoping boom. Both types are rated for capacity and the manufacturers specifications should be consulted prior to use.

For both types of forklifts, the basic lifting considerations discussed previously must be determined prior to selection and use. Load weight, lift height, distance, and setup location must match the forklift's structural capacity and stability. As with other lifting equipment, as size increases, so does lifting capacity. For telescoping boom forklifts, as size increases so does the vertical reach and horizontal operating range.

Vertical mast forklifts require a relatively stable and flat lifting surface. Basically, the boom goes straight up with little tilting ability. Because of its limited tilting range, the forklift must be able to get close to the structure on which the load is to be placed. The forklift positions perpendicular to the building after transporting the load to the placement location. The load is raised with the mast tilted back to the cab. When the load is high enough to be set down, the mast is tilted forward or the forklift is pulled closer to the placement surface and the mast tilted forward. When positioned, the load is lowered and placed. Typically wood runners or shims are placed under the load so the forks can be removed when the load is secure. Once the load is clear, the mast is tilted back to the cab and the forklift backs up while continuing to lower the mast. The cage over the operator should be adequate to deflect a load should it shift or fall.

Telescoping boom forklifts are typically designed for more irregular terrain and greater lifting distances. The load placement process is very similar to a telescoping boom crane. The boom can extend horizontally while lifting. The fork apparatus can tilt and push forward when in placing position. A major advantage is that this lift does not have to be up next to the structure in order to place the load. Some models of telescoping boom forklifts have outriggers or stabilizers.

Forklift production is determined by how long it takes to secure, transport, and unload the load at the placement location. This is influenced by the distance and speed of transport. The speed of transport is influenced by congestion, the travel surface, and the amount of maneuvering that is required.

### 4.6.3 Personnel Lifts

Personnel lifts play a supporting role and are ideal for hoisting workers, tools, and equipment into position to secure structural components or materials in locations unreachable by a ladder or other means. They are typically tired and self-propelled and require a stable operating surface. Personnel lifts come in many operation types and lifting and height capacities. Most have the operation controls on the work platform so height and platform location can be adjusted from the work platform while in use. There are two basic types of personnel lifts found on most construction jobs. One has a scissor type lifting mechanism and a work platform, and the other has a telescoping or retractable boom and a work platform. The same lifting limitations that apply to cranes apply to thesc lifts as well. A tipping condition will result with too much weight, whether people, materials or tools. Lift users should be aware of these operating restrictions prior to use. Both types are rated for capacity, and the manufacturers specifications should be consulted prior to use.

Figure 4.32 shows two scissor lifts and a telescoping boom lift working together. The name scissor lift comes from the configuration of the lifting apparatus. Note the small

**FIGURE 4.32** Scissor and Telescoping Boom Personnel Lifts.

diameter solid rubber tires. This lift is typically used for interior work on a flat stable work surface. They are ideal for MEP rough-in and trim out and sheetrock, ceiling, or soffit installation. Scissor lifts can reach up to 60′ and accommodate a load up to a ton of load. Larger work platforms require larger and more powerful base units. As the required work height increases, the less stable the scissor lift becomes. This is their greatest limitation.

The lifting mechanism for telescoping/retractable boom personnel-lift works the same way as a telescoping boom crane. Work platforms used on these types of lifts are typically smaller than those that can be used on scissor lifts; however, the reach can be much greater. Telescoping boom personnel lifts can reach up to 150′. The work platform swivels and levels as the boom positions. These lifts are ideal as support equipment for securing structural steel components together and exterior caulking and cleaning.

## REFERENCES

[1] Burch, D. (1997). *Estimating Excavation*. New York: Craftsman Book Company.

[2] Schaufelberger, J.E. (1999). *Construction Equipment Management*. Upper Saddle River: Prentice-Hall, Inc.

[3] Caterpillar Inc. (2017). *Caterpillar Performance Handbook*. 47th ed. Peoria: Caterpillar Inc.

[4] Capachi, N. (1987). *Excavation and Grading Handbook*. New York: Craftsman Book Company.

[5] Brown, D.C. (2005). Graders Shift into High Gear. *Grading & Excavation Contractor*, 7(1), pp. 24–32.

[6] Peurifoy, R.L., Schexnayder, C.J., Schmitt, R. and Shapira, A. (2018). *Construction Planning, Equipment and Methods*. 9th ed. New York: McGraw Hill.

[7] Gransberg, D.D. (1996). Optimizing Haul Unit Size and Number Based on Loading Facility Characteristics. *Journal of Construction Engineering and Management*, 122(3), pp. 248–253.

[8] Power Crane and Shovel Association (PCSA). (2019). Milwaukee, Wisconsin. www.aem.org/groups/product-specific/power-crane-shovel-association-pcsa/. (Accessed November 27, 2019).

[9] Infrastructure Health and Safety Association. (2019). Mobile Crane Manual. Etobicoke, Ontario, Canada. www.ihsa.ca/products/MC001. (Accessed November 27, 2019).

## CHAPTER PROBLEMS

1. Calculate the tractive effort generated by a 92,000 lb loaded scraper traveling on a maintained dirt haul route where the tires sink about 2″ into the travel surface.
2. Calculate the total resistance if the scraper in the previous example must haul up a 3% grade on part of the haul route.
3. A bulldozer with a 13′ wide blade is to be used to excavate and push fairly loose material. According to the soils report the material to be moved has a 16% swell factor. When the dozer arrives at the jobsite, a series of test blade loads are excavated to estimate a typical load. The average $H = 5'$, the average load width in front of the blade = 6.7′, and the load length = 10′. The load time suggested by the manufacturer is 0.10 min. Once the blade goes through the cut, the haul push is 100′ with an average speed of 2.5 mph. Backtrack distance is 120′, and the dozer will travel at a speed of about 3.5 mph. Once back to the hole, the dozer takes about 0.07 min to reposition. The ownership and operating (O&O) cost is $70/hour. The operator costs $32/hour including fringes and benefits. Answer the following questions:
   a. How much material (lcy) can be moved in one production cycle?
   b. How much material (bcy) can be moved in one production cycle? The quantity take-off is in bcy so the lcy load must be converted to bcy.
   c. What is the cycle time for one production cycle?

d. What is the work hour productivity if the operator works 48 min per 60-min hour?
e. How long will it take to move the 2,500 bcy?
f. How much will it cost?

4. A wheeled front-end loader has a 1.5 lcy bucket and a 0.2 min cycle time is to be used to load a dump truck with a heaped capacity of 16.0 lcy. It takes the truck 1.0 min to dump its load. Answer the following questions:
   a. How long does it take to load the truck?
   b. How much material can be hauled in a working hour?
   c. How many trucks should be assigned to this equipment package?
5. Conduct a "Lift Check" on the Grove RT650E series– 105 ft. Main Boom for the following situation. The placement point for a load is 130′ above the ground. The load is 6′ tall. The rigging is 8′ and the boom clearance is about 16′. The height above ground should be about 160′. The distance from centerline of rotation is 100′.
   a. Determine a minimum boom length
   b. Determine the lift capacity

# 5 Estimating Construction Equipment Productivity

## 5.0 INTRODUCTION

The 21st century is an era of accelerating technology, which has seen and will see technological advancements in the development of larger, faster, and more productive construction machinery. Increased machine productivity has resulted in an increase in overall project size. These factors have combined to produce a capital-intensive environment in which the equipment fleet managers must operate. The risk they must bear is further increased by volatile economic trends. As a result, the construction industry has been forced to search for methods to reduce the high level of risk. Historically, the lowest cost method for reducing risk has been to provide detailed estimating and planning prior to undertaking an equipment-intensive project and solid management throughout the course of the project. Estimating and planning involves the judicious selection of equipment, the careful scheduling of time and resources, and the accurate determination of expected system productivity and cost. Management involves putting the plan into action. The key management ingredient is having predetermined standards by which actual system outputs can be measured and upon which future decisions can be based.

Even a seemingly straightforward operation such as earthmoving is a highly dynamic system. A hauling operation contains several components that interact in a very complex manner. Analytical methods, based on engineering fundamentals, have been developed to solve the problem of bringing these components together in a logical manner. These methods mathematically model hauling systems. Their solutions are numerical results that may be used in the decision-making process of estimating, planning, and managing an equipment-intensive project.

## 5.1 BACKGROUND

Early methods made the somewhat naive assumption that optimizing productivity based on physical constraints of the environment would in turn minimize the overall production cost. Therefore, no effort was made to include cost or profit variables in those mathematical models. The early models developed by Gates and Scarpa [1] were the first to recognize the importance of the cost function in overall system optimization. Many methods currently in use do not adequately model physical conditions, depending instead on the judgment and experience of the user.

## 5.2 THE PEURIFOY METHOD OF OPTIMIZING PRODUCTIVITY

The first author to propose a method to optimize the productivity of construction equipment systems was Peurifoy [2]. His method involves determining all the physical constraints on the hauling system and evaluating them to determine the system's ultimate performance. The constraints are as follows:

1. Haul road rolling resistance: The haul road is broken down to segments of like road materials (i.e., asphalt, rutted earth, etc.), and rolling resistance in pounds per

ton is assigned to each segment. These assigned values are then used as part of an equation to determine the maximum velocities of haul units.
2. Haul road grades: The route is evaluated to determine the grade of the haul road for use in the velocity calculation.
3. Haul unit horsepower: This value is used to determine the maximum amount of rimpull, which can be developed by the haul unit. It is then used to determine the maximum velocity attainable by the haul unit in a loaded and unloaded condition.
4. Haul unit loaded and empty weight: The weights are used to determine first whether sufficient power is available to move the vehicle and secondly in the velocity calculation.
5. Haul unit transmission characteristics: These characteristics are used to determine the amount of time required to accelerate to top speed.
6. Haul unit loading time: The loading time is necessary to determine both cycle time and the optimum number of haul units.
7. Haul unit travel time: Travel time is one of the parts of cycle time.
8. Haul unit delay time: This consists of all times encountered in the cycle time except travel and loading time.
9. Altitude of project site: Altitude affects engine performance and thereby alters the engine's ability to produce rimpull.

### 5.2.1 Rimpull

The first concept that must be understood is rimpull. Rimpull is defined as the tractive force between the driving wheels and the surface on which they travel. If the coefficient of traction is high enough that the tires do not slip, maximum rimpull is a function of the power of the engine and the gear ratio between the engine and the driving wheels. The following equation can be used to determine maximum rimpull.

$$\mathrm{RP} = \frac{375(\mathrm{HP})(e)}{V} \qquad (5.1)$$

where:

RP = maximum rimpull (lbs)
HP = horsepower of engine
$e$ = efficiency of engine (decimals)
$V$ = velocity (miles per hour, mph)

The rimpull required to overcome grade and rolling resistance is given by the following formula.

$$\mathrm{RP_R} = W(\mathrm{RR} + 20(+S) \qquad (5.2)$$

where:

$\mathrm{RP_R}$ = rimpull required (lbs)
$W$ = weight of vehicle (tons)
RR = rolling resistance (lbs/ton)
$S$ = slope of grade (%)

The difference between the maximum rimpull and the required rimpull equals the amount of force available to accelerate the vehicle to top speed. The acceleration in miles per hour per minute is as follows:

$$a = \frac{0.66(\mathrm{RP_a})}{W} \tag{5.3}$$

where:

$a$ = acceleration (mph/min)

$\mathrm{RP_a}$ = available rimpull (i.e. $\mathrm{RP_a} = \mathrm{RP} - \mathrm{RP_R}$)

Thus, if the maximum speeds in each gear are known, the time to accelerate to top speed and that speed can be determined. Example 5.1 demonstrates how the above equations are used.

**Example 5.1:** If a truck with a 150 horsepower engine with an efficiency of 0.81*, weighs 38,000 pounds fully loaded and has maximum speeds of 3.0, 5.2, 9.2, 16.8, and 27.7 mph in first through fifth gears, respectively, the top speed and time to reach that speed on a level road with a rolling resistance of 60 lbs/ton can be found as follows.

*Note: $e$ = 0.81 is an average value for engine efficiency. This can vary from 0.60 when a truck is cruising empty in high gear to 0.92 when a loaded truck is climbing a grade in low gear.

Subtracting Equation 5.2 from Equation 5.1 yields:

In 1st gear:

$$\begin{aligned}\mathrm{RP}_a &= \frac{375(\mathrm{HP})(e)}{V} - W(\mathrm{RR} + 20(\pm S))\\ &= \frac{375(150)(0.81)}{3.0} - \frac{38000(60+20(0))}{2000}\\ &= 15,187.5 - 1,140 = 14,047.5\ \mathrm{lbs}\end{aligned}$$

$$\text{Maximum available rimpull per ton} = \frac{\mathrm{RP_a}}{W} = \frac{14,047.5}{(38,000/2000)} = 739.34\ \mathrm{lbs/ton}$$

As the maximum is often not reached due to lack of driver courage and mechanical losses in the gears, this value is reduced to 300 lbs/ton, the maximum achievable value cited by Peurifoy [2].

Then from Equation 5.3:

$a$ = 0.66 (300) = 198 mph/min

And the time to accelerate from 0 to 3 mph will equal:

$$\text{time} = \frac{3.0}{198} = 0.015\ \text{minutes}$$

The same set of calculations is then made for each of the five gears keeping the 300 lbs/ton as the maximum in mind. The results are as follows:

Acceleration time 2nd gear = 0.011 minutes

Acceleration time 3rd gear = 0.030 minutes

Acceleration time 4th gear = 0.139 minutes

Acceleration time 5th gear = 0.622 minutes

Assuming 4 seconds per gear change, the total time to accelerate to a top speed of 27.7 mph equals 1.069 minutes.

### 5.2.2 Cycle Time and Optimum Number of Units

At this point, the foregoing calculations must be made for the empty weight of the truck so that the time for the return trip may also be determined. Once this is done, the top speeds found are used to determine the travel times. The time to load and discharge material must also be estimated. The total cycle time can now be calculated.

$$C = L + T + D \tag{5.4}$$

where: $C$ = cycle time (minutes)
$L$ = loading time (minutes)
$T$ = travel time (minutes)
$D$ = discharge time plus time for other delays for turns, maneuvering, acceleration, etc. (minutes)

The optimum number of haul units ($N$) is determined as follows:

$$N = C/L \tag{5.5}$$

The productivity can be estimated with the following equation:

$$P = \frac{60N(S_H)}{C} \tag{5.6}$$

where: $P$ = productivity (tons or cubic yards per hour)
$S_H$ = size of hauling unit (tons or cubic yards)
60 = conversion factor from minutes to hours

Example 5.2 illustrates how to compute productivity in this fashion.

**Example 5.2:** An 18 cubic yard dump truck has a loading time equals 3 minutes, a travel time of 7 minutes, and the dumping and delay times equal 5 minutes. Calculate the cycle time, optimum number of hauling units and productivity.

$C = 3 + 7 + 5 = 15$ minutes
Therefore: $N = 15/3 = 5$ units

and $P = \frac{60(5)(18)}{15} = 360$ cubic yards per hour

Peurifoy's techniques allow the engineer to relate the hauling process to engineering fundamentals and make estimates of system productivity based on these fundamentals. The primary weakness with this model is that it does not include cost factors, and it causes the estimator to base estimates on instantaneous production rather than sustained production. Instantaneous production is the maximum theoretical production achievable at any given instant. Sustained production is the average realistic production achievable throughout the course of the project that considers hard to quantify factors for human frailty, equipment reliability, and environmental instability. Peurifoy's calculations tend to become rather complex and have provided a basis upon which subsequent authors have expanded.

## 5.3 THE PHELPS' METHOD

Phelps [3] takes Peurifoy's method and carries it a step farther by introducing a factor of realism to the computations. This method strives to estimate the production that can realistically be achieved in a given period of time. Phelps defines this as sustained production. To do this, the amount of time that is wasted due to human weakness and imperfect management is apportioned to each cycle. In industry, equipment managers sometimes try to compensate for the human factor by using a 45- to 50-minute productive hour. This does not give an accurate estimate because operations that have long cycle times allow less opportunity to waste time than ones with short cycle times. For example, the average truck driver is more likely to waste time using the restroom or getting a drink when being loaded or waiting to be loaded than when engaged in the haul, dump, and return portion of the cycle. Thus, on projects with longer haul distances, the total amount of time wasted is less than on shorter hauls because a greater portion of the cycle time is spent actively engaged in operating the vehicle.

### 5.3.1 Fixed Time

Phelps breaks the cycle into three parts: fixed time, variable time, and loading time. The fixed time consists of the delays that are built into the system due to mechanical constraints and the human factor. These include times to accelerate, decelerate, turn, dump, and waste (i.e., nonproductive times). The acceleration and deceleration can be estimated by using empirical values [3].

$$\text{Total acceleration time} = 0.3x + 0.2y \tag{5.7}$$

$$\text{Total deceleration time} = 0.02(x + y) \tag{5.8}$$

where:

$x$ = number of accelerations while loaded
$y$ = number of accelerations while empty

The total fixed time ($F$) can be estimated by using the following empirical values shown in Table 5.1, which were established from actual project information.

### 5.3.2 Variable Time

The loading time ($L$) is estimated from the production characteristics of the loader given by the manufacturer. The variable time ($V$) is calculated using the following equations:

$$V_H = \frac{375(\text{HP})(e)}{W_F(\text{RR} + 20(\pm S))} \tag{5.9}$$

**TABLE 5.1**
**Phelps Method Fixed Time (*F*) Values When Loading Time (*L*) Is Given [3]**

| Haul Type | Distance (ft) | Fixed Time Formula |
|---|---|---|
| Short haul | 200′–1,200′ | $F$ = 4.5 minutes + $L$ |
| Medium haul | 1,200′–5,000′ | $F$ = 4.0 minutes + $L$ |
| Long haul | 5,000′–9,600′ | $F$ = 3.5 minutes + $L$ |

*Note*: These values contain one acceleration and deceleration for each haul and return trip. Therefore, if intermediate stops occur, this value should be increased appropriately.

$$V_R = \frac{375(\text{HP})(e)}{W_E(\text{RR} + 20(\pm S))} \tag{5.10}$$

$$V = \frac{60d}{V_H} + \frac{60d}{V_R} \tag{5.11}$$

where: $V_H$ = velocity of haul direction (i.e., while loaded) (mph)
$V_R$ = velocity of return direction (i.e., while empty) (mph)
$W_F$ = weight fully loaded (tons)
$W_E$ = weight empty (tons)
$V$ = variable time (minutes)
$d$ = haul distance (miles)

### 5.3.3 Instantaneous and Sustained Cycle Times

With the above information, the instantaneous cycle time ($C$) can be calculated.

$$C = F_i + V + L \tag{5.12}$$

where: $F_i$ = instantaneous fixed time (minutes) [i.e., the sum of all fixed time components except waste time ($W$)] or

$$F_i = F - W \tag{5.12a}$$

The number of units can be calculated using Equation 5.5. The total wasted time for the entire project is estimated and apportioned to each cycle to determine the waste time per cycle ($W$). With this, the sustained cycle time ($C_s$) is calculated. The sustained productivity ($P_s$) can also be computed. Example 5.3 illustrates this process.

$$C_s = C + W \tag{5.13}$$

$$P_s = \frac{60(N)(S_H)(H)}{C_s}$$

$$P_s = \frac{60(N)(S_H)(H)}{C_s} \tag{5.14}$$

where: $S_H$ = capacity of haul unit (tons or cubic yards)
$H$ = shift length (hours)

**Example 5.3:** Given a haul length of 1,300 ft, a loading time ($L$) of 3.0 minutes, a variable time ($V$) of 4.0 minutes, compute the sustained cycle time, the optimum number of hauling units ($N$), and sustained production rate. The hauler has a capacity of 20 bank cubic yards. The shift is 8 hours long, and waste time ($W$) is 2.0 minutes per cycle.

Using the empirical estimates for fixed time ($F$) shown in Table 5.1,
$F$ = 4.0 + $L$ = 4.0 + 3 = 7.0 minutes
and from Equation 5.12a
$F_i$ = 7.0–2.0 = 5.0 minutes
From Equations 5.12 and 5.5, respectively,
$C$ = 5.0 + 4.0 + 3.0 = 12 minutes

and

$$N = \frac{12.0}{3.0} = 4 \text{ units}$$

From Equation 5.13:

$C_s = 12 + 2 = 14$ minutes

From Equation 5.14:

$$P_s = \frac{60(4)(20)(8)}{14} = 2743 \text{ cubic yards per shift}$$

It should be noted that Phelps does not fix the physical constraints, which can be varied. For instance, the poor haul road maintenance can cause the rolling resistance to markedly increase, which decreases the achievable speeds. This causes an increase in the variable times of the hauling units. As any component of the sustained cycle time increases, the optimum number of hauling units' changes, and the system's ability to maintain the calculated sustained productivity begins to fail. Therefore, the use of this method should include an analysis of changing physical constraints to determine the most economical situation. Thus, the maximum achievable production can be determined in context with the appropriate equipment mix, inherent physical conditions, and ancillary requirements such as haul road maintenance. The result is a fully optimized system within the physical constraints of the project environment.

## 5.4 OPTIMIZING THE HAULING SYSTEM BASED ON LOADING FACILITY CHARACTERISTICS

Arriving at an optimum equipment fleet for a given hauling task necessarily involves relating the two major types of equipment in the system to one another. This can be done by mathematically characterizing the operational characteristics of the loading facility to the mathematical description of the hauling unit using a load growth curve.

### 5.4.1 Load Growth Curve Construction

An earthmoving system's productivity is limited by the production of the loading facility. In other words, regardless of the size, number, and speeds of the hauling units, the ability of the loading facility to load the haul units will determine the maximum productivity of the system. As a result, the loading facility, characteristics must be carefully considered in the planning and subsequent steps of a hauling operation. Most models do include some function describing the loading facility, such as loading time or loader productivity. Generally, loading time is derived by dividing the haul unit capacity by the equipment manufacturer's figure for loader productivity. This does not consider the fact that the size of the haul unit may not be an even multiple of the loader bucket capacity. For example, if a front loader with a 1.5 cubic yard bucket is loading a 10.0 cubic yard dump truck, it would require 6.67 buckets to fill the truck. As it takes virtually the same amount of time for a loader to load two-thirds of a bucket as it does to load a full bucket, the theoretical productivity is not achieved. Additionally, legal haul restrictions and material weight must play a part in the selection of an optimum mix of loader and hauling unit. Therefore, improvements to existing methods must be made to more adequately consider the characteristics of a loading facility.

The *Caterpillar Performance Handbook* [4] contains load growth curves for bottom-loaded earthmovers. Experience with this management tool in the field has shown it to be valuable in modeling actual occurrences. The same concept can be applied to top-loading operations. To construct a load growth curve, the unit of hauler capacity is plotted against the loading time. A given loading facility's loading cycle must first be separated into its various elements. These elements are then divided into productive and nonproductive categories. The physical act of placing material into a haul unit is considered productive. Other elements such as filling the bucket, maneuvering, and movement are considered nonproductive in this application. Productive elements are plotted as sloping vertical deflections, and nonproductive elements are plotted as horizontal displacements.

**Example 5.4:** A front loader with a 1.5 bank cubic yard bucket has the following cycle elements:

| | |
|---|---|
| Move to stockpile | 0.05 minutes |
| Fill bucket | 0.10 minutes |
| Move to truck and maneuver to load | 0.15 minutes |
| Dump loaded bucket | 0.10 minutes |
| Total cycle time | 0.40 minutes |

The constructed load growth curve is shown in Figure 5.1. Note that there are a total of 0.3 minutes of nonproductive time and 0.1 minutes of productive time.

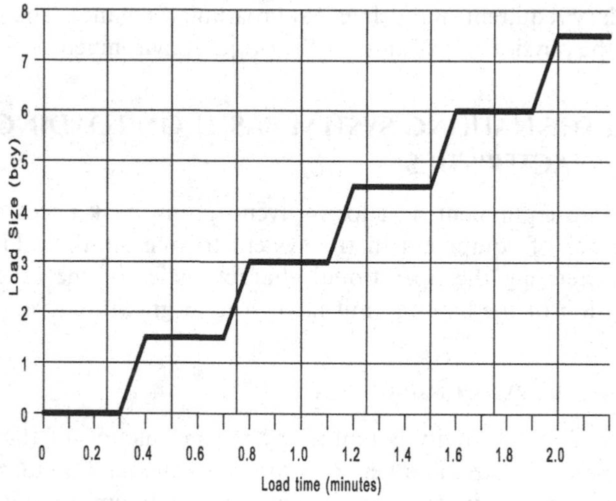


**FIGURE 5.1** Load Growth Curve for Bucket Loader.

### 5.4.2 Belt Conveyor Load Growth Curve

The same theory can be applied to all types of loading facilities. It should be noted that the load growth curve for a belt conveyor is parabolic until it reaches its top operating speed, where it then becomes a straight line. Thus, it has two elements of cycle time: accelerate to operating speed and operate at that speed until the haul unit is full. Both of these elements are productive. This can be simplified as a straight line by decreasing the slope of the steady state line to compensate for the initial acceleration time. The next set of examples illustrates the construction of load growth curves for a belt conveyor and a discharge hopper.

**Example 5.5:** A belt conveyor has a theoretical productivity of 2,000 tons per hour. The time to accelerate to operating speed is 0.1 minute. Construct a simplified load growth curve for this machine.

$$\text{Steady State Slope} = \frac{2000\ \text{TPH}}{60\ \text{min}/\text{hr}} = 33.33\ \text{tons/minute}$$

Assume average loading duration = 3.0 minutes
Therefore: percent slope reduction = 0.1/3.0 = 0.03 or 3.0%
Thus the slope for design purposes = (1.0 − 0.03) (33.33) = 32.33 tons/minute

Figure 5.2 is the load growth curve for the belt conveyor.

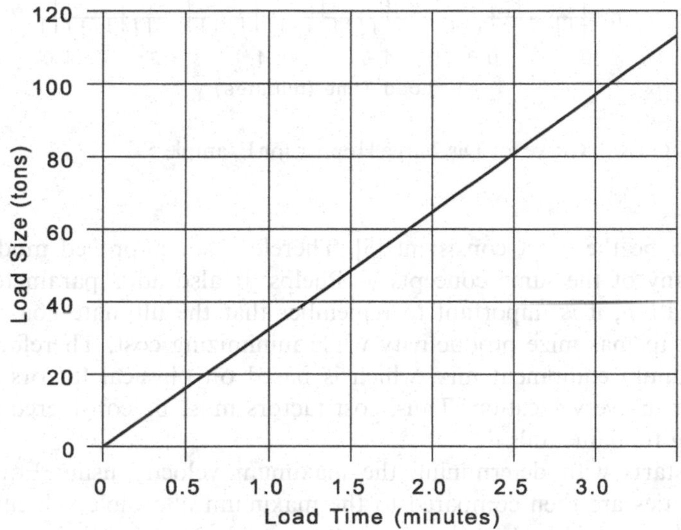


**FIGURE 5.2** Belt Conveyor Load Growth Curve for Example 5.5.

**Example 5.6:** A 10.0 bank cubic yard discharge hopper is filled by a belt conveyor, which is loaded by a 5 BCY bucket loader. The productivity of the conveyor is greater than the productivity of the loader. Therefore, as the conveyor's productivity is limited by the productivity of its loading facility, its theoretical productivity is of no significance. The hopper's loading cycle can be broken into two elements:

Fill hopper = 0.7 minutes
and
Discharge load into haul units = 0.1 minutes.

Figure 5.3 is the load growth curve for this situation. It looks much like the bucket loader load growth curve shown in Figure 5.1. This is because the bucket loader in this example is controlling system productivity.

### 5.4.3 Determining Optimum Number of Haul Units

The next model takes the best characteristics from the Phelps method and combines them with load growth curve information to determine the optimum number of haul units. A comparison of five optimization methods with actual data gathered in the field found the

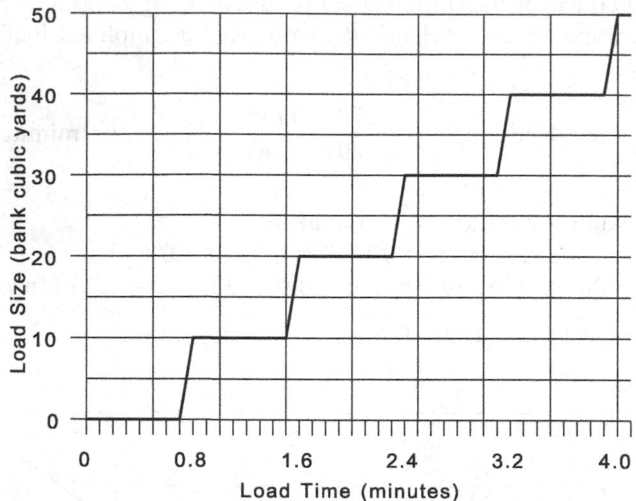


**FIGURE 5.3** Load Growth Curve for Discharge Hopper for Example 5.6.

Phelps method to be the most consistent [5]. Therefore, an improved model was devised, which utilizes many of the same concepts as Phelps. It also adds parameters for cost. As costs vary by location, it is important to remember that the ultimate goal of optimizing a hauling system is to maximize productivity while minimizing cost. Therefore, it is conceivable that an optimum equipment mix, which is based on physical factors alone, may not minimize the cost in every location. Thus, cost factors must be considered equally important to engineering fundamentals.

The analysis starts with determining the maximum velocity using Equations 5.9 and 5.10. These velocities are then compared to the maximum allowable velocity (i.e., the legal speed limit or other restriction) to determine the actual velocities to be used in the travel time ($T$) calculation.

$$T = \frac{d}{88}\left(\frac{1}{V_H} + \frac{1}{V_R}\right) \tag{5.15}$$

The loading time ($L$) is then taken off the load growth curve constructed for the given loading facility. The delay time ($D$) along the route is estimated. These are then added to the travel time to calculate the instantaneous cycle time from Equation 5.4 and the optimum numbers of haul units from Equation 5.5.

$N$ is usually not a whole number and must therefore be rounded. The rounding decision is of great importance because it will ultimately determine the maximum productivity of the hauling system. Two analytical methods are available to make this decision.

### 5.4.4 Rounding Based on Productivity

The decision of whether to round the optimum number of haul units up or down can have a marked effect on the system's productivity. Rounding the number up maximizes the loading facility productivity. Rounding the number down maximizes haul unit productivity. Therefore it is logical to check both productivities and select the higher of the two. This process is shown by Example 5.7.

**Example 5.7:** A 1.5 cubic yard front-end loader is going to load dump trucks with a capacity of 9.0 cubic yards. The loader takes 0.4 minutes to fill and load one bucket. The travel time in the haul is 4.0 minutes. Dump and delay times are 2.5 minutes combined.

$$L = \frac{9(0.4)}{1.5} = 2.4$$

$$C = 4.0 + 2.5 + 2.4 = 8.9 \text{ minutes}$$

$$\text{and } N = \frac{8.9}{2.4} = 3.71 \text{ haul units}$$

Rounding down will maximize haul unit productivity. In other words, the haul units will not have to wait to be loaded, but the loader will be idle during a portion of each cycle. Therefore:

$$\text{Productivity of 3 haul units} = \frac{9(3)(60)}{8.9} = 182 \text{ cubic yards per hour}$$

Rounding up will maximize loader productivity with the haul units having to wait for a portion of each cycle. This assumes that there will always be a truck waiting to be loaded as the loader finishes loading the previous truck. Therefore:

$$\text{Loader productivity} = \frac{1.5(60)}{0.4} = 225 \text{ cubic yards per hour}$$

This number can be checked by calculating the productivity of four haul units. The additional time each truck spends waiting to be loaded ($A$) can be calculated as follows:

$$A = N(L) - C \tag{5.16}$$

In this case: $A$ = 4(2.4) – 8.9 = 0.7 minutes per cycle
Thus: actual cycle time = 8.9 + 0.7 = 9.6 minutes per cycle

$$\text{and : Productivity of four haul units} = \frac{9(4)(60)}{9.6} = 225 \text{ cubic yards per hour}$$

This is equal to the productivity of the loader. Therefore, it checks. When comparing the two possible productivities, it appears that it is best to round up in this case. Thus, four haul units are selected. This decision also makes intuitive sense. No matter how many trucks were added to the system, they could never haul more material than the loader could load. The only way that a higher level of productivity could be achieved in this example is to add another loader or use a larger loader.

### 5.4.5 Rounding Based on Profit Differential

Another philosophy on rounding the optimum number of haul units involves analyzing both cases to determine which would yield the greatest amount of profit. The aim is to find the best trade-off between the added cost of an extra vehicle and the benefit of having or not having that vehicle, as shown in Example 5.8.

**Example 5.8:** A 1.5 cubic yard front-end loader in has an hourly cost ($C_L$) of $150.00 with operator. This figure includes jobsite fixed costs such as supervision, etc. The hourly cost of a dump truck ($C_t$) is $50.00 per hour with a driver. The instantaneous cycle time ($C$) is 8.0 minutes, and the loading time ($L$) is 1.5 minutes per truck. The size of the truck ($S_H$) is 10 cubic yards. The project quantity ($M$) is a total of 10,000 cubic yards of material that requires hauling, and the bid unit price is $2.00 per cubic yard.

$$N = \frac{8.0}{1.5} = 5.33 \text{ haul units}$$

The total cost (TC) to complete the project can be described by the following equation:

$$TC = \frac{M(C)(C_t(N) + C_L)}{N(S_H)(60)} \tag{5.17}$$

Therefore, the total cost if $N$ is rounded down to five units is:

$$\text{TC}_5 = \frac{10,000(8)(50(5) + 150)}{5(10)(60)} = \$10,667$$

The total cost if $N$ is rounded up to six units is:

$$\text{TC}_6 = \frac{10,000(9)(50(6) + 150)}{6(10)(60)} = \$11,250$$

$$\begin{aligned}&\text{The total revenue for the project} = 2.00(10,000) = \$20,000\\&\text{Then : Profit with 5 trucks} = \$9,333\\&\text{Profit with 6 trucks} = \$8,750\end{aligned}$$

In this case it is better to round down, as greater profit is realized.

In practice, an old rule of thumb should be considered when making rounding decisions: "Always round down, as it is easier to add another truck when necessary than to delete one that is not required." The simple logic of this rule speaks for itself. The manager should never make this decision arbitrarily. Factors such as time, equipment, and labor constraints must be considered before the decision is made. Finally, the experience of the decision maker must ultimately be relied upon to determine the most advantageous situation.

### 5.4.6 Optimizing with Cost Index Number

Once the rounding decision has been made, the sustained cycle time ($C_s$) can be calculated. If $N$ is rounded down, $C_s$ is found directly from Equation 5.13 because the productivity of the haul units are controlling system productivity. If $N$ is rounded up to allow the productivity of the loading facility to control, $C_s$ is found by adding Equations 5.13 and 5.16. The result is shown below.

$$C_s = C + W + A \tag{5.18}$$

The total time (TT) to complete the haul of a given amount of material and the system's cost index number (CIN) can be computed as follows:

$$\mathrm{TT} = \frac{M(C_s)}{60(N)(S_H)} \tag{5.19}$$

$$\mathrm{CIN} = \frac{\mathrm{TT}(N(\mathrm{EOC} + \mathrm{MOC} + \mathrm{OC}) + \mathrm{IC})}{M} \tag{5.20}$$

where: TT = total time to complete haul (hours)
$S_H$ = size of haul unit (tons or cubic yards)
$M$ = amount of material (tons or cubic yards to match $S_H$)
$N$ = optimum number of haul units
CIN = cost index number
EOC = equipment ownership cost ($ per hour)
MOC = maintenance and operating cost ($ per hour)
OC = operator cost ($ per hour)

### 5.4.7 Selecting Optimum Haul Unit Size

In most situations, a construction contractor will not be constrained to the size of the haul unit that must be used prior to bidding on a project. In many cases, trucks will be rented for the duration of the project, either directly or via a subcontract. Therefore, it is very important to select the equipment mix which best satisfies the physical constraints of the actual project environment. The above model can be used to do just that. The process is illustrated in the next example.

**Example 5.9:** The front-end loader from Example 5.4 with a bucket size of 1.5 loose cubic yards will be used to load material from a stockpile. Its load growth curve is shown in Figure 5.1. 10,000 loose cubic yards of materials are to be hauled to complete this project. Three sizes of haul units are available to the equipment manager. Their details are shown in Table 5.2. Project costs, which are independent of haul unit size selection, are estimated to be $300 per hour. The material must be hauled over a haul road that has a one-way length of 5,000 ft, 60 pounds per ton rolling resistance, and an average slope of +2.0% in the haul direction. The unit weight of the material is 3,000 pounds per loose cubic yard. The speed limit of the haul road is 35 mph, and the cost for a truck driver is $15 per hour.

**TABLE 5.2**
**Specifications for Haul Units in Example 5.9**

| Item | Haul Unit A | Haul Unit B | Haul Unit C |
|---|---|---|---|
| Capacity (lcy) | 6–8 | 12–14 | 15 -17 |
| Horsepower | 109 | 260 | 260 |
| Efficiency | 0.80 | 0.80 | 0.80 |
| Weight empty (tons) | 8.7 | 18.4 | 18.7 |
| Weight full (tons) | 17.7 | 36.4 | 41.2 |
| EOC ($/hr) | 8.96 | 11.18 | 13.52 |
| MOC ($/hr) | 6.04 | 6.20 | 7.94 |
| Labor ($/hr) | 15.00 | 15.00 | 15.00 |

From Equations 5.9 and 5.10 for haul unit A:

$$V_H = \frac{375(109)(0.8)}{17.7(60 + 20(+2))} = 18.47 \text{ mph}$$

and

$$V_R = \frac{375(109)(0.8)}{8.7(60 + 20(-2))} = 187.93 \text{ mph}$$

Then comparing $V_{max}$ = 35 mph
$V_H$ = 18 mph
$V_R$ = 35 mph
From Equation 5.15:

$$T = \frac{5000}{88}\left(\frac{1}{18} + \frac{1}{35}\right) = 4.78 \text{ minutes; use 4.8 minutes}$$

From Figure 5.1: Entering the load growth curve on the *Y*-axis at 6.0 cubic yards, the loading time (*L*) for haul unit A is found to be 1.6 minutes.

The delay times are estimated as follows [3]:

| | |
|---|---|
| Accelerate after load | – 0.3 minutes/cycle |
| Decelerate to dump | – 0.2 minutes /cycle |
| Maneuver and dump | – 1.0 minutes/cycle |
| Accelerate empty | – 0.2 minutes/cycle |
| Decelerate | – 0.2 minutes/cycle |
| Total | 1.9 minutes/cycle Therefore: $D$ = 1.9 minutes |

Then from Equation 5.4:
$C$ = 1.6 + 4.8 + 1.9 = 8.3 minutes
From Equation 5.5:

$$N = \frac{8.3}{1.6} = 5.19 \text{ units}$$

As the maximum achievable system productivity is the productivity of the loader, this number will be rounded up to 6 units. Thus each truck will have an additional time waiting to load each cycle. From Equation 5.16:

$A$ = 6(1.6) − 8.3 = 1.3 minutes
and
Driver waste time ($W$) is estimated to be 2.0 minutes/cycle.
Therefore, from Equation 5.13:
$C_s$ = 8.3 + 1.3 + 2.0 = 11.6 minutes per cycle
From Equations 5.19 and 5.20, respectively:

$$\text{TT} = \frac{10,000(11.6)}{60(6)(6)} = 53.7 \text{ hours of hauling}$$

and

$$\text{CIN} = \frac{53.7((8.96 + 6.04 + 15)(6) + 300)}{10,000} = 2.58$$

Repeating the above calculations for haul units B and C yields the following numbers:
$N_B = 3$ and $CIN_B = 2.13$; $N_C = 3$ and $CIN_C = 2.12$

Assuming that the addition of sideboards would allow one more bucket of material to be loaded per cycle, the following numbers of haul units and CIN's are found:

$N_A'$ (optimum number of type A haul units with sideboards, 7.5 loose cubic yards) = 5
$CIN_A' = 2.40$
$N_B'(13.5 \text{ lcy}) = 3$
$CIN_B' = 2.09$
$N_C'(16.5 \text{ lcy}) = 2$
$CIN_C' = 2.47$

Plotting CIN versus size in loose cubic yards yields Figure 5.4. This shows that the use of three (3) type B (12 lcy basic size) with sideboards provides the minimum CIN, and therefore, this is the optimum size and number of hauling units for this project.

### 5.4.8 Optimizing the System with a Belt Conveyor

As previously mentioned, the logic shown in Example 5.9 can be used for any type of loading facility. The next example demonstrates the use of this model with a belt conveyor. This loading facility has the advantage of allowing a variable amount of material to be loaded. Thus the project manager can analyze several differently sized loads for each available type of haul unit. This allows the project manager to more closely optimize the load in relation to rolling resistance and horsepower.

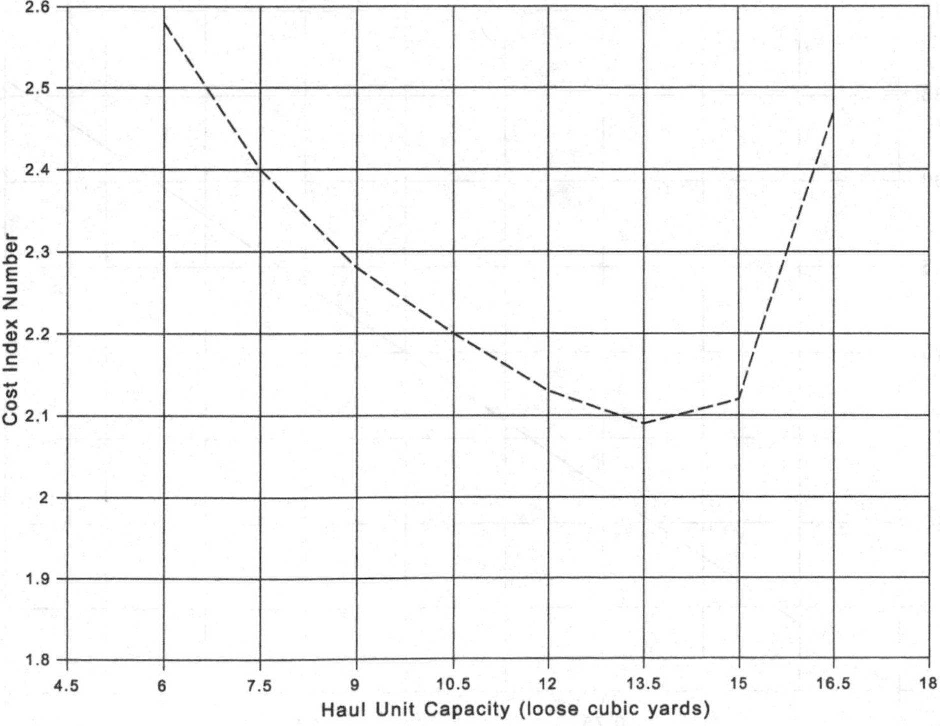


**FIGURE 5.4** CIN Comparison for Example 5.9.

**Example 5.10:** The project used in Example 5.9 will be accomplished using a belt conveyor that is buried in a stockpile. Therefore, the conveyor can be assumed to be continuously loaded so that its productivity will control system productivity. The conveyor has a theoretical productivity of 1,000 tons per hour. Its load growth curve is shown in Figure 5.5.

After performing the same set of calculations as in Example 5.9, the results of the nine possible combinations of loads on the three different haul units are as follows:

Load Haul Unit A with: 6 lcy: $N = 12$ and CIN = 1.09
7 lcy: $N = 11$ and CIN = 1.03
8 lcy: $N = 10$ and CIN = 0.98
Load haul unit B with: 12 lcy: $N = 6$ and CIN = 0.84
13 lcy: $N = 6$ and CIN = 0.79
14 lcy: $N = 5$ and CIN = 0.85
Load haul unit C with: 15 lcy: $N = 5$ and CIN = 0.86
16 lcy: $N = 5$ and CIN = 0.82
17 lcy: $N = 5$ and CIN = 0.80

Comparing CIN's, it is found that using six type B haul units loaded to 13 loose cubic yards minimizes the CIN. The use of five type C haul units loaded to 17 loose cubic yards yields a CIN very close to the optimum. However, the problem of vehicle reliability should be considered in the final decision. If one of the six type B units breaks down, there would be a 17 % loss in production until it is returned to service. On the other hand, if one of the five type C units is lost, the system suffers a 20% production drop. Therefore, the use of six type B haul units is the best solution.

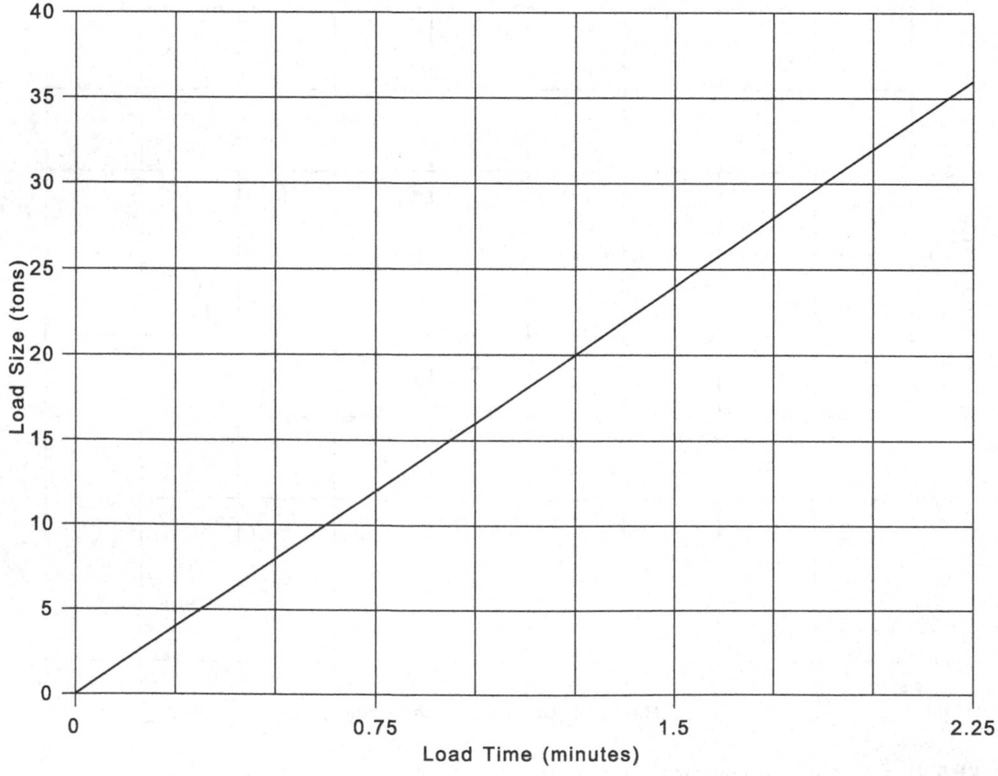


**FIGURE 5.5** Load Growth Curve for Example 5.10.

### 5.4.9 Selecting the Optimum Size Loading Facility

All the discussion to this point has centered on selecting the optimum size and number of haul units given a particular loading facility. There are times when just the opposite decision must be made. The previous model can be adapted to pick the optimum size loading facility when the size and maximum number of haul units are fixed.

**Example 5.11:** Using the project information from Example 5.9, a project manager has ten type C haul units available and a choice of three bucket loaders to rent. The characteristics of each loader are shown in Table 5.3.

The load growth curve for Loader I is shown in Figure 5.1. Corresponding load growth curves would be constructed for Loaders II and III. It is poor practice to load less than a full bucket. Therefore, each loader should be analyzed before loading the given haul unit with all feasible combinations of full buckets. In other words, Loader I can load the type C haul unit with either 15 loose cubic yards (10 full buckets) or 16.5 loose cubic yards (11 full buckets).

The results of the calculations are shown below.

Loader I: 15.0 lcy load: $N = 2$ and CIN = 2.18

16.5 lcy load: $N = 2$ and CIN = 2.09

Loader II: 16.0 lcy load: $N = 2$ and CIN = 2.00

Loader III: 15.0 lcy load $N = 3$ and CIN = 1.48

From these calculations, Loader III with the 2.5 loose cubic yard bucket should be chosen. It should load 15 loose cubic yards (six full buckets) on the type C haul unit.

**TABLE 5.3**
**Loader Characteristics for Example 5.11**

| **Loader** | **I** | **II** | **III** |
|---|---|---|---|
| Bucket size (lcy) | 1.5 | 2.0 | 2.5 |
| **Cycle elements** | | | |
| Move to pile (minutes) | 0.05 | 0.05 | 0.05 |
| Fill bucket (minutes) | 0.10 | 0.13 | 0.17 |
| Maneuver to load (minutes) | 0.15 | 0.15 | 0.15 |
| Load truck (minutes) | 0.10 | 0.13 | 0.17 |
| Total load time (minutes) | 0.40 | 0.46 | 0.54 |

## 5.5 OPTIMIZING SPECIAL PURPOSE EQUIPMENT

The principles detailed in the previous sections are applicable to a variety of typical construction equipment optimization problems that utilize special purpose construction machinery. While it is beyond the scope of this book to provide a comprehensive set of examples, this section will demonstrate how one common construction project can have its fully optimize the equipment required to complete a highway chip seal project.

### 5.5.1 Chip Seal Equipment Requirements

Chip seals consist of two materials: asphalt binder and aggregate (chips). They are constructed by spraying the binder on the surface of an existing pavement using an asphalt distributor and the covering with aggregate using a self-propelled chip spreader or a dump truck. The aggregate is then embedded using rollers. Both steel wheel and pneumatic rollers have been used, but the pneumatic roller is the most comment. Steel wheel roller can only be used on aggregate that is very hard and not subject to crushing. The rolling process reorients the aggregate chips so that their center of gravity is in its lowest position. Figure 5.6 illustrates both the cross-sectional design of a chip seal and the center of gravity issue.

Chip seals are usually constructed under traffic, which makes optimizing cycle times with quality control requirements paramount to completing the project and expeditiously opening it to traffic [6]. In this type of project, the key decision criterion is to minimize traffic disruption because the longer the normal flow of traffic is impacted, the higher the probability of a traffic accident in the construction work zone. This consideration must be balanced against conducting the necessary construction activities required to properly place the new chip seal surfacing. Early chip seal failure is characterized by the loss of aggregate due to failure of the contractor to adequately embed it in the bituminous binder [7]. In the chip seal paving train, the pneumatic roller is the tool used to achieve embedment. It is also the last link in the equipment package, and its purpose is embedment, a factor which cannot reliably be measured like compaction. These two factors combine with the pressure to release the newly sealed road to traffic and create a situation where the rollers might be taken off the surface before proper embedment is achieved [8].

Since, chip seals are among the least complex of construction processes, the design process is minimal, consisting of selecting an appropriate binder and aggregate gradation. Binder and aggregate application rates shown on contract documents are average rates application rates are commonly adjusted in the field to account for the relative surface condition of the pavement being sealed. The relative success for all types of chip seals is more a function of the quality of the construction procedures than the details of the design [7]. Thus, as a result, construction means and methods are key to achieving required chip seal performance. In other words, chip seal construction quality is driven by the equipment used to construct it. The following are the usual pieces of equipment used on seal coat projects:

- Asphalt distributors
- Chip spreaders
- Dump trucks
- Rollers
- Brooms
- Bucket loaders

Of these, the distributor, the chip spreader, and the rollers determine the sustained production of the chip seal paving train shown in Figure 5.7. The distributor is the controlling link in the production system because no other piece of equipment can begin to produce its

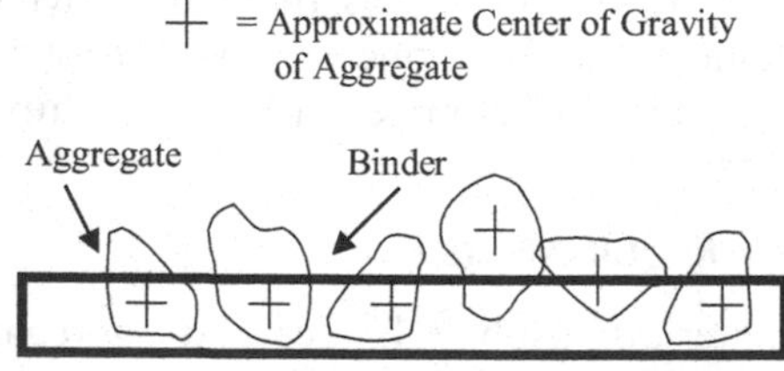


**FIGURE 5.6** Chip Seal Cross-Section and Aggregate Center of Gravity – Embedment Relationship.

**FIGURE 5.7** Chip Seal Paving Train.

function until the distributor has applied the binder to the surface. Thus, the production rates of the chip spreader and the rollers must equal or exceed the distributor's sustained production to maximize production. Therefore, to maintain the required quality, other equipment in the train must be able to keep up with the production of the distributor. When rollers' production rate is less than the distributor, overall system production will decrease, or the roller operators will sacrifice quality by speeding up to maintain the distributor's pace.

The primary measurement of quality for the roller team is called linger time. This is the total time spent rolling a specific area of freshly chipped surface. Contract requirements vary according to specifications from 1,000 square yards per hour to 5,000 square yards per hour [8]. Thus, linger time becomes the rollers' minimum allowable production rate. This rate must be matched to the production rate of the distributor if the system is to achieve the distributor's maximum sustained production.

## 5.5.2 Matching Roller and Distributor Production

A typical equipment package for a standard US highway with a 12-foot lane includes between two and four pneumatic rollers depending on the actual equipment selected. The typical asphalt distributor would be set up to make a 12-foot wide shot (the length of road it takes to empty the distributor's tank) to match the width of one lane, allowing the opposing lane to permit traffic flow through the work zone. In practice, the production rates of the chip spreader and asphalt distributor are roughly equal, making the combined production rate of the system equal to the distributor's sustained production rate. Thus, the roller team's production rate becomes the link in question that needs to be evaluated, and the required number of rollers needed to sustain the distributor's production is the key to optimizing the chip seal equipment system.

The roadway is a two-lane highway with the dimensions shown in Figure 5.8.

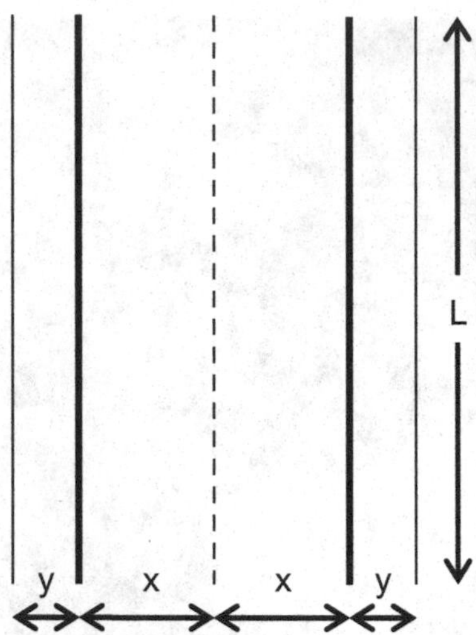


**FIGURE 5.8** Typical Chip Seal Roadway Geometry.

Where $x$ = lane width
$y$ = shoulder width
$L$ = length of the chip seal shot

If the linger time is one hour per contract specified area ($A$), the required rolling time ($t$) for a specified shot length ($L$) depends on the shot width (in this analysis the lane width, $x$) and the roller speed ($V_r$). Equations 5.20 and 5.21 algebraically describe the relationships.

$$A = xL \tag{5.20}$$

$$t = \frac{L}{V_r} \tag{5.21}$$

The rolling time requirement (1 hour/A) implies that roller in the package will move at a fixed speed and linger on the sealed area for the period of time specified as the contract. If the shot width is greater than the actual rolling width of a single roller, then more than one roller will be required. The rolling time ($t$) required to travel a distance of $L$ at the traveling speed of $V_r$ permits the total time spent by all rollers per each pass, $T$ (the total rolling time), to be calculated by $t$ by the number of rollers, $N$, in the package as shown in Equation 5.22.

$$T = tN \tag{5.22}$$

The required number of passes (NP) can then be calculated once the time per one pass is determined. This calculation is formulated in Equation 5.23.

$$\text{NP} = \frac{H}{T} \tag{5.23}$$

Where: $H$ is 1 hour as in 1 hour per 5,000 yd$^2$ rolling time.

Roller production, $P_r$, is defined as the linear miles of chip seal section rolled per hour. This isa function of the number of passes and shown in Equation 5.24.

$$P_r = \frac{V_r}{\text{NP}} \tag{5.24}$$

Combining Equations 5.20, 5.21, 5.22, 5.23, and 5.24 allows the roller production rate, $P_r$, to be expressed in terms of the number of rollers, $N$, as shown in Equation 5.25.

From Equations 5.24 $P_r = \frac{V_r}{\text{NP}}$ and 5.23 $\text{NP} = \frac{H}{T}$

$\therefore P_r = \frac{V_r T}{H}$ and Equation 5.22 $T = tN$

$\therefore P_r = \frac{V_r tN}{H}$ and Equation 5.21 $t = \frac{L}{V_r}$

$\therefore P_r = \frac{V_r LN}{HV_r} = \frac{LN}{H}$ and Equation 5.20 $L = \frac{A}{x}$

$\therefore$

$$P_r = \frac{AN}{xH} \tag{5.25}$$

The distributor speed to produce the required asphalt binder rate can be calculated from Equation 5.26. It must be noted that the distributor's spray bar output is dependent on the type of the binder sprayer used. $W$ is the width of the shot and is used interchangeably with $x$ value in this calculation as $x = W$.

$$S_f = \frac{9G_t}{WR} \tag{5.26}$$

where $S_f$ = distributor speed (ft/min)
$G_t$ = spray bar output (gal/min)
$W$ = sprayed width (ft)
$R$ = rate of binder application (gal/sy)
9 = conversion factor from sy to sf

Distributor speed, $S_f$, can be modeled as the distributor production rate in lineal miles per hour, allowing the roller production rates to be computed in this same unit to determine system production. If the chip spreader's and roller team's production is less than the production of the distributor, the distributor will have to reduce its production, or the other equipment will not be able to produce the required level of quality.

### 5.5.3 Maximizing Chip Seal System Production

As previously stated, the production rate of all other links in the equipment system had to be greater than or equal to the production of the controlling piece to permit that piece to maximize its sustained production. Equation 5.27 describes the instantaneous production of the rollers in relation to the instantaneous production of the distributor.

$$P_r \geq P_D \tag{5.27}$$

When production is measured in lineal miles per hour for a given shot width, distributor production, $P_D$, equals distributor speed, $S_f$. One could also compare sustained production rates. However, it is unnecessary since the object is merely matching the production of one of the system's components to the production of another. Additionally, instantancous production as the parameter of comparison yields a conservative solution by ensuring that the "best" production of the roller is greater than or equal to the "best" production of the distributor (See Example 5.12).

**Example 5.12:** A contractor owns several Dynapac CP 132 nine-wheel pneumatic tire rollers. The roller has two forward gears that two speeds of 6.2 mph (10 km/hr) and 12.4 mph (20 km/hr). Measured across the outside tires, the actual rolling width is 69.3 inches (176 cm). The lower speed of 6.2 mph (10 km/hr) is used in the calculations because it reduces the number of rollers needed to meet the minimum rolling time requirement.

$V_r$ = 6.2 mph (10 km/hr)

Other factors are as follows:

- The road has the following dimensions: $x$ = 12 ft. and $y$ = 10 ft.
- The contract specifies a minimum of one hour of rolling time per 5,000 square yards of sealed surface.
- The contractor intends to use three rollers if adequate.
- The distributor's production rate is 90 gal/min
- The specified binder application rate is 0.33 gal/sy
- The shot length is 1,250 yd.

1. The distributor's production rate of 90 gpm must be converted to lmph which equals the distributor speed from Equation 5.26:

$$S_f = \frac{9G_t}{WR} = \frac{9 \text{ sf/sy}(90 \text{ gpm})(60 \text{ min/hr})}{(12 \text{ ft})(0.33 \text{ gal/sy})(5290 \text{ ft/mi})} = 2.32 \text{ lmph}$$

2. Calculate the number of passes using Equations 5.20, 5.21, 5.22, and 5.23:

$$A = xL = 12 \text{ ft}(0.33 \text{ yd/ft})(1{,}250 \text{ yd}) = 4{,}950 \text{ sy}$$

$$\text{t} = \frac{L}{V_r} \frac{1{,}250\text{yd}}{6.2\text{mi/hr}(1{,}760 \text{ yd/mi})} = 0.115 \text{ hr}$$

$$T = tN = 0.115 \text{ hr}(3 \text{ rollers}) = 0.345 \text{ hr}$$

$$\text{NP} = \frac{\text{H}}{\text{T}} = \frac{1 \text{ hr}}{0.345 \text{ hr}} = 2.91 \text{ passes}$$

Therefore, the actual NP must become 3 because the system must always have an odd number of complete passes. In this case, the rollers would follow the chip spreader on their first pass, reverse course and roll backwards to the start of the shot on the second pass and return to catch back up with the spreader on the third pass.

3. Calculate the production of the roller spread is calculated using Equation 5.25

$$P_r = \frac{AN}{xH} = \frac{4{,}950 \text{ sy}(3 \text{ rollers})}{12 \text{ ft}\left(0.33\frac{\text{yd}}{\text{ft}}\right)(1 \text{ hr})} = 2.07 \text{ lmph}$$

At this production rate, 3 rollers making 3 passes will achieve the required 1 hour of rolling time per 5,000 square yards (4,180 m$^2$) of surface area. Comparing the production rate of the distributor at 2.33 lmph to that of the rollers at 2.07 lmph, one finds that either:

- The rollers will not be able to achieve the rolling time requirement if distributor production is maximized, or
- The distributor production will be reduced if the rolling time is strictly enforced.

4. Quantify the impact of this production mismatch.With the rollers moving at 0.26 lmph slower than the distributor, the three rollers would lag the distributor by about a quarter of a mile every hour if they scrupulously maintained the minimum rolling time requirement. At the end of a typical 12-hour day, the rollers would be 3 miles (4.8 km) behind the rest of the equipment and would need to continue to roll for another 1.5 hours to finish the remaining newly sealed surface. This creates a different construction quality problem because as the binder would be cold and the ability the rollers would be able to embed the aggregate per specifications. Hence, quality suffers either by failing to meet the specified linger time or by needing to roll cold binder. Thus, a team of three rollers of this size and type is not optimum.
5. Taking Equation 5.25, algebraically isolating the number of rollers, $N$, and changing $P_r$ to the distributor's production ($P_d$) as the minimum required roller production, allow one to solve for the required number of rollers ($N'$) to match or exceed the distributor's production ($P_d$).

$$\text{Equation 5.25 } P_r = \frac{AN}{xH} \text{ becomes } N' = \frac{P_d x H}{A}$$

$$N' = \frac{2.33\ \text{lmph}(4\text{yd})(1,760\ \text{yd/mi})}{5,000\ \text{sq yd/hr}} = 3.28\ \text{rollers} \rightarrow 4\ \text{rollers}$$

Thus, four rollers would be required to keep up with the distributor and achieve the required roller linger time.

### 5.5.4 Determining Roller Coverage and Rolling Patterns

To meet the required chip seal performance standards, the rollers must also evenly cover the entire sealed surface. Thus, once the minimum number of rollers is determined, it must next be tested to ensure that the roller team will be able to completely cover the sealed surface. Commonly, rollers are employed in a staggered pattern with one roller on the centerline, one roller on the outside edge, and the third bringing up the center. Using the Dynapac CP132 from the previous examples, the resultant roller pattern for three rollers in a staggered formation is shown in Figure 5.9. It shows that there are three areas that receive three passes. Two are roughly 37 inches wide on the outsides of the lane, and the third is about five inches wide between the wheel paths. The wheel paths receive six passes due to the overlap between two rollers. Research has shown that loss of aggregate is most prevalent between and outside of the wheel paths where the least amount of rolling occurs with this pattern [8]. On the other hand, the four rollers in a diagonal pattern shown in Figure 5.9 provide six passes across most of the land width. Thus, adding a fourth roller to the chip seal paving train not only allows the contractor to maximize the production of the distributor but also reduces the probability of early chip seal failure due to improper aggregate embedment.

## 5.6 COMMENTS ON OPTIMIZING EQUIPMENT FLEETS

The examples discussed in this chapter clearly demonstrate the relative ease and objectivity with which construction equipment fleet composition decisions can be made using the salient physical parameters of a given project. The great danger that is faced by both project managers and estimators is the bias toward using equipment that is currently in the company's inventory without regard to the potential impact on project productivity. As a

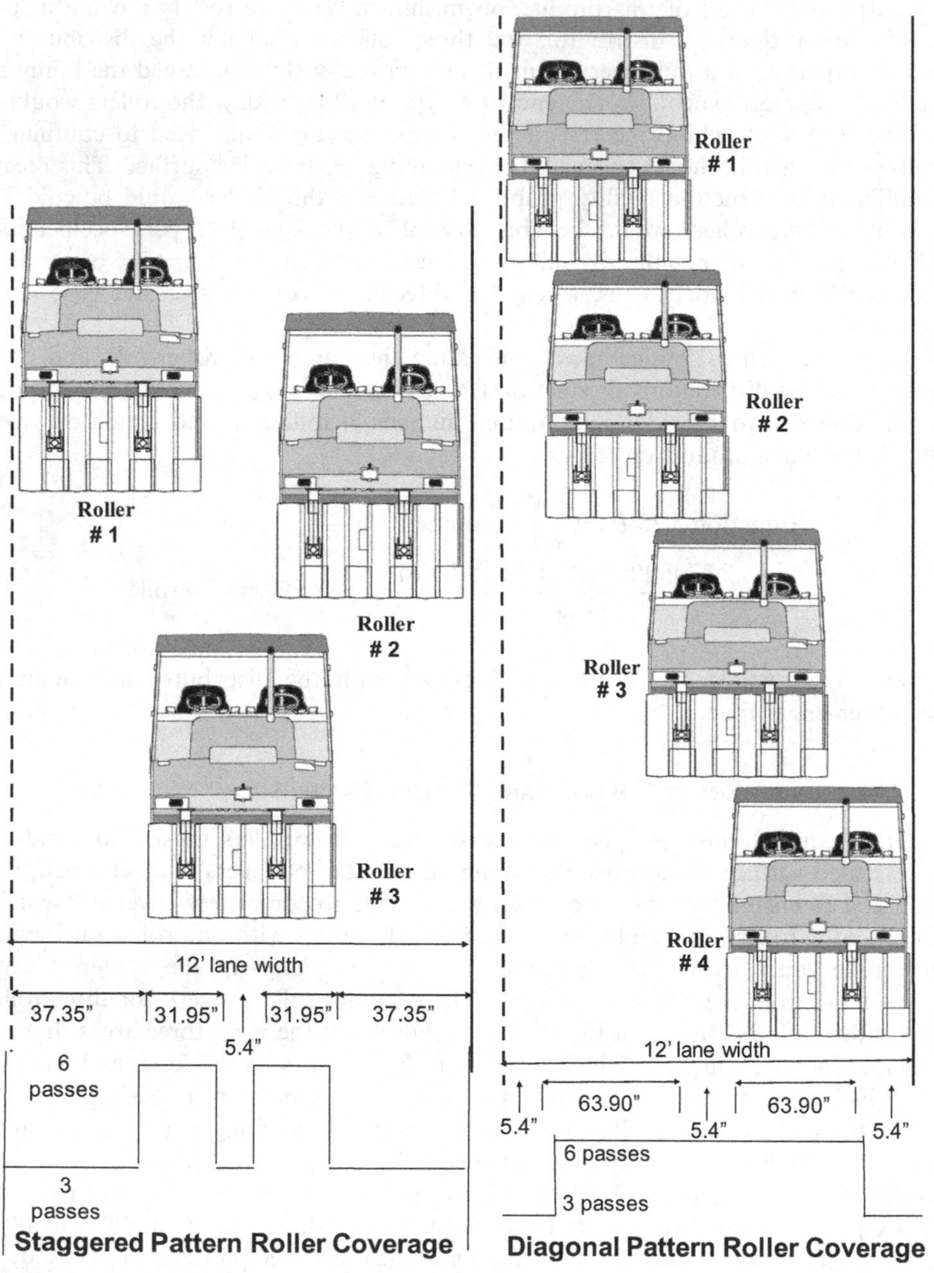


**FIGURE 5.9** Comparison of Rolling Pattern Coverage.

minimum, the option of renting an optimized equipment mix should be evaluated against using current equipment. In this analysis, the cost of idle equipment should be factored into the final result to allow management to select the lowest cost solution. If renting an optimized equipment mix is selected, the bid price should include the cost of idle equipment to allow the organization to recover those costs as well as the actual rental costs.

Rounding is another decision that has been shown to be very important. One option not analyzed in this chapter is to round the number of haul units up, and use one of the units as a standby vehicle. In other words, if the optimum number of haul units was rounded up to six, five of the trucks would be put into production with drivers, and the sixth vehicle would be brought on site for use if a production vehicle were to break down. The broken unit would then become the standby unit once it is repaired. Another method would be to rotate the standby unit every day and utilize the time a vehicle is out of production to perform preventive maintenance. Fluid level can be checked. Worn out tires can be replaced, and minor adjustments to major assemblies such as clutches and brakes can be made. This management technique not only maximizes equipment availability but also reduces overall maintenance and repair costs as well adjusted and lubricated assemblies fail at a much lower rate. Additionally, there are unquantifiable savings due to the psychological attitude of the operator. Those who have worked in the construction industry can verify that an operator who is sitting in a clean, well-maintained vehicle tends to operate that vehicle with more confidence and care and thereby achieves higher production. Thus, a program of regular rotation of operational vehicle for on-site preventive maintenance reduces the amount of equipment time lost to unscheduled breakdowns.

## REFERENCES

[1] Gates, M. and Scarpa, A. (1975). Optimum Size of Hauling Units. *Journal of the Construction Division*, 101(CO4), Proceedings Paper 11771, pp. 853–860.

[2] Peurifoy, R.L., Schexnayder, C.J., Schmitt, R. and Shapira, A. (2018). *Construction Planning, Equipment and Methods.* 9th ed. New York: McGraw Hill.

[3] Phelps, R.E. (1997). Equipment Costs, Unpublished Working Paper, Oregon State University, Corvallis, Oregon.

[4] Gransberg, D.D. (1996). Optimizing Haul Unit Size and Number Based on Loading Facility Characteristics. *Journal of Construction Engineering and Management*, 122(3), pp. 248–253.

[5] Caterpillar Inc. (2017). *Caterpillar Performance Handbook*. 47th ed. Peoria: Caterpillar Inc.

[6] Gransberg, D.D., Senadheera, S. and Karaca, I. (2004). Calculating Roller Requirements for Chip Seal Projects. *Journal of Construction Engineering and Management*, 130(3), June, pp. 378–384.

[7] Gransberg, D.D. (2006). Correlating Chip Seal Performance and Construction Methods. *Transportation Research Record, Journal of the Transportation Research Board, No. 1958*, National Academies, pp.54–58.

[8] Gransberg, D.D., Senadheera, S. and Karaca, I. (1999). *Analysis of Statewide Seal Coat Constructability Review.* Texas Department of Transportation Research Report TX- 98/0-1787-1R. Lubbock, TX: Texas Tech University.

## CHAPTER PROBLEMS

1. If a truck with a 249 horsepower engine with an efficiency of 0.81*, weighs 80,000 pounds fully loaded and has maximum speeds of 2.0, 4.8.2, 8.2, 18.2, and 41.2 mph in first through fifth gears, respectively, calculate the top speed and time to reach that speed on a level road with a rolling resistance of 45 lbs/ton.
2. A 24 cubic yard dump truck has a loading time that equals 4 minutes, a travel time of 12 minutes, and the dumping and delay times equal 4 minutes. Calculate the cycle time, optimum number of hauling units and productivity.
3. Given a haul length of 2,500 ft, a loading time ($L$) of 3.5 minutes, a variable time ($V$) of 8.0 minutes, compute the sustained cycle time, the optimum number of hauling units ($N$), and sustained production rate using the Phelps Method. The hauler has a capacity of 17.5 loose cubic yards. The shift is 10 hours long, and waste time ($W$) is 2.0 minutes per cycle.
4. Calculate the optimum number of trucks for the following situation. A front loader with a 3.0 loose cubic yard bucket has the following cycle elements:

- Move to stockpile 0.05 minutes
- Fill bucket 0.20 minutes
- Move to truck and maneuver to load 0.15 minutes
- Dump loaded bucket 0.15 minutes

The loader will service a fleet of dump trucks with a capacity of 20 loose cubic yards. The trucks have the following cycle elements:

- Maneuver to loading position 0.5 minutes
- Travel to dump site 10.0 minutes
- Spot, maneuver, and dump 0.5 minutes
- Dump load 0.5 minutes

5. A 3.0 cubic yard front-end loader has an hourly cost ($C_L$) of $175.00 with operator. This figure includes jobsite fixed costs such as supervision, etc. The hourly cost of a dump truck ($C_t$) is $80.00 with a driver. The instantaneous cycle time ($C$) is 12.0 minutes, and the loading time ($L$) is 1.5 minutes per truck. The size of the truck ($S_H$) is 20 cubic yards. The project quantity ($M$) is a total of 100,000 cubic yards of material that requires hauling, and the bid unit price is $1.00 per cubic yard. Determine the correct number of dump trucks by rounding:

   a. Based on productivity
   b. Based on profit differential

6. An asphalt distributor has a production rate of 100 gal/min. It will shoot chip seal binder on the 9-foot wide shoulders of a road using an application rate of 0.30 gal/sy. What is the distributor's speed in lineal miles per hour?

# 6 Stochastic Methods for Estimating Productivity

## 6.0 INTRODUCTION

Chapter 5 furnished several different methods for calculating the productivity of a given piece of equipment. Each of these relied on fixed estimates of time and cost. The concept of a sustained versus an instantaneous cycle time was introduced, and this allows the estimator to compensate for the unforeseen interruptions in a typical equipment production cycle. The sustained cycle time, which might also be called the average daily cycle time, recognizes that equipment production systems are indeed variable even though the mathematical model that was used to estimate production is deterministic.

## 6.1 BACKGROUND

The next step is to recognize that each input variable in the production rate estimate has its own characteristic variability. For instance, in a loader/dump truck production system, the variability of the loader's cycle time is normally less than that of the trucks for no other reason than that there is typically only one loader and hence one operator. Whereas, there will be more than one dump truck, and the truck drivers will have their own individual ability to efficiently operate the machine. Thus, while the loader will cycle within a tight range of instantaneous cycle times, the trucks' times will experience greater variability across the course of the workday. If the haul is over a public highway, variation in cycle times will increase by another increment as the truck drivers have to deal with different traffic situations on each cycle, some of which will cause big delays, such as a traffic jam, and others that will improve the cycle time for a given cycle where the truck happened to hit all green lights as it proceeded down the road without having to stop.

Accounting for this variation would seem to be an impossibly complex mathematical situation as the influence of traffic on a public road is infinitely random and infinitely variable. It would be in the deterministic models illustrated in Chapter 5. However, the laws of probability and statistics were developed to specifically give the analyst tools to able to account for systems that encounter a measurable degree of variation in their normal circumstances. Thus, given some background information about the inherent variation in each input variable, equipment production can be estimated using a stochastic model that includes all the normal variation and furnishes output that shows the expected range of productivity that can be achieved by the equipment package under analysis. Typically, this range will be displayed as a best possible case, a worst possible case, and a most likely case.

Given this type of information, the estimator can then get a better feeling for what is realistically achievable under the project's conditions and can adjust the final estimated production rate for a given crew accordingly. Thus, the accuracy of the final estimate should be enhanced. Additionally, the project manager can then take the information to the field and better manage the actual construction, ensuring that the actual production rates do not leave the range in the estimate. As a result, stochastic estimates of productivity furnish not only a more precise pre-project estimate but also a mechanism that can transcend the office and take the estimating assumptions to the field and enact them [1].

This chapter will discuss the simple rules of probability and statistics that are used to develop a stochastic mathematical model of a typical equipment production system, building on the deterministic models shown in Chapter 5. It will discuss the use of standard commercial simulation software packages for solving equipment production problems. Finally, it will spend some time discussing the elements required to validate a simulation model to ensure that it is adequately predictive.

## 6.2 DEVELOPING MATHEMATICAL MODELS

Chapter 3 discussed the development of a mathematical model for determining optimum equipment replacement timing and strategy. The rules shown for that problem are no different than the rules for modeling equipment production on a stochastic basis. Essentially, there are three major rules. First, the analyst must be able to mathematically describe the system under analysis in terms of interrelated equations based on the physical constraints of the given system. Next, the variables used in the equations must have values that can be used to solve them. In the deterministic model, each variable has a single "best" value that is used. In a stochastic model, a range of possible values for each variable is used and each value range has an associated probability distribution function (PDF). Finally, there must be a clearly defined decision criterion that mathematically describes the final solution (i.e. "maximize earthmoving crew production").

With these in hand, the model can be developed. Often the model starts out as a deterministic one and is then modified to permit a range of input variable values rather than a single value. Thus, the equations that describe the deterministic model can be transferred to a computer program, which, in many cases, is merely a standard commercial spreadsheet program. There are other modeling programs available that interact with the model expressed in the spreadsheet that add the stochastic dimension to the process and change the output from being deterministic to stochastic. Several of these will be discussed and demonstrated in Section 6.3.

At this point, a short discussion of the application of probability and statistics to the calculation of equipment productivity is warranted. The following sections will merely highlight the important concepts as applied to equipment production models. For further detail, the reader is referred to the sources shown in the references at the end of the chapter.

### 6.2.1 Probability Theory

In construction equipment production system analysis, probability theory is used to account for the fact that the real cycle times will vary from cycle to cycle depending on many circumstances that are far too complex for the estimator to model. For instance, the cycle time of a push-loaded scraper is dependent on the cycle time of the pusher dozer. If on a given cycle, the dozer operator takes a short break between scrapers to take a drink of water or use the bathroom, the next scraper will probably end up waiting for several seconds or minutes longer than normal to begin loading. These types of delays are predictable and the deterministic production model accounts for them using the concept of sustained cycle time rather than making all the calculations using the instantaneous cycle time.

Nevertheless, there are other delays that are not included in the sustained cycle time. An example of this type would be the production time lost for the pusher dozer to refuel at mid-morning because the operator forgot to top off the machine's fuel tank the night before. On the other hand, there will be cycles during the normal day that take less than the sustained cycle time because the system can actually achieve its computed instantaneous cycle times. Thus, using the sustained cycle time as a sort of average, an observer with a stop watch who recorded the actual cycle time of a given scraper would record times that are both less than and greater than the sustained cycle time that was used in the estimate. A table or graph of these times would be called a "frequency distribution" because it shows the number of ("how many") times in a given period a specific cycle time was observed.

Example 5.3 computed the sustained cycle time for the loader-truck hauling system to be 14 minutes per cycle. Figure 6.1 is a graphical depiction of a hypothetical frequency distribution of actual cycle times on 30-second increments for that system. One can see that this system's cycle times ranged from a low of 10 minutes to a high of 20 minutes, with the 14-minute sustained cycle time being the time that was observed most often. Also looking at this chart, one can see that there seems to be a greater number of actual cycle times that are more than the sustained cycle time of 14 minutes than there are times less than 14 minutes. In fact, the average observed cycle time is 14.9 minutes, nearly a full minute greater than that computed by the estimator in this example. This indicates that the actual production is going to be less than the estimated production. Rerunning that example with a 15-minute sustained cycle time gives a production rate of 2,560 cubic yards per day, which is roughly nine truckloads of earth less than estimated. So, one can see that if the estimator had known how the actual cycle times would be distributed in relation to the computed sustained cycle time, the bid could have been adjusted to account for the variation in the field. This is exactly what a stochastic production estimating model provides.

For the next project, the Example 5.3 estimator now has a historical record of actual cycle times and could develop a stochastic model to estimate future production rates based on this data. The software that is used to run the stochastic model usually requires the estimator to identify the type of frequency distribution that applies to a given variable. This can be done in any one of two ways. First, if the estimator does not have historical data like that shown in Figure 6.1, an assumption for the type of distribution can be made. The second way, which requires historical data, is to use curve-fitting software to assist in defining the appropriate distribution based on the available data. There are quite many frequency distributions that are used in statistics. However, the four most common for the types of simulations that are used in equipment production modeling are the following:

- Triangular distribution (Figure 6.2): This one is selected when the estimator can only estimate the lowest possible value, the highest possible value, and the most likely value.
- Poisson distribution (Figure 6.3): This distribution is appropriate when the variations in values under analysis are random, such as the arrival of public traffic at

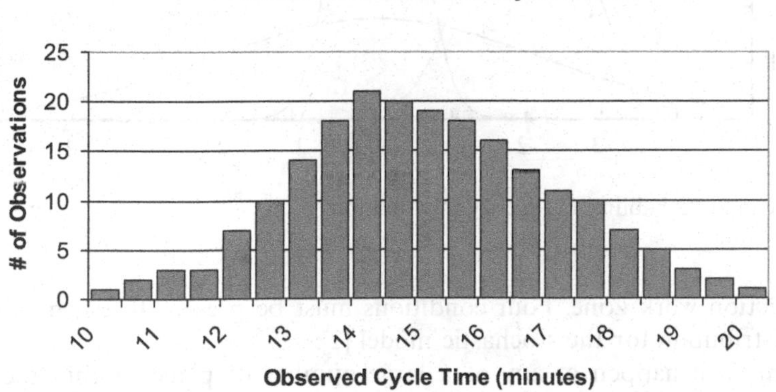


**FIGURE 6.1** Hypothetical Frequency Distribution for Example 5.3 Loader-Truck Haul Crew.

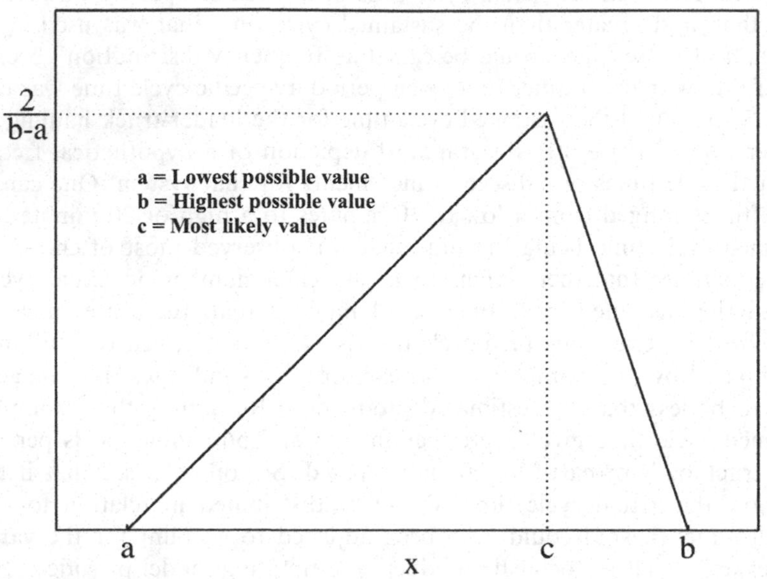


FIGURE 6.2 Triangular Probability Frequency Distribution.

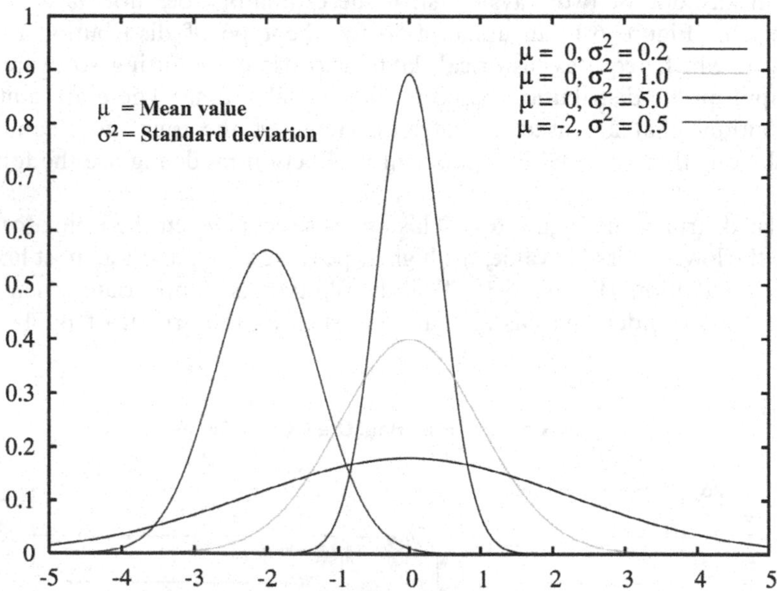


FIGURE 6.3 Normal Probability Frequency Distribution.

a construction work zone. Four conditions must be present to permit the selection of this distribution for the stochastic model [2]:

- "Events can happen at any of a large number of places within the unit of measurement and along a continuum.
- At any specific point, the probability of an event is small.

  - Events happen independently of other events.
  - The average number of events over a unit of measure is constant" [2].
- Normal distribution (Figure 6.4): This is the classic "bell curve." This is used when the uncertain value is subject to "many different sources of uncertainty or error" [2]. The following are some notable qualities of the normal distribution:
  - The density function is symmetric about its mean value.
  - The mean is also its mode and median.
  - 68.27% of the area under the curve is within one standard deviation of the mean. 95.45% of the area is within two standard deviations. 99.73% of the area is within three standard deviations.
- Uniform distribution (Figure 6.5): This is used when the high and low values are known and the chance of observing any value that range is equal to all other chances.

Thus, to be able to move a deterministic equipment production model to a stochastic one, the estimator will need to do the following:

- Assemble the variables that are going to be part of the model.
- Develop the mathematical relationships for each variable.
- Identify the possible range in values for each variable.
- Associate a probability frequency distribution with each variable that is going to be allowed to vary in the impending simulation.
- Develop the remaining input that is required by the simulation software that will be used.

### 6.2.2 Statistical Analysis

Statistical analysis is fundamental to the preceeding discussion. In estimating equipment productivity, it is common to use the "average" times or loads. The mathematical average is

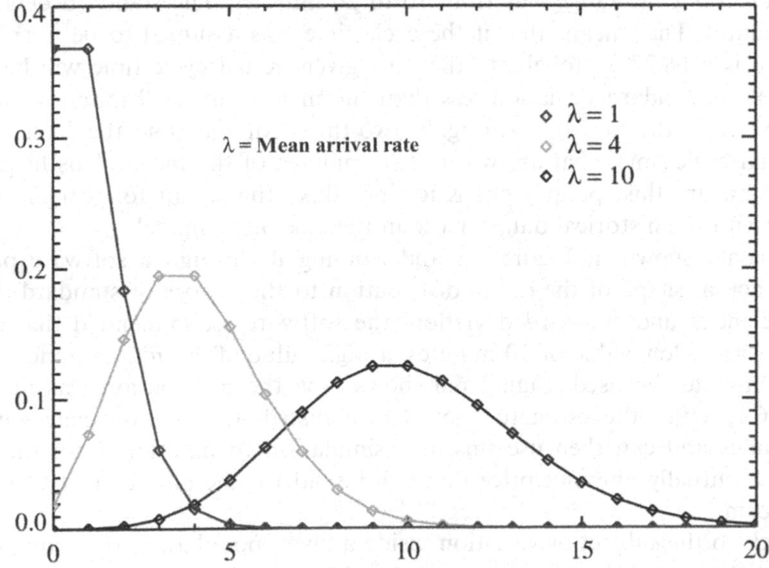


**FIGURE 6.4** Poisson Probability Frequency Distribution.

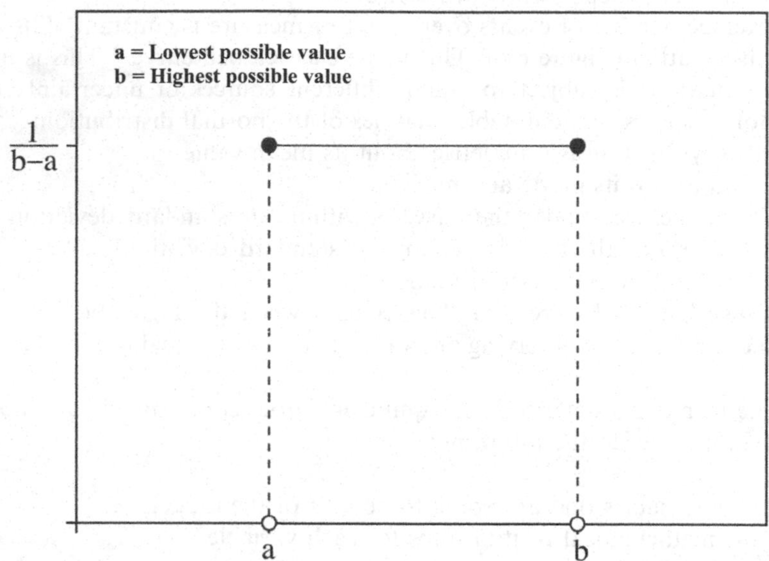


**FIGURE 6.5** Uniform Probability Frequency Distribution.

a statistical measure. It is more properly called the "mean." Along with the mean comes the "median," which is literally the middle value without regard to frequency, and the "mode," which is the most frequent value in the population. These are all called measures of "central tendency," which means that they describe the axis about which the actual observed values are oriented. The next concept that must be understood to utilize a stochastic equipment production model is the idea of variation within a given sample population.

Figure 6.1 showed the variation of cycle times for the hypothetical haul crew. The mean was 14.9 minutes, but the range was from 10 to 20 minutes. The standard deviation of the data is 2.0 minutes. That means that if the cycle time was assumed to be normally distributed that there is a 68.27% probability that any given actual cycle time will be somewhere between 12.9 (one standard deviation less than the mean) and 16.9 minutes (one standard deviation more than the mean); so roughly two-thirds of the time the haul crew will be achieving actual cycle times that are within two minutes of the mean. This helps define the natural variation in this parameter. Knowing this, the estimator can now associate a distribution with the historical data for use in the stochastic model.

Taking the data shown in Figure 6.1 and running it through a software package that compares the actual shape of the data's distribution to the shapes of standard distributions with the same mean and standard deviation, the software recommended that a triangular distribution with the low value of 10 minutes, a high value of 20 minutes, and a most likely value of 14.5 minutes be used. Figure 6.6 shows how the distribution was fit to the data from Figure 6.1. Thus, the estimator now has a distribution to associate with the haul crew's cycle times and can then use this in a simulation to fine-tune both the production estimates and eventually, the unit price that will be bid for the pay item that this particular crew will perform.

Getting back to the subject of variation inside a given population, the estimator can also utilize the statistical analysis of construction functions to temper its judgment regarding how conservative the estimated production rates should be to cover the unknowns discussed

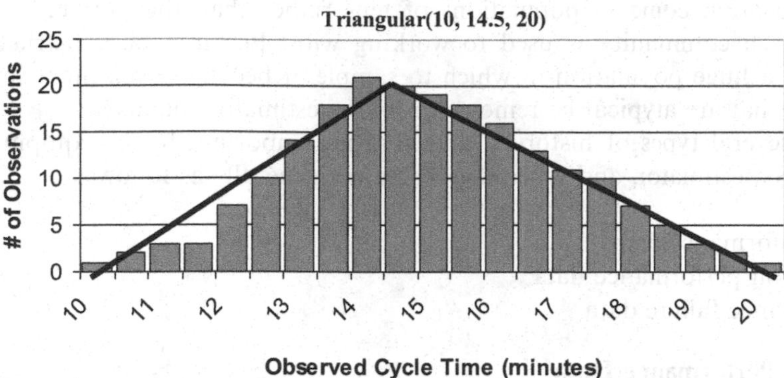


**FIGURE 6.6** Triangular Distribution Fit to Figure 6.1 Data.

in the first paragraph of this section. Given a mean and a standard deviation for a parameter like haul cycle time, the measures of central tendency and dispersion about the mean allow one to make the following judgments:

- If the standard deviation is relatively small as compared to the mean, then it indicates that variation in this parameter value will be correspondingly small. Thus, this situation would allow the estimator to use the mean as the predicted value for the parameter with good confidence.
- If the standard deviation is relatively large as compared to the mean, then it indicates that variation in this particular parameter value will be correspondingly great, and the estimator will need to use a value that is greater than the mean as the predicted value for the parameter in order to increase the confidence that the actual average value will be less than or equal to the value used in the estimate.

### 6.2.3 Historical Data

To conduct the types of statistical analyses discussed in the preceding section, one must have historical data on which to operate. Obtaining, reducing, and maintaining historical data is a difficult and time-consuming process. As a result, many equipment managers and estimators merely use the information from the last project and assume that it will be close enough to the upcoming project. This is not always the case. The heart of good estimating is a robust database that springs from actual costs and production rates that were achieved on a representative sample of past projects. The term "representative sample" is very important to understand.

Using all the past projects in an organization's files would mean including projects that had anomalies in them. For instance, a given project might have achieved highly weekly production rates because there was no adverse weather during the entire job. On the other hand, a project that was plagued with bad weather and muddy working conditions would experience below-normal production rates. Therefore, it is important to eliminate data from projects that were not typical from the production estimating database.

In statistical terms, this is called "removing outliers" and is one of the most controversial phases of a rigorous engineering research methodologies. There are accepted statistical methods for doing this, but they were designed for use on large sample sizes and would probably not be usable in this application. This is because construction projects for the average organization come in populations of tens rather than the hundreds or thousands that the research community is used to working with. Because the construction industry does not have a huge population in which to sample, it becomes even more important that those projects that are atypical be removed from the estimating database.

There are several types of historical data that are important to the equipment manager, project manager, estimator, and scheduler. These are generally as follows:

- Cost performance data
- Production performance data
- Maintenance failure data

#### 6.2.3.1 Cost Performance Data

Cost performance data is different from purely cost data in that it not only records the actual costs that were incurred but it also relates the actual costs to the estimated costs. It can be expressed as a ratio and described by the following equation:

$$\text{Cost performance} = \frac{\text{Actual cost for a given item}}{\text{Estimated cost for that item}} \tag{6.1}$$

Ideally, the ratio will be unity. However, if the ratio is greater than one, the estimator should investigate to find out what was different and adjust the estimator factors for future projects if necessary. The same thing should happen if it is less than one because this indicates that the estimate was too conservative and bidding too high often leads to not winning the project. In reality, there will be a range, say 0.95 to 1.05, within which the estimator will feel comfortable leaving the estimating factors alone.

The difference between cost performance data and cost data is an important distinction because cost estimates can be impacted by many different factors. Some of the factors can be controlled during project execution through diligent project management, but some of the factors are totally outside the ability of the project personnel to influence. A controllable factor would be the amount of overtime that is worked during a project, whereas an uncontrollable factor would be an unexpected raise in the rate of the overtime premium pay during project execution due to a newly negotiated union labor agreement. In the first case, the cost performance can be controlled by estimating a given amount of overtime work as a means of controlling cost risk and putting that cost in the bid. Whereas, in the second case, a change in the overtime premium rate is impossible to predict and, therefore, cannot be accounted for in the bid. Thus, in the first case, the estimator can check the performance of labor costs versus the estimate and determine if the amount of overtime estimated for that project was adequate and adjust that rate if it was too high or too low. In the second case, the estimator can only change the formulas for estimating total labor costs to reflect the newly negotiated labor rate. The cost performance in the second case becomes an anomaly and should be eliminated from the estimating database, as it is clearly atypical.

The second issue related to cost performance is the impact of inflation on the actual costs of labor and materials. A prudent estimator will normally factor in some inflation, particularly if the period of performance is quite long. Nevertheless, the market is full of surprises, and those contractors who had bid 2003 structural steel prices for projects to be constructed in 2004 got the surprise of their lives when those prices inflated by 40% to

60% [3]. This was an unexpected development and caused so much pain in the construction industry that many public owners negotiated price increases to critical construction contracts rather than bankrupt their contractors and have their projects go unfinished. Nevertheless, this is a useful example on cost performance. Projects that were underway when the steel price hike happened become anomalies because their ratio of actual costs to estimated costs will be quite high and not representative of future projects that will be bid using the higher steel prices.

The final issue with cost performance data is the age of the data in the database. At some point in time, the prices of old projects are no longer valid and only serve to arithmetically reduce the average values, which leads to cost performance ratios that are greater than one and worse, to reduced profitability. Thus, some mechanism for defining when a project's data has become dated and needs to be removed should be in place. A common system is to use a 24 to 36 month moving average, which automatically drops projects from the database when they become dated. When making these adjustments, the estimator must remember that cost performance is a direct function of production performance and ensure that the production rates achieved, which are reflected in the cost data, are both reasonable and representative of expectations for the given project environment.

### 6.2.3.2 Production Performance Data

Understanding how the actual production related to the estimated production is essential for any equipment-intensive project. These projects are by definition production-driven. The reader will be able to see just how important this is when reading Chapter 7 on scheduling. Thus, capturing production performance data is essential. Perhaps more important than capturing the cost performance data because the production data is related to the physical conditions in which the project was undertaken. The price of diesel fuel could double on a project, as happened in the summer of 2005, making the cost performance look awful, but if the production performance was close to unity, the estimator would know that the unexpected change in the price of fuel was the cause for missing the target profit, not the ability of the crews to construct the project as planned. Production performance can be expressed in an equation similar to cost performance:

$$\text{Production performance} = \frac{\text{Actual production for a given item}}{\text{Estimated production for that item}} \tag{6.2}$$

Typical production data should be collected for every major crew or equipment resource package. Tracking the actual production achieved during construction furnishes a good project control metric and assists the project manager in identifying those crews that require assistance or need to work extra hours early in the game. The one major failing of production performance data is that the estimator often assumes that all the machinery will be available all the time during the project. This is a poor assumption. Construction equipment gets hard usage and requires an aggressive preventive maintenance and repair program to maximize its availability. Thus, it is equally important to collect maintenance failure data.

### 6.2.3.3 Maintenance Failure Data

This data needs to be collected for each type of machine that is typically used in the course of the types of projects that a given organization will build. The purpose of this data is to give the estimator and scheduler the information necessary to adjust production assumptions to deal with this reality. This idea is best explained by Example 6.1.

**Example 6.1:** A bulldozer normally is used for five 10-hour shifts per week for 50 weeks each year. In the past three years, it has been unavailable because of breakdown, routine maintenance, and servicing 44 hours in year 1, 150 hours in year 2, and 163 hours in year 3. The earthmoving crew to which it is assigned has a sustained production rate of 2,000 cubic yards per day assuming 100% availability. Find the estimated production rate for a project that will last one year based on the data in this problem.

$$\text{Maximum availability} = 10\ \text{hr/day}(5\ \text{days/week})(50\ \text{weeks/year}) = 2,500\ \text{hours}$$
$$\text{Average yearly down time} = (44 + 150 + 163\ \text{hours/year})/3\ \text{years} = \text{hr/year}$$
$$\text{Adjusted production rate} = (1-(119/2,500))(2,000\ \text{CY/day}) = 1,905\ \text{CY/day}$$

At this point, having collected all the data that is required, the estimator can move on and develop the simulation model to allow the stochastic calculation of equipment production rates. It must be remembered that the purpose of taking this analysis to the next level is to get a better handle on the risks due to uncertainty. Understanding the credible range in possible production rates for every equipment resourced crew in the project increases the amount of analytic information available to the estimator and to the individual that will be responsible for making the final business decision of exactly how much the final bid price will be for the equipment-intensive project, or if the organization is a public agency, exactly 'how much funding will be authorized to self-perform the equipment-intensive project with in-house construction forces.

## 6.3 SIMULATIONS

Thirty years ago, simulations were only used by academics that had easy access to great amounts of computing power. As a result, the technique has gotten a bad name in the construction industry as a tool that only adds value to academic research. That stigma needs to be removed as the power of the personal computer has increased and the availability of inexpensive simulation software packages that operate from the standard commercial spreadsheets that are ubiquitous throughout the construction industry. As a result, the power of the enhanced information that can be obtained through the use of simulations is readily available to the average project manager, estimator, or scheduler if they are willing to take a few hours to learn how to make the simulation software operate with the spreadsheet program that they already know well.

There are several well-known simulations that have been used specifically for the purpose of analyzing equipment production. The major benefit derived is the ability to be able to inexpensively compare possible alternatives to see which one furnishes the maximum production or the minimum unit cost. For example, making the decision as to how many dozer-scraper teams to use on a given earthmoving project can determine the ultimate profitability of that job. It is economically impossible to go out to the field and physically run experiments with real equipment teams to determine which one optimizes the team's production. However, using a computer simulation, the estimator can "try" all the combinations and permutations of dozers and scraper that make sense for a given task and compare their outcomes mathematically. If the computer simulation is run thousands of times, the picture of the expected outcomes becomes more clear and the level of uncertainty that is associated with using a deterministic value for the parameters in the model is greatly reduced because eventually all possible outcomes will be found. The graph of the set of simulation outcomes becomes an approximation of the probability distribution for the different alternatives that can then guide the decision-maker to the final decisions.

### 6.3.1 Monte Carlo Simulation Theory

Most of the common commercial simulation software programs are based on Monte Carlo game theory. This type of simulation essentially uses a random-number generator to randomly select possible values for each of the variables in the model that have an associated probability distribution, and then it calculates the outcome. It then repeats this sequence many times and then accumulates each iteration's output to form an approximate probability distribution for the expected outcome. In equipment production simulation, those outcomes could be an expected daily production rate, an expected unit cost, or an expected length of schedule for the project. The estimator or project manager can then take that output and use it to adjust the bid pricing that is based on the deterministic solution to the same problem.

#### 6.3.1.1 Developing Monte Carlo Simulation Input

Input for a standard Monte Carlo simulation is described in detail in the preceding section. Developing the input is mainly a function of satisfying the input requirements of the simulation software that is being used. This will be shown by an example (Example 6.2) using a commercial software package called *@Risk* ® [4]. The information from Example 5.3 will be used.

**Example 6.2:** A 1.5 cubic yard front-end loader is going to load dump trucks with a capacity of 9.0 cubic yards. The loader takes 0.4 minutes to fill and load one bucket. The travel time in the haul is 4.0 minutes. Dump and delay times are 2.5 minutes combined.

$$L = 2.4 \text{ minutes}; C = 8.9 \text{ minutes and } N = \frac{8.9}{2.4} = 3.71 \text{ haul units}$$

Rounding down will maximize haul unit productivity. In other words, the haul units will not have to wait to be loaded, but the loader will be idle during a portion of each cycle. Rounding up will maximize loader productivity with the haul units having to wait for a portion of each cycle. This assumes that there will always be a truck waiting to be loaded as the loader finishes loading the previous truck.

$$\text{Productivity of 3 haul units} = 182 \text{ CY/hr}$$

$$\text{Productivity of 4 haul units} = \text{Loader productivity} = 225 \text{ CY/hr}$$

Example 6.2 is the deterministic solution to the problem of estimating the haul crew's production rates and making the decision of how many trucks should be assigned to this crew. Based on the assumption that the project manager would want to maximize production, then the crew would have a loader and four trucks assigned to it.

However, when the estimator collected the cycle time data, it was found that three of the variables had substantial ranges of actual times rather than specifically achieve the same time in every observation. First, the time it took the loader to load a single bucket varied between 0.3 and 0.5 minutes, with 0.4 minutes being the most likely time. Because the haul was over a public road, the trucks travel time also varied due to interference of public traffic as follows: 3.5 minutes if there was no traffic interference to 5.0 minutes if the truck got stuck behind a slow-moving vehicle with 4.0 minutes being the average time observed. Finally, the delay time was a function of the truck's ability to pass through a traffic light and make a left turn across traffic. It then varied between 1.5 and 3.0 minutes, with each 0.5-minute increment being as likely as the other because both possible delays were essentially random.

Based on the equations given in Chapter 5 for Example 5.3, a computer spreadsheet was developed based on the deterministic values given in the original problem statement. The @Risk ® software was then activated and three input variables were designated along with their associated probability distributions which are shown in Figure 6.7:

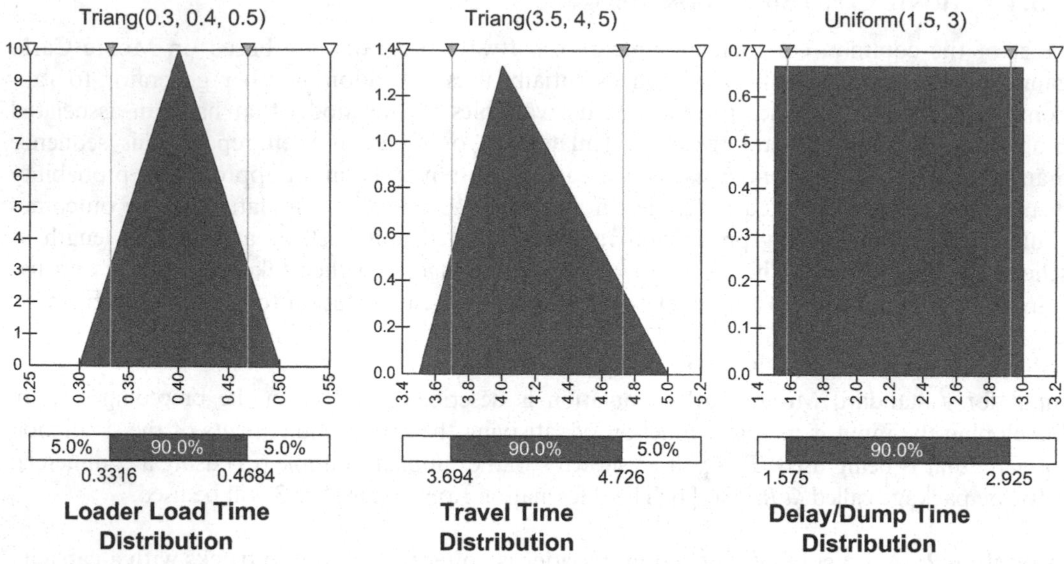


**FIGURE 6.7** Probability Distributions for Three Input Variables in Example 6.2.

- Loader load time: Assigned a triangular distribution with 0.3 minutes as the lowest possible value, 0.5 minutes as the highest possible value, and 0.4 minutes being the most likely time. *Triang(0.3,0.4,0.5)*
- Truck travel time: Also assigned a triangular distribution with 3.5 minutes as the lowest possible value, 5.0 minutes as the highest possible value and 4.0 minutes being the most likely time. *Triang(3.5,4.0,.5.0)*
- Delay/dump time: As this variable had the same probability for all possible times, it was assigned a uniform distribution with 1.5 minutes as the lowest possible value and 3.0 minutes as the highest possible value. *Uniform(1.5,3.0)*

Once all the input has been properly recorded in the simulation software, the simulation is ready to be run, and its output can be analyzed.

### 6.3.1.2 Analyzing Monte Carlo Simulation Output

The process of simulation output analysis will also be described using Example 6.2. After running 100 iterations of the Monte Carlo simulation, the output shown in Table 6.1 was produced. These numbers can easily be interpreted to assist the production-based equipment fleet size and composition decisions.

Looking at Table 6.1, two alternatives are shown. The first are the predicted hourly production rates if the crew is resourced with a 1.5-cubic yard front-end loader and four 9-cubic yard dump trucks and shown as simulation output 1. The second shown as simulation output 2 are the predicted rates if the crew is given only three trucks. As in the deterministic solution to this problem, the 4-truck crew has the higher predicted productivity. However, when one compares the mean predicted values to the values calculated using the deterministic model, it can be seen that both are higher, and the 3-truck crew is predicted to produce 13 cubic yards per more than was calculated deterministically. This is a 7% increase in that average production for that crew.

One can also see that the ranges in production for the two alternative crews overlap indicating that it is possible to achieve production rates greater than or equal to the 4-truck crew with the 3-truck crew. Thus, the potential for using three instead of four trucks and

**TABLE 6.1**
**Example 6.2 Simulation Output**

| Simulation Value | Variable Name | Deterministic Value | Minimum Value | Mean Value | Maximum Value | 5th Percentile Value | 95th Percentile Value |
|---|---|---|---|---|---|---|---|
| Output 1 | Production N rounded up to 4 trucks (CY/hr) | 225 | 180 | 227 | 288 | 192 | 269 |
| Output 2 | Production N rounded down to 3 trucks (CY/hr) | 182 | 130 | 195 | 265 | 164 | 245 |
| Input 1 | Loader load time (min) | 0.4 | 0.3 | 0.4 | 0.5 | 0.3 | 0.5 |
| Input 2 | Travel time (min) | 4.0 | 3.5 | 4.2 | 4.9 | 3.7 | 4.7 |
| Input 3 | Delay/dump time (min) | 2.5 | 1.5 | 2.2 | 3.0 | 1.6 | 2.9 |

saving the cost of a truck and driver for the entire length of the project should be investigated. Whereas in the deterministic analysis, it was immediately rejected as the estimated production of the 3-truck crew was substantially less than the 4-truck crew.

To assist in making this decision, another function of the simulation output can be consulted. Probability distributions for the output variables can be drawn and manipulated to gain additional information about each alternative. Figure 6.8 is the distribution for the 3-truck and 4-truck crews.

Making the assumption that the project manager believes that the crew needs to have a minimum hourly production rate of at least the 225 cubic yards per hour that was selected in Example 5.3, one can enter Figure 6.8 and find that there is a 15% probability that three trucks will produce at least 225 cubic yards per hour. Thus, there is an 85% chance that on any given hour, that production quota will be missed. The 4-truck crew has a probability of

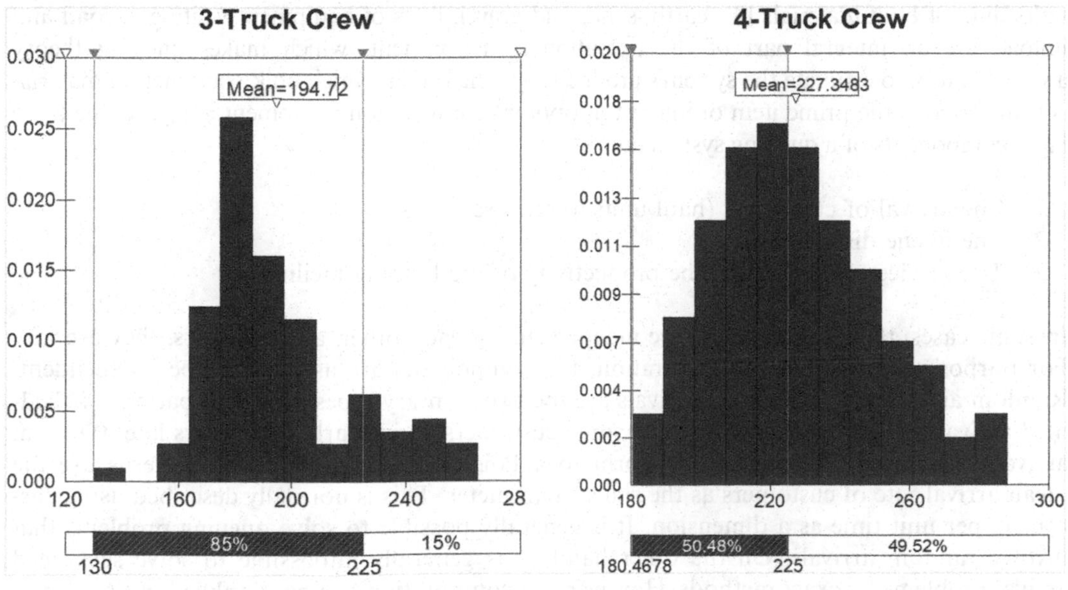


**FIGURE 6.8** Probability Distribution for 3-Truck and 4-Truck Crews in Example 6.2.

roughly 50% of meeting or exceeding the 225 cubic yards per hour production objective. This gives the decision-maker a way to quantify the risk of using only three trucks on this job rather than the four that would have been indicated from the deterministic analysis.

### 6.3.2 Other Simulations

There are two main simulation strategies: "Process Interaction" and "Activity Scanning." A third approach is called "Event Scheduling" and is sometimes combined with Process Interaction [5]. The software packages that utilize these simulation strategies generally require that their users be able to perform computer programming functions to input model characteristics and mathematics. These programs are very powerful and produce very specific information based on the quality of the input.

In the 1990s, equipment simulations were developed based on specific code using the approaches discussed in the preceding paragraph. CYCLONE is probably the oldest general construction simulation program in use. CYCLONE is an activity scanning-based simulation where the activity cycle diagram is the simulation model itself and not just a plan for a simulation program. STROBOSCOPE (an acronym for STate and ResOurce Based Simulation of Construction ProcEsses) is one that was designed specifically for the modeling of construction operations [5]. It is a simulation programming language based on the activity scanning simulation paradigm and activity cycle diagrams. The reader is directed to an excellent paper by Martinez and Ioannou [6] that contains a detailed description and comparison of these two construction simulation packages. The increase in the computing power and flexibility of commercial spreadsheets has made specific simulation programs obsolete. Hence, current practice combines commercial spreadsheets with open-ware or proprietary Monte Carlo add-ons to permit the analyst to develop individual simulations tailored to the specific problem at hand.

Another type of simulation that is generally applicable to construction equipment simulations is the program that is based on "queuing theory." In French, the word *queue* means a line. Queuing theory deals with the formation and operation of lines. By using the fundamental laws of probability and statistics, the planner can model almost any system in which a line is formed by using queuing theory. As the typical construction equipment project involves the loading and unloading of bulk materials like earth, sand, and gravel, lines of equipment waiting to load and unload are an integral part of the operational environment, which makes queuing theory a valuable tool to estimate the system's productivity. The rate at which bulk construction material is transferred is the prime item of interest in optimal construction equipment analysis. The three basic components of a queuing system are:

1. The arrival of customers (haul units, trucks, scrapers),
2. The queue discipline, and
3. The service of customers (the productivity of the loading facility).

In some cases, these components are independent of each other, and in others, they are not. For purposes of analysis and illustration, the components are assumed to be independent. Random arrivals and scheduled arrivals are the two primary types of arrival patterns. Scheduled arrivals include patterns in which some customers arrive early and others late. Random arrival patterns are assumed to conform to a Poisson distribution. Both patterns use the mean arrival rate of customers as the salient parameter. This is normally described using customers per unit time as a dimension. It is generally possible to solve queuing problems that involve random arrivals. On the other hand, it is generally impossible to solve scheduled arrival problems by exact methods. However, the computation can be simplified in the case of a construction hauling system by assuming that the haul units will arrive one at a time and that the time interval between arrivals will be a function of cycle time plus delays.

One simple commercial spreadsheet add-on that is available for this type of analysis is QUEUE.XLA [7]. This is an open-ware discreet event simulation and uses queuing theory to simulate the motion of customers through lines and predicts estimated waiting times for different combinations of customer service facilities. In construction, the service facilities would be the piece of equipment responsible for loading another piece of equipment, which in turn would haul, dump, and return to enter the queue and be serviced once again. Thus, this type of simulation software would be useful to compare alternative hauling crew sizes to reach a point where the waiting time is minimized. This is important because when a piece of construction equipment, be it loader or hauler, it is not productive, yet it costs money for the equipment and its operator to wait for its turn to become productive. General formulae for these answers for single channel systems are as follows:

$$W_q = \frac{\lambda}{\mu(\mu - \lambda)} \tag{6.3}$$

$$P_o = 1 - \lambda/\mu \tag{6.4}$$

where $W_q$ = mean waiting time
$P_o$ = probability that the facility is idle
$\lambda$ = mean arrival rate (haul units/unit time)
$\mu$ = mean service rate (haul units/unit time)

Example 6.3 shows how the simulation software based on this theory operates on a typical construction problem. The deterministic method in the example will obviously need to be developed for input into the stochastic simulation in much the same way as the Monte Carlo simulation example of the preceeding section.

**Example 6.3:** A ready-mix concrete company is going to replace its fleet of mixer trucks and is seeking to select the optimum number of trucks in relation to the service rate of its batch plant and the rate at which orders are received. The concrete batch plant loads mixer trucks with a maximum capacity of 12 cubic yards. Orders are filled on a first come–first served basis and are received on a random basis throughout the day. Some orders are placed in advance (i.e., scheduled), but the time of day at which concrete is required on an advance order is sufficiently spread across the course of an average day that all orders can be assumed to be random and conform to a Poisson distribution. On an average day, the company will fill orders totaling 480 cubic yards of concrete, with the average load being 9 cubic yards. The batch plant can discharge concrete into the mixer trucks at a rate of 2 cubic yards per minute. The total ownership and operating cost of the batch plant including labor and overhead is $1,500 per hour. The same cost for the trucks is $80 per hour. As the batch plant is the loading facility, this situation can be modeled as a single channel queuing problem with N =1.

$$\text{The arrival rate, } \lambda = \frac{480\,\text{CY/day}}{9\,\text{CY/loads}(8\,\text{hr/day})} = 6.67\,\text{loads/hr or } 53.33\,\text{loads/day}$$

$$\text{The service rate, } \mu = \frac{2\,\text{CY/min}(60\,\text{min/hr})}{9\,\text{CY/load}} = 13.3\,\text{loads/hr}$$

$$\text{From Equation 6.3: } W_q = \frac{6.67}{13.3(13.3 - 6.67)} = 0.08\,\text{hr/load} = 4.8\,\text{minutes/load}$$

and from Equation 6.4 : $P_o = 1 - \frac{6.67}{13.3} = 0.50$

Average total daily waiting time for the fleet = 0.08 hr/load (53.33 loads/day)
= 4.26 hr/day

Average daily loading facility idle time = 0.50 (8)= 4 hrs/day

At the optimum number of trucks [8]:

Cost of waiting trucks = Cost of facility idle time

Therefore,

Cost of waiting trucks = 4.26 hrs/day($80/hr) (N)

and Cost of idle facility = 4 hrs/day ($1, 500/hr)(1)

Setting these equations equal to each other and solving for N, the optimum number of trucks is 17.6. A rounding decision must now be made. If the number is rounded up to 18, each truck would average three loads per day. It is easy to see how queuing simulations could then be developed into stochastic simulations to enhance the amount of information that is available to the equipment manager.

#### 6.3.2.1 Developing Input

Input development for each of these other simulations will be specific to the software package itself. As a result, it is difficult to generalize the requirements at this writing. Nevertheless, the estimator will need to come prepared with as much historical data on the operations that are to be modeled as possible. As a minimum, the probable range of potential values for each variable and parameter in the model must be known to give the simulation program the ability to use realistic numbers. The old "Garbage In-Garbage Out" cliché is very applicable to the subject of equipment simulations.

#### 6.3.2.2 Analyzing Output

Again, the output will be specific to the program used. To analyze its meaning, the estimator will need to have a set of clearly written decision criteria that can be used to measure the relative success or failure of each alternative under analysis to satisfy the specific objectives of the simulation. Such criteria may look as follows:

- Minimize waiting time of haul units
- Maximize daily production rate
- Minimize unit costs

## 6.4 EXPECTED PRODUCTION

The purpose of a stochastic equipment production simulation is to compute expected production rates based on the constraints inherent to the project itself. The expected production output from these simulations is then utilized in two ways. First, these rates are adjusted as required using the estimator's professional judgment and used to compute the unit costs of equipment-intensive tasks on the project for use in bid preparation. In this application, the simulation gives the estimator the ability to quantify the uncertainty associated with each possible alternative and allows a decision as what each crew will cost based on their ability to produce. Secondly, the expected production rates that are used in the estimate become project control metrics that the project manager can use in the field to manage the job and measure project progress.

### 6.4.1 Cost Estimating Factors

To develop the estimating factors from the simulation output, the estimator will take the following steps:

- Select the crew size and composition.
- Select the expected production rate that flows from the crew size and composition.
- Gather equipment ownership and operating costs data including labor rates for the given crew.
- Apply those to the crew's expected production rate to arrive at an hourly crew cost.
- Given the hourly crew costs and the quantity of work to be performed by the crew, develop a unit cost for each pay item that the crew will accomplish.

This process will be illustrated by continuing with the haul crew developed in Example 6.2 with associated costs data.

**Example 6.4:** Calculate the unit cost for the crew given that the cost of for the 4-truck haul crew is as follows:

$$\begin{aligned} \text{Loader with operator} &= \$77.45/\text{hr} \\ \text{Truck with operator} &= \$83.75/\text{hr} \\ \text{Hourly crew cost} &= \$77.45 + 4(83.75) = \$412.45/\text{hr} \end{aligned}$$

From Table 6.1, the expected production = 227 CY/hr

Therefore: $\text{Unit cost} = \dfrac{\$412.45/\text{hr}}{227\text{CY}/\text{hr}} = \$1.82/\text{CY}$

It should be noted that this figure includes only the direct cost of labor and equipment and would then be marked up as appropriate to cover indirect costs, overhead, and profit, if applicable.

### 6.4.2 Production Management Factors

Production management factors are project control metrics that can be used in the field to make routine daily equipment management decisions and to measure project progress. For the haul crew in Example 6.4, the expected production rate was 227 cubic yards per hour. By definition, this rate must be understood as an average *sustained* production rate. Therefore, it serves as a basis to be expanded to a series of larger units of time to allow the project manager to measure the crew's actual productivity against its planned production. Example 6.5 shows how these are used.

**Example 6.5:** The project that the 4-truck haul crew will work on has a total of 175,000 cubic yards of aggregate to be hauled by the crew. The crew will work a standard 40-hour work-week. Calculate the daily and weekly minimum production rate based on the expected production rate from the simulation and determine how many weeks this project should last.

$$\text{Expected daily production} = 227\ \text{CY/hr}(8\text{hr/day}) = \underline{\underline{1,816\ \text{CY/day}}}$$

$$\text{Expected weekly production} = 227\ \text{CY/hr}(40\text{hr/week}) = \underline{\underline{9,080\ \text{CY/week}}}$$

$$\text{Expected project duration} = 175,000\ \text{CY}/9,080\ \text{CY/week} = \underline{\underline{19.3\ \text{weeks}}}$$

Given the above production factors, the project manager can now measure actual production against planned production and use the comparison to make equipment management decisions. For instance, if actual daily production has been exceeding the expected daily production and one of the trucks breaks down, the project manager can look at the accumulated excess production and determine whether or not to schedule the haul crew to work overtime to make up for the loss of one truck for a given period of time.

## 6.5 VALIDATING SIMULATION MODELS

The final requirement in utilizing a simulation model is to validate its accuracy by comparing predicted results with observed results. To do this, a validation data collection plan must be developed so that field personnel can obtain the necessary information that will allow the estimator to refine the model's assumptions and mathematical relationships. Each project that is simulated based on historical data will be somewhat different that the projects from which the historical data was drawn. Thus, recalibrating the simulation model is essential to preserving its authority for future estimates. The validation process essentially consists of verifying the model's assumptions and fundamental input data. Then a sensitivity analysis is conducted to identify those input parameters that have the greatest effect on the model's output.

### 6.5.1 Verifying Assumptions and Inputs

A good estimator always documents the assumptions that were used in the estimate. This rule applies to equipment-intensive project estimates as well as others. Once the simulations have been run and the estimating factors have been determined, it is worthwhile to review the initial set of assumptions that were used to develop the model to ensure that they have not been unintentionally changed during the analysis. One good method for doing this is the use of a trial set of input values that correspond with the assumptions that are being checked for which the estimator already knows the answer. These are fed into a fresh copy of the simulation, run through the model, and the output derived from this exercise is checked against the known values. If the answers are roughly the same, then the model development process did not unintentionally alter the initial set of assumptions.

The reader should remember that since most simulations use random values taken from each variables probability distribution, it is unlikely that the answers will be exactly the same as each simulation will use a somewhat different set of random values. If the output of the known sample does not match, then the estimator needs to thoroughly check the simulation model and identify where the error lies and correct it before moving on with the estimate using the simulations derived factors.

The next check could be best described as the "reality check." Regardless of the level of experience that the estimator has in the construction industry, all the salient assumptions and input values for production rates, crew sizes, and other issues that will eventually drive the bid's bottom-line should be presented to another knowledgeable and experienced member of the organization to see if they seem realistic at face value. This individual should preferably be one who will be responsible for eventually building the project the bid price. Any changes or adjustments that come out of this process should be made, and then the final simulation can be run to produce the production estimating factors.

### 6.5.2 Sensitivity Analysis

The type of sensitivity analysis that was described in Section 3.3 is also applicable to simulations as well. In essence, a sensitivity analysis is being performed as the simulation is run with the software randomly selecting possible values from within the specified range of the probability distribution for each variable. Commercial simulation software packages like @Risk® often conduct the sensitivity analysis in conjunction with the simulation and report the results in various forms in the simulation output. For the Example 6.2 simulation, tornado diagrams were produced for both alternatives and the one for the 4-truck haul crew that was selected, as is shown in Figure 6.9.

Looking at this figure, it shows that the production rate of the crew is extremely sensitive to the time it takes for the loader to load a single bucket. The reader will remember that the range of possible times for the loader was 0.3 to 0.5 minutes per bucket (18 to 30 seconds). That is a very narrow range. Each truck will require six buckets to fill it before it can leave to complete the haul. Thus, knowing this, the project manager needs to set up the area in which the loader will load the trucks in a fashion that will minimize, if not eliminate, any and all distractions to the loader operator. Other managers that will be on the project will need to be

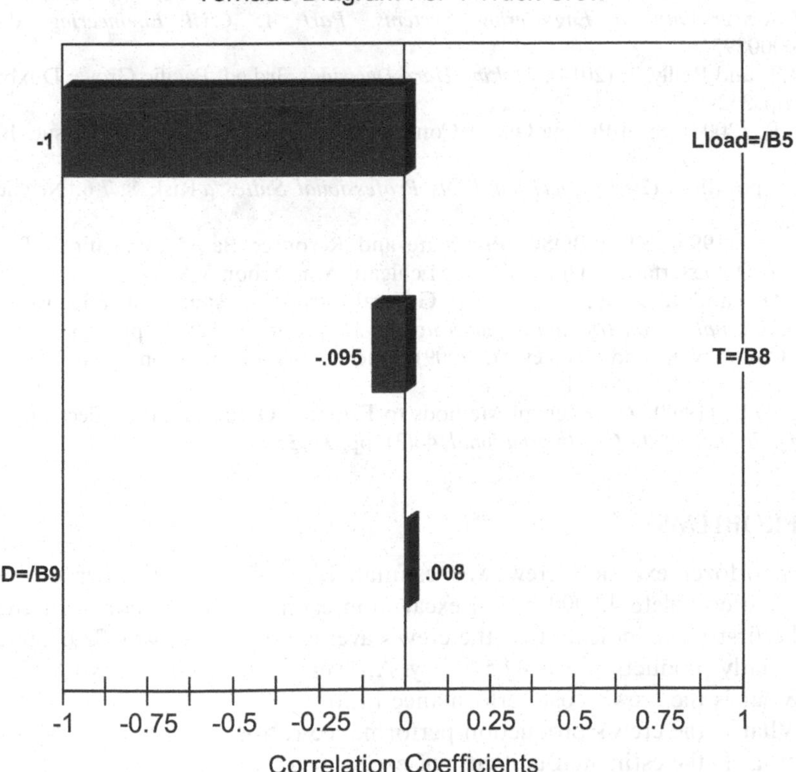


**FIGURE 6.9** Tornado Diagram for 4-Truck Haul Crew Sensitivity Analysis.

told to leave the loader alone and not ask for it to be pulled off the material haul for short periods of time to accomplish other minor tasks. Indeed, the very success of this task is dependent on the loader being able to maintain the cycle time that is going into the estimate. Since the model is extremely sensitive to this one variable, the range of values should probably be physically checked in the field before the bid is submitted.

This example furnishes an excellent illustration of how much additional valuable information can be derived at a very little cost from a simple Monte Carlo simulation. When equipment simulations first came on the scene in the days of the mainframe computer, they were merely an interesting academic exercise that took too long and cost too much to be of much added value to the estimator on an equipment-intensive project. However, with the advent of powerful personal computers combined with commercial simulation software packages that work with most common spreadsheet programs, they have become a mechanism for quantifying the uncertainty that is inherent to the estimating and bidding process. Simulation can give the estimator and project manager information that cannot be determined through simple deterministic spreadsheet calculations. Their greatest benefit is in their ability to more closely model the reality of a construction project where not every cycle is equal to or less than the cycle time used in the bid. By allowing experienced construction management professionals to see things in a different quantitative fashion, simulations allow those persons to temper their professional judgment with hard facts and numbers, creating a means to better manage the risks inherent to any equipment-intensive project.

## REFERENCES

[1] Gardner, B.J., Gransberg, D.D. and Rueda-Benavides, J.A. (2016). Stochastic Conceptual Cost Estimating of Highway Projects to Communicate Uncertainty Using Bootstrap Sampling. *Journal of Risk and Uncertainty in Engineering Systems, Part A: Civil Engineering*. doi: 10.1061/AJRUA6.0000895.

[2] Clemen, R.T. and Reilly, T. (2014). *Making Hard Decisions.* 3rd ed. Pacific Grove: Duxbury, Thomas Learning. pp. 352–371.

[3] Boyken, D.R. (2004). Steel Pricing Out of Control. *Developments.* Spring 2004 Issue. Boyken International, p. 6.

[4] Palisade Corporation. (2019). *Decision Tools Professional Suite.* @Risk ® 7.6. Newfield: Palisade Corp.

[5] Martinez, J.C. (1996). STROBOSCOPE: State and Resource Based Simulation of Construction Processes. Ph.D. Dissertation, University of Michigan, Ann Arbor, MI.

[6] Martinez, J.C. and Ioannou, P.G. (1999). General-Purpose Systems for Effective Construction Simulation. *Journal of Construction Engineering and Management*, 125(4), pp. 265–274.

[7] Morris, J., Connolly, K. and Hershey, W. (2009). Fine-Tuning a Federation of Models–The Quest for a Discrete Event Module. *Fuel*, 1, pp. 1–8.

[8] Gransberg, D.D. (1998). A Review of Methods to Estimate Optimum Haul Fleet Production. *Cost Engineering, Journal of AACE, International*, 40(3), pp. 31–35.

## CHAPTER PROBLEMS

1. A scraper/dozer exaction crew was estimated to cost $28,000 per day and was required to complete 40,000 bcy of excavation each day. The records for the project after the first week indicate that the crew's average daily cost was $29,750 and their average daily production was 42,570 bcy. Answer the following questions:
    a. What is the crew's cost performance ratio?
    b. What is the crew's production performance ratio?
    c. What is the estimated unit cost of excavation?
    d. What is the actual unit cost of excavation?

2. Given the probability density function shown in Figure 6.8, answer the following questions:
   a. What is the approximate probability that a 3-truck crew will exceed 160 cubic yards per hour production?
   b. What is the approximate probability that a 4-truck crew will exceed 160 cubic yards per hour production?
   c. What is the approximate probability that a 3-truck crew will exceed 260 cubic yards per hour production?
   d. What is the approximate probability that a 4-truck crew will exceed 260 cubic yards per hour production?
3. A hot-mix asphalt batch plant loads 40-ton bottom-dump trucks. Orders are filled on a first come–first served basis and are received on a random basis throughout the day. The orders can be assumed to be random and conform to a Poisson distribution. On an average day, the company will fill orders totaling 2,000 tons with the average load being 38 tons. The batch plant can discharge hot-mix asphalt into the trucks at a rate of 4 tons per minute. The total ownership and operating cost of the batch plant including labor and overhead is $8,000 per hour. The same cost for the trucks is $180 per hour. Modeling this system as a single channel queuing problem with N =1, answer the following questions.
   a. Arrival rate, $\lambda$.
   b. Service rate, $\mu$.
   c. Mean waiting time, $W_q$.
   d. Probability that the facility is idle, $P_o$.
   e. The optimum number of trucks,
   f. Discuss the implications on production and cost of the rounding decision.

# 7 Scheduling Equipment-Intensive Horizontal Construction Projects

## 7.0 INTRODUCTION

Projects that require large numbers of machines present their own challenges to construction schedulers. Activities such as preventive maintenance, stand-by vehicles, and multiple shift operation require special knowledge to integrate into the normal project schedule. Most, if not all, construction projects have contractual completion dates that, if missed, significantly alter the financial profitability of the project through either liquidated damages or late-finish penalties. Even in the absence of these, it hurts the contractor's overall profitability because keeping equipment committed to a given project longer than originally planned prevents it from working on other projects where it could have earned the company additional profit in that fiscal year. Thus, understanding the scheduling dynamics of an equipment intensive project is key to achieving target profit margins for the equipment's owner.

From a scheduling point of view, equipment-intensive projects are defined as those projects where the production of the equipment is significantly more important than the production of the labor. For instance, a highway construction project may have 30 employees working on it each day in a variety of trades. However, the great majority of those people will be operating pieces of construction machinery, with only a few being assigned jobs as common laborers using hand tools. Conversely, labor-intensive projects typically have a few pieces of equipment that are used for localized tasks and most of the workers are in trades that require hand tools like cement finishers, carpenters, and common laborers. Thus, the scheduler of an equipment-intensive project is primarily concerned with managing the work site in a manner that eliminates conflicts between the equipment; whereas, the scheduler of a labor-intensive project is primarily concerned with managing the work site in a manner that eliminates conflicts between the trades. The scheduler of an equipment-intensive project establishes the durations of each activity in the schedule based on shift length and equipment package production rates, and the scheduler of a labor-intensive project establishes the duration of each activity in the schedule based on shift length and trade crew production rates.

In each case, the scheduler's task is to devise a sequence of work that permits the project to be finished before its contract completion date. However, in the equipment-intensive project, the scheduler typically is dealing with large quantities of materials that must be moved, processed, and/or installed, and a minor error in the assumed production rate of a given equipment package can be translated into a major time or cost problem. As a result, scheduling an equipment-intensive project requires an in-depth understanding of how production rates of interdependent equipment packages impact the schedule.

## 7.1 BACKGROUND

At this point, the reader should note that the following is a brief description of the mechanics of schedule development and is intended as a "refresher" for a reader who has knowledge of the subject. It is not intended to be a comprehensive review of scheduling methods. If the reader needs further detail on this subject, the references listed at the end of this chapter should be consulted [1–4].

There are two different methods that are used to mathematically model the construction schedule and deconflict the schedule for simultaneous activities, which in turn produces a schedule that reasonably estimates whether the project can be completed in the time allotted. The first method is called Critical Path Method (CPM). There are two different CPM models: Activity-on-Arrow (AOA) and Activity-on-Node (AON) [1]. Figure 7.1 shows the notional concepts of each. AOA only permits a relationship between activities (called a precedence relationship) where one activity must finish before subsequent activities can start (called a Finish-to-Start (FS) relationship). As a result, it is very cumbersome to work with when modeling large projects with thousands of activities. AON, on the other hand, allows the following four different precedent relationships to be modeled, which are shown in Figure 7.2:

1. Finish-to-Start (FS): The preceding activity must finish before subsequent activities can start.
2. Start-to-Start (SS): The preceding activity must start before subsequent activities can start.
3. Finish-to-Finish (FF): The preceding activity must finish before subsequent activities can finish.
4. Start-to-Finish (SF): The preceding activity must start before subsequent activities can finish.

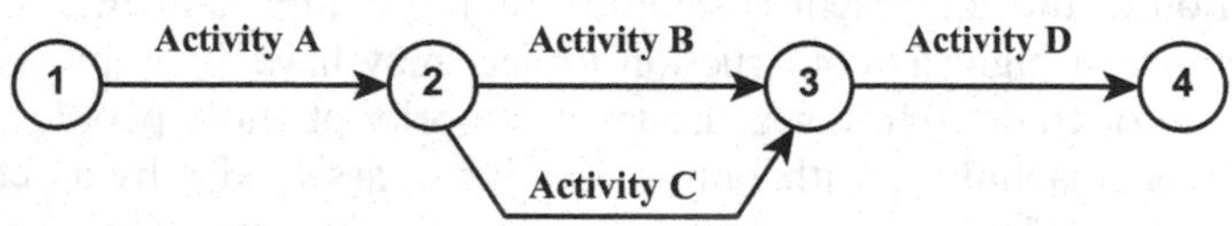


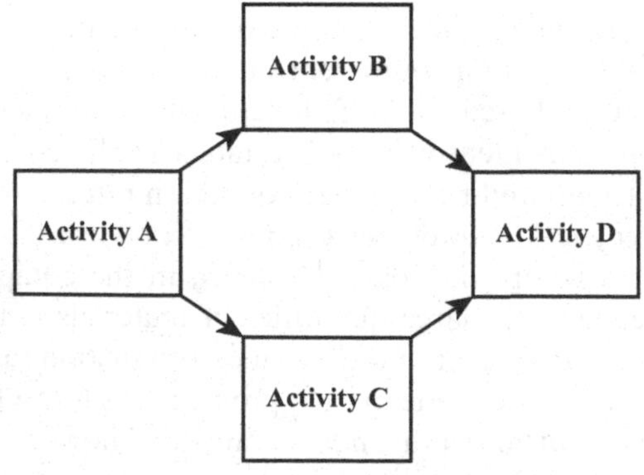


**FIGURE 7.1** Notional Concepts of the AOA and AON Networking Methods.

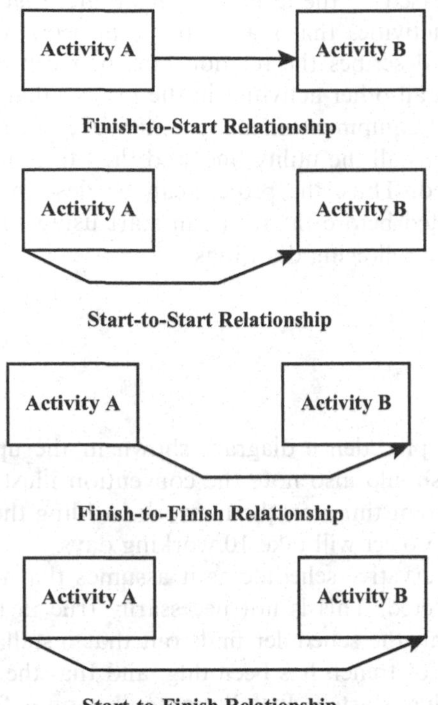


**FIGURE 7.2** PDM Precedence Relationships.

The AON method has been adopted by most current commercial computer scheduling packages. Section 7.2 will briefly explain the AON network described as the Precedence Diagramming Method (PDM). CPM's primary goal is to identify the sequence of work that is most critical to completing the project on time and calls this the "critical path." All other related work sequences have float or slack time in their path, and thus, the scheduler can concentrate on managing the work sequences that are critical.

The second common scheduling technique is called Linear Scheduling (LS). Linear scheduling is production-based rather than activity-based like PDM. Although some writers have described ways to determine the critical path of a linear schedule, its primary goal is to maximize the production of all equipment resource packages (more commonly called "crews") by ensuring that one activity's production rate does not unintentionally control the production of another one. This is done using a graphical approach rather than the networking approach used in CPM. The graph has two axes: time is on the "Y" axis, and location is on the "X" axis. Thus, LS not only tracks the project in time, but it also ensures that there are no conflicts between crews on the actual ground. Section 7.3 explains how this is accomplished.

The remainder of this chapter will briefly describe the internal algorithms of each scheduling method and then show how linear scheduling can be used to plan the work sequence for the major features of work and then converted to a PDM schedule where all other activities can be added in a comprehensive schedule for the entire project.

## 7.2 PRECEDENCE DIAGRAMMING METHOD

PDM is nothing more than AON with a few extra rules to provide continuity in the process of developing a logic diagram. This diagram takes the form of a network that logically lays out the sequence of work based on the technical constraints that define the precedence relationships between all the activities that make up the project. A precedence relationship is a technical constraint that describes the relationships between either the start or finish of the activity in question and all other activities in the project that are related to it.

For example, in a typical equipment-intensive utility project, the contractor must dig the utility trench before it can install the utility line, and the utility line must be installed before the trench can be backfilled. Thus, the project can be described as three linear activities where one must be completed before the next can start using Finish-to-Start (FS) relationships. The activities have the following durations:

- Trenching: 3 days
- Installation: 5 days
- Backfill: 2 days

With this information, the precedence diagram shown in the upper portion of Figure 7.3 can be drawn. The reader should also note the convention illustrated in this figure for displaying the early and late event times on the network. Adding the cumulative durations, the scheduler can see that this project will take 10 working days.

Obviously, this is a conservative schedule as it assumes that no activity can start before the preceding one is completed. This is not necessarily true in this case. After discussions with the field superintendent, the scheduler finds out that installation of the utility line can begin after one day's worth of trench has been dug, and that the backfilling can begin three days after the installation has started. Installation still cannot finish until the trenching is finished, and the same finish relationship holds true between installation and backfilling. Thus, the scheduler can revise the schedule, as shown in the lower part of Figure 7.3, and sequence the activities in parallel. In other words, the diagram now shows that on the second day of the project, both the trenching and installation crews will be working. The scheduler does this by using SS and FF relationships and showing the delay between the start of trenching and the start of installation by the small box on that arrow with a number showing how many days of "Lag" exists on that path. In other words, the start relationship between the two activities can now be articulated as follows: *Installation cannot start until 1 day AFTER Trenching starts.* Thus, this is a SS relationship with one day of lag.

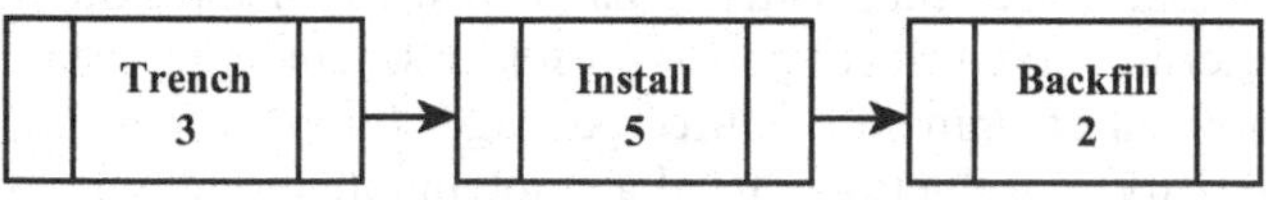


**Utility project with linear work sequence**

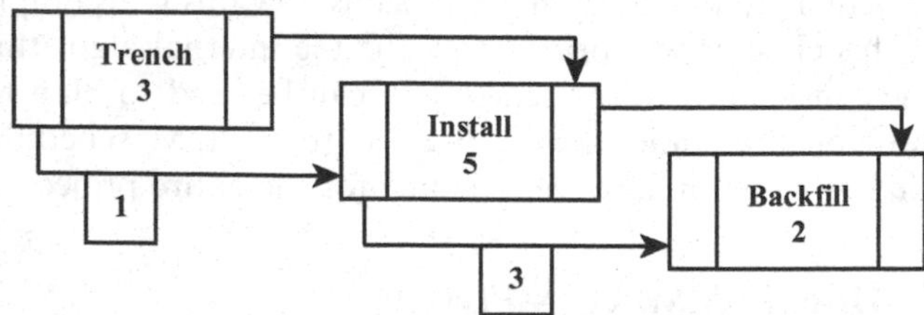


**Utility project with parallel work sequence**

**FIGURE 7.3** PDM for Linear and Parallel Work Sequences.

### 7.2.1 Determining the Critical Path

To determine how long a project will take, a series of calculations must be made on the PDM network. The following are three discreet sets of calculations:

- Forward pass calculation: This process determines the early event times (the earliest each activity can begin and end) for the project.
- Backward pass calculation: This process determines the late event times (the latest each activity can begin and end) for the project.
- Determination of float: Float is the amount of time an activity can start after its early start time without creating a delay for the entire project.

These calculations are best understood by demonstration. The two networks shown in Figure 7.3 will furnish the input for this demonstration.

#### 7.2.1.1 Forward Pass Calculation

Looking first at the network describing the linear work sequence, one can conduct the forward pass calculations using the following formulas:

$$EF = ES + d \tag{7.1}$$

Where, EF = Early finish time
ES = Early start time
d = Duration of the activity,

and

$$ES(\text{succeeding activity}) = EF(\text{preceding activity}) \tag{7.2}$$

Thus, looking at the first activity "Trench" with its duration of 3 days, its ES would be 0, then:

$$EF_{\text{Trench}} = 0 + 3 = 3 \text{ days}$$

Moving on to the next activity, "Install," it cannot start until "Trench" is finished so the ES for "Install" (the succeeding activity) would equal the EF of "Trench" (the preceding activity) or 3 days. Next, adding its 5-day duration to its ES, we get:

$$EF_{\text{Install}} = 3 + 5 = 8 \text{ days}$$

Using the same calculation, we can then calculate the EF of "Backfill" to be 10 days, and this concludes the forward pass for this network. The results are displayed in Figure 7.4 for both networks. Finally, if more than one arrow leads into a start of an activity from two or more preceding activities, the highest of the EF values of the preceding activities should be selected as ES for the activity in question.

Next, one can see that the parallel work sequence allows the contractor to complete the project 4 days earlier than the linear work sequence.

#### 7.2.1.2 Backward Pass Calculation

Next, the late event times need to be calculated using the backward pass that is based on Equations 7.3 and 7.4:

$$LS = LF - d \tag{7.3}$$

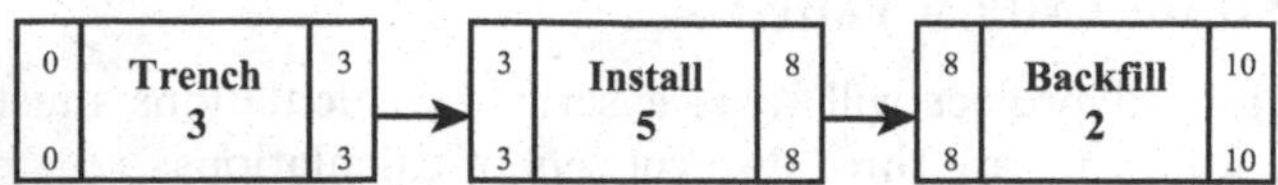


Utility project with linear work sequence

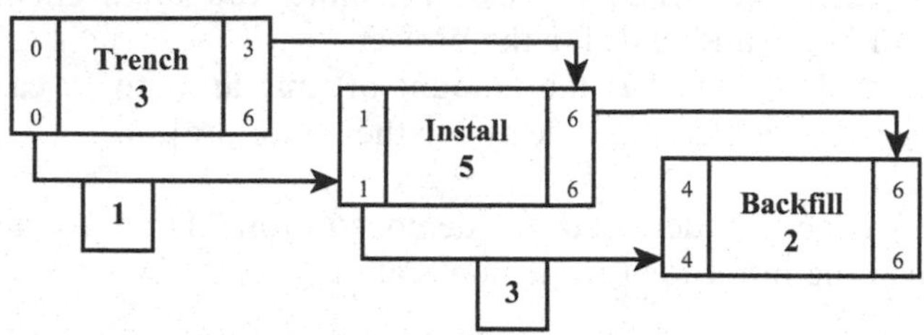


Utility project with parallel work sequence

**FIGURE 7.4** PDM Forward and Backward Pass Calculations.

Where, LS = Late start time
LF = Late finish time
d = Duration of the activity,

and

$$\text{LF(preceding activity)} = \text{LS(succeeding activity) if FS} \tag{7.4a}$$

or

$$= \text{LF(succeeding activity) if FF} \tag{7.4b}$$

If both FS and FF exist, then:

$$\text{LF(preceding activity)} = \text{the lowest value of all relationships} \tag{7.4c}$$

As the linear work sequence network contains only FS relationships, and there is only one path through the network from beginning to end, the late event times found in the backward pass calculation are equal to the early event times for every activity. Therefore, the backward pass will be illustrated using the parallel work sequence network shown in Figure 7.4. Starting with the last activity "Backfill" and assuming that the project must be completed as quickly as possible, the LF will equal the EF or 6. We then execute the calculations shown in Table 7.1 and can arrive at the late event times for each activity. With the early and late event times calculated, the scheduler can now calculate the float for the two networks.

#### 7.2.1.3 Calculating Float

In PDM, there are three types of float and the equations for calculating each are as follows:

- Start Float (SF): Float associated with an activity's start.

$$\text{SF} = \text{EF} - \text{ES} \tag{7.5}$$

**TABLE 7.1**
**Backward Pass Calculations**

| Act | Dur | LF | Calculation | LS |
|---|---|---|---|---|
| Backfill | 2 | 6 | LS = 6–2 | 4 |
| Install | 5 | =LF(Backfill) = 6 | LS = 6–5 = 1 | |
| | | OR | LS = LS(Backfill)–Lag<br>= 4–3 = 1 | 1 |
| Trench | 3 | =LF(Install) = 6 | LS = 6–3 = 3 | |
| | | OR | LS = LS(Install)–Lag<br>= 1–1 = 0 | Select the smaller of 3 or 0<br>Therefore, LS = 0 |

- Finish Float (FnF): Float associated with an activity's finish.

$$FnF = LF - LS \quad (7.6)$$

- Total Float (TF): Float associated with the path in which an activity falls.

$$TF = LF - ES - d \quad (7.7)$$

It is possible for an activity to have zero SF or FF and still have a quantity of TF. This means quite literally only the start or finish of that activity has no float, but the path (sequence of work) in which it falls does have float. The scheduler must calculate the float for every path in the network to determine the critical path. The critical path is the longest path through the network and is determined by the sequence of activities in which the float is zero. Table 7.2 is the matrix associated with the parallel work sequence shown in Figure 7.4 with the associated float calculations completed.

Looking at the float in the network, one can see that everything except the finish of "Trench" has no float and is therefore critical. This table shows us that we must start "Trench" on its early start date but could finish it 3 days late and still be able to finish the project by the end of day 6. However, this is not the case with the other two activities, which must start and finish on their early dates. Any increase in their actual durations will cause the project to finish later than the end of day 6. Thus, it is important to ensure that those two activities are assigned sufficient resources to ensure their timely start and completion. This then leads to the discussion of critical resource identification.

### 7.2.2 Critical Resource Identification

It is imperative that the project manager knows which resources are assigned to the project can directly impact on the project's timely completion. It is also essential that the

**TABLE 7.2**
**Float Calculations for Parallel Work Sequence**

| Activity | d | ES | EF | LS | LF | SF | FnF | TF |
|---|---|---|---|---|---|---|---|---|
| Trench | 3 | 0 | 3 | 0 | 6 | 0 | 3 | 3 |
| Install | 5 | 1 | 6 | 1 | 6 | 0 | 0 | 0 |
| Backfill | 2 | 4 | 6 | 4 | 6 | 0 | 0 | 0 |

estimator who bids the job has a pretty good idea as to which resources will become critical.

There are two common ways to identify critical resources. The first is to look at the float available in each activity in the schedule and declare that the resources associated with each critical activity automatically become critical resources. Thus, in the above example, if the "Install" activity required a backhoe, a pipe truck, and an air compressor plus the workers to operate them, these machines with their operators and any other associated labor would be coded as critical resources. Measures would then be taken to ensure that these resources would be assigned to the critical activities as prescribed by the schedule before allowing them to be used on other less critical tasks. This is a good project control technique, and it highlights the connection between the estimated time of performance and the labor and equipment costs contained in the cost estimate because it links crews to the production activities on which they will be used.

The second method takes a less theoretical and more pragmatic approach to this issue. It recognizes that the schedule is just an estimate of the time it will take to complete the project and sees the activity durations as targets rather than absolute values. The durations are a function of the production rates that the estimator applied to each crew and can be changed by adding or subtracting resources as required to meet the activity completion targets (also referred to as "milestones"). Thus, it looks at the project manager's ability to acquire more equipment and labor on short notice to differentiate between critical and noncritical resources. This method then defines a critical resource as one that physically or technically cannot be increased within a specific period of time and then seeks to "hand-manage" all critical resources letting the noncritical ones rise and fall based on the needs of the project. The best example of a critical resource in this approach is a tower crane on a building project. Typically, the project will erect one tower crane to service the project from beginning to end. Thus, the scheduler must ensure that if there is only a single crane on a given project that no two activities that require the crane are scheduled to happen at the same time. Often this will put activities that require the critical resource on the critical path because they must be scheduled in series as they cannot be scheduled in parallel.

The major advantage of the second approach is that it greatly reduces the number of critical resources that must be managed during project execution. This allows the project manager to maintain a keen focus on those resources that most greatly affect the actual duration of the project. The danger that comes from selecting this approach over the first one is that it tends to uncouple the equipment resource package assumptions and their planned production rates that were made during the estimate from the actual execution of the project by allowing the project manager to increase the allocation of noncritical resources to accommodate the short-term needs of the project. If this is done with great discipline, its attendant increase in actual cost can be controlled. However, one must be careful not to give in to the temptation to retain the larger size equipment resource package as insurance against unknown future delays. Thus, to mitigate the risk that the actual cost of production will exceed the estimated cost, the scheduler then needs to resource load the schedule.

### 7.2.3 Resource Loading the Schedule

Resource loaded schedules are typically done using commercial construction scheduling software [4]. They can be quite arcane and complicated. The aim of resource loading is to ensure that the schedule is indeed realistic within the constraints of the available resources and their associated production rates. The purpose of this section is not to "teach" the reader the mathematical mechanics of resource-loaded schedules. It is rather to highlight the benefits that can be achieved by utilizing this powerful project control tool, and show through a simple example of how this technique can be used to increase the accuracy of the estimate for an equipment-intensive project.

Resource loading has two objectives:

- To permit the accumulation of resource requirement data across the life of the project, which then permits the project manager to plan the hiring of labor and the acquisition of equipment for the project.
- To ensure that the durations for those resources that have been designated as critical are realistic for the production rates that can be reasonably be achieved by the given crew.

Resource loading is accomplished by developing a resource dictionary in the software package that is being used by the scheduler. This task essentially consists of identifying the equipment and labor requirements need for each crew and then associating them with the activities to which they will be assigned. The software then produces a histogram for each individual trade and specific piece of equipment, which then allows the project manager to procure these resources, as they are needed in the project. The following discussion demonstrates what happens inside the software when a project's schedule is resource loaded using the information contained in Table 7.3.

Figure 7.5 below shows the resource histograms for the backhoe requirements for the two potential work sequences in the Figure 7.4 example. One can see that by choosing to complete each activity before starting the next activity, the project can be completed with only one backhoe at the cost of 10 days worth of backhoe rental. However, if the parallel work sequence is selected, the project will need two backhoes on days 2, 3, 5, and 6. Because the second backhoe will be idle on day 4, the estimator will need to include 11 days of backhoe cost to the estimate unless the project manager has the flexibility to return the backhoe for one day to the rental company or use it on another project. This simple example shows how important it is to plan the details of construction equipment utilization during the bidding process to ensure that the final cost estimate reflects the actual work sequence that will be used on the job, which quite naturally leads to the discussion of cost loading the schedule.

### 7.2.4 Cost Loading the Schedule

Cost loaded schedules are commonplace in the commercial building construction industry. The American Institute of Architects standard contract between the owner and the general contractor [5] contains a provision requiring the submission of a "schedule of values" against which the contractor will be paid for satisfactory progress each pay period. To do so requires the estimator to accumulate all the direct costs associated with each feature of work that is listed in the project schedule of values, assign indirect costs and profit margins, and furnish a lump sum value for each feature of work. After this, the owner would pay

**TABLE 7.3**
**Resource Requirements for Utility Project**

| Act | d | Backhoe | Pipe Truck | Air Compressor | Tamping Machine | Operator | Laborer |
|---|---|---|---|---|---|---|---|
| Trench | 3 | 1 | | | | 1 | 1 |
| Install | 5 | 1 | 1 | 1 | | 1 | 5 |
| Backfill | 2 | 1 | | | 1 | 1 | 2 |

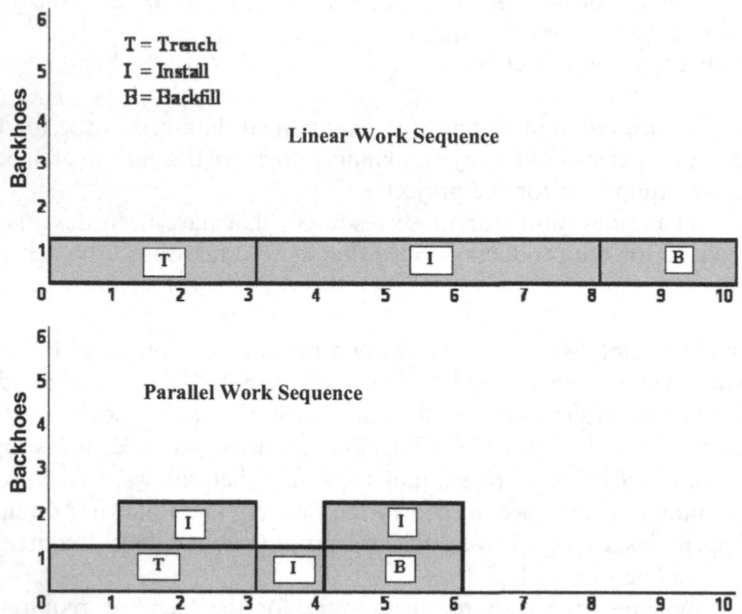


**FIGURE 7.5** Resource Histogram for Utility Project Options.

a portion of that lump sum commensurate with the item's current completion percentage. In other words, if a work item has a value of $100,000 and is 50% complete, the owner would then pay $50,000 less any retainage that might be appropriate.

Many, if not most, equipment-intensive projects utilize unit price rather than lump sum contracts. Thus, the unit price for each pay item reflects its individual cost and markups. Some unit price contracts have hundreds of individual pay items, making them difficult to use the pay items themselves as a cost control measure during construction as it is extremely complicated to relate them directly to an activity-based schedule. Thus, the estimator must roll-up the various pay items and precisely synchronize them with the work described for each activity in the schedule. This creates a schedule of values for the project and the necessary input to allow the project manager to cost load the schedule. Table 7.4 shows the costs associated with the example utility project.

Figure 7.6 shows the utility project parallel work sequence schedule as a cost loaded bar chart. One can see that when displayed in this manner, it shows the expected cash flow for the life of the project. This is sometimes called "earned value" because once the contractor has completed an activity, the activity's value has been "earned" or in other words, the contractor

**TABLE 7.4**
**Cost Loading Input for Utility Project**

| Activity | Duration | Cost ($) | Daily Cost ($/day) |
|---|---|---|---|
| Trench | 3 | $36,000 | $12,000 |
| Install | 5 | $150,000 | $30,000 |
| Backfill | 2 | $28,000 | $14,000 |

*NOTE:* the above costs were contrived for example purposes only and do not reflect any attempt to achieve realism.

| | Day | 1 | 2 | 3 | 4 | 5 | 6 |
|---|---|---|---|---|---|---|---|
| Activity | d | | | | | | |
| Trench | 3 | 12K | 12K | 12K | | | |
| Install Pipe | 5 | | 30K | 30K | 30k | 30K | 30K |
| Backfill | 2 | | | | | 14K | 14K |
| Daily Cost | | 12K | 42k | 42k | 30k | 34k | 34k |
| Cumulative Cost | | 12k | 54k | 96k | 126k | 170k | 214k |

**FIGURE 7.6** Cost Loaded Bar Chart for Utility Project.

can apply to be paid for that activity. For instance, by the end of day 3, this contractor would have expected to earn $96,000 because activity "Trench" would be complete, and two days worth of "Install Pipe" would also have been finished. Cost loaded schedules allow one to track the financial completion of the project, along with its physical completion.

Figure 7.7 shows the daily cost histogram and the cumulative cost curve for the project. These two graphics can be very useful project control measures. Looking at this at a more

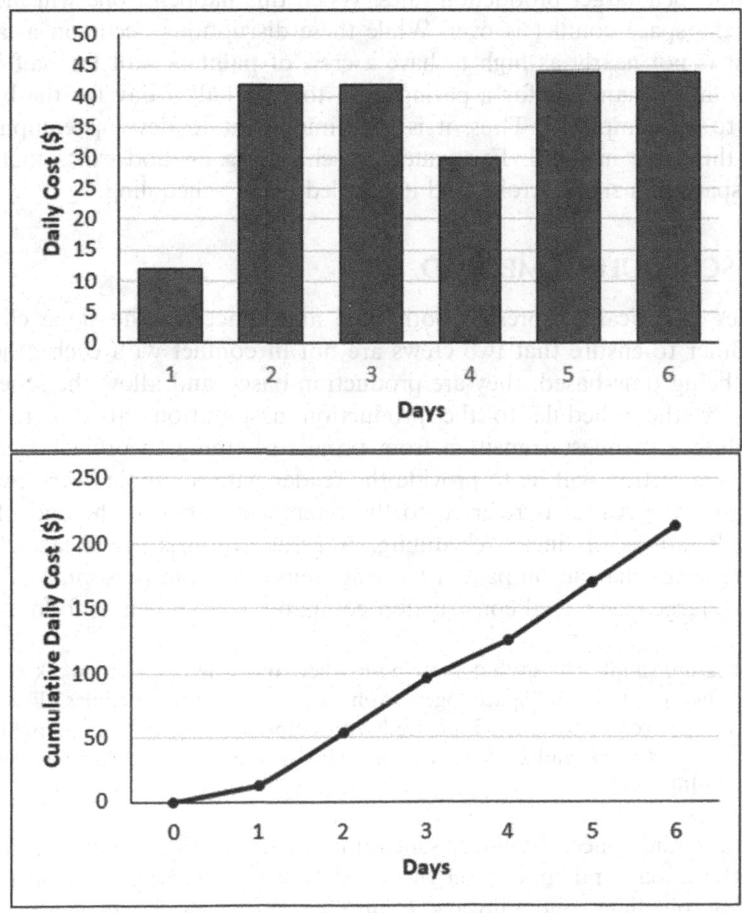


**FIGURE 7.7** Daily Cost Histogram and Cumulative Daily Cost Curve for Utility Project.

global level than shown by this extremely simple example, in order to finish as planned, a project must be properly financed, and inadequate cash flow is one of the major causes of contractor bankruptcy [5]. Thus, taking the project plan consisting of the estimate and the schedule and reducing it to the visual form illustrated in Figure 7.7 can give a project manager a very powerful tool with which to develop the project's financing plan.

For instance, if the example utility project could only receive $35,000 per day of financing, the daily cost histogram shows there are four days where this plan will exceed that limit. Therefore, even though there seems to be no technical reason why the project cannot be finished in six days, the financial constraints will make it impossible. This would force the project manager to seek a different schedule that fit the financial constraint just described.

To briefly summarize, the PDM of scheduling is an activity-based methodology that permits the scheduling of equipment-intensive projects. Its output can be structured to furnish powerful project control tools that are useful in executing the project plan in a manner that fits both cost and time constraints. The major assumption that the estimator must make in using this method is that the durations derived from the production rates for the crews associated with each activity will not be hindered or conflicted by the work going in other parallel activities. This is because this scheduling method has no inherent algorithmic mechanism to manage both space and time simultaneously. Thus, the danger in the field is that two sets of equipment and their crews will converge on the ground and need to use the same space in order to maintain their target production rates. When this happens, one will inevitably delay the other until the space conflict is over. While these disruptions occur on a labor-intensive project, the cost is not nearly as high to have a crew of painters wait for half a day for the drywall crew to finish than it is for a paving train to wait half a day for the base on which they will pave to be completed. Thus, it is very important to develop equipment-intensive schedules with this issue in mind. Fortunately, a scheduling method exists that can manage both time and space in a single stroke, and it is called linear scheduling.

## 7.3 LINEAR SCHEDULING METHOD

Linear schedules graphically represent both time and space on the same chart and thus allow the scheduler to ensure that two crews are not in conflict with each other. Additionally, instead of being time-based, they are production-based and allow the scheduler to precisely synchronize the schedule to the production assumptions that were used in the estimate, providing a seamless transition from project planning to project execution. Thus, the purpose of this section will be to provide the reader with a simple overview of this technique. Once again, the reader is referred to the references listed at the end of the chapter for detailed explanations of linear scheduling. A great synopsis on linear scheduling was published in the notes that accompany a training course for the personnel of Peter Kiewit and Sons, Inc., a large heavy civil construction company. It is shown as follows:

> Linear schedules are simple charts that show both when and where a given work activity will take place. Because they put time and space together on one chart, linear schedules allow us to see how the pieces of the project fit together. Enhanced with color, varying shades, or patterns, they also communicate types of work and crew movement. This is something neither bar charts nor CPM schedules can do [6].

To track both time and space, the linear scheduling method utilizes a standard graph where location is on the x-axis and time is on the y-axis. Thus, it is best used on linear projects such as highways, pipelines, and railways. Figure 7.8 shows the concept for a road project that is 1,000 feet (10 stations) long. In this example, there are three activities:

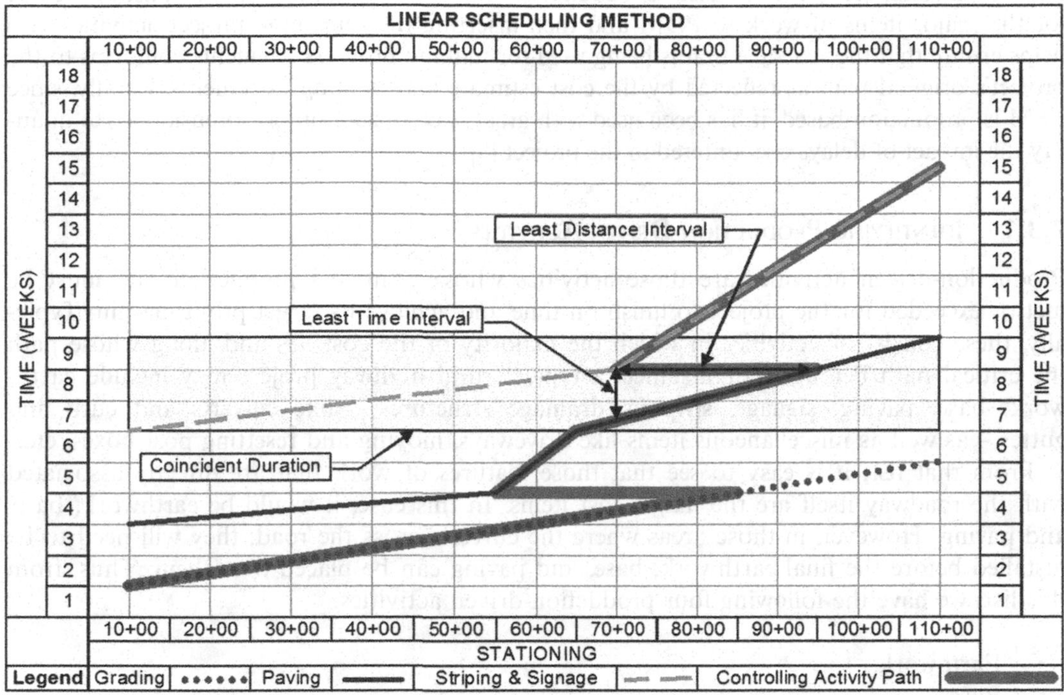


**FIGURE 7.8** Example Linear Schedule for Rural Highway Rehabilitation Project.

1. Grading
2. Paving
3. Striping and Signage

In this figure, the contractor has decided to start at the east end of the project (STA 1+00) where it has an area to store its equipment and then work on the westbound lane toward the west (STA 0+00) completing subgrade preparation and the stabilized base. It will then do the same activities on the eastbound lane working back to the east end (STA 10+00). Because the asphalt batch plant is located west of the project, it will reverse the work process and pave from west to east and back. Figure 7.8 shows the final schedule for this project. The steps used to arrive at this will be explained in the upcoming sections of this chapter. This scenario was selected to demonstrate the flexibility inherent with linear scheduling.

Linear scheduling is best used in the planning process of the project [3,6,7]. While it is possible to display every single feature of work on the chart, it is often not valuable to do so. Thus, most practitioners that use this technique use it to plan the "big items of work" and then convert the result to a PDM network adding the remainder of the project's less important activities at that stage. This approach has three advantages. First, it allows the scheduler to utilize the estimator's cardinal production rates for the major features of construction to ensure that those items that are critical to project profitability are scheduled in a manner where the planned production rates can be realistically achieved. Next, it creates a focus on those activities that will generate the majority of the cash flow and keeps them from being lost in the minutiae of the complete construction schedule for a major project. It does this by forcing the scheduler to "plug" the minor items of work into the schedule around the major items and prevents them from unintentionally controlling the overall pace of construction as could happen when PDM is used alone. As many owners require the use of a construction schedule configured in a specified commercial PDM-based scheduling software package, converting the linear schedule for the major items of work to PDM and then inserting the remaining project activities also helps ensure that the schedule that is being used for making progress payments conforms to the project's original plan as reflected by the cost estimate and bidding documents. Finally, since LSM is production-based, it has been used with great success as a simple forensic tool to quantify the impact of delays encountered in the project [8].

### 7.3.1 Identifying Production-Driven Activities

Production-driven activities are those activities whose estimated production rate must be met or exceeded for the project to finish on time and achieve its target profit margin. Typically, these consist of activities in which the majority of the cost lies and along whose path the critical path belongs. For instance, a typical rural highway project may include earthwork, base, paving, signage, striping, drainage structures – like culverts and curb and gutter – as well as miscellaneous items like driveways, moving and resetting post boxes, etc.

From that list, it is easy to see that those features of work that are directly associated with the roadway itself are the major cost items. In this case, it would be earthwork, base, and paving. However, in those areas where the culverts cross the road, they will need to be installed before the final earthwork, base, and paving can be placed over them. Thus, from this list, we have the following four production-driven activities:

- Earthwork,
- Base,
- Paving, and
- Culverts.

The remainder of the work can be scheduled as required in a manner that will not interfere with the production-driven activities. Additionally, depending on the project site, it may be possible to construct the culverts at the same time as roadway activities are underway as long as the culverts are completed before the other crews physically reach the locations of the culverts. Thus, the initial focus of the linear schedule will be to deconflict the space between the culvert construction crews and the other three roadway construction crews in a manner that permits all four to achieve the production rates assumed in the project estimate. Once the production-driven activities are identified, their requisite production rates can be calculated.

### 7.3.2 Establishing Production Rates

As discussed in Chapter 5, there are two forms in which production can be expressed:

- Instantaneous production: the production rate of a single cycle.
- Sustained production: the average rate of multiple cycles over a protracted period.

To this list, the required production rate should be added. The required production rate has little to do with the numbers or sizes of equipment in the resource package. It is project dependent and is the number of units of work that must be accomplished divided by the amount of time the contract allows for that work to be completed. It can be expressed by the following equation:

$$P_r = \frac{M}{CT} \tag{7.8}$$

Where, $P_r$ = Required daily production rate (units per day)
M = Total units that need to be processed or moved (units)
CT = Contract time allowed (days)

This is the starting point for the production rate calculations that are made in conjunction with the development of the linear schedule. The idea is to ensure that adequate resources are allocated to each crew to ensure that the project's contractual requirements are met. This is best explained by Example 7.1.

**Example 7.1:** A project involves hauling 100,000 cubic yards of aggregate to a stockpile at an asphalt batch plant. The contract allows 10 working days in which to build the stockpile. Using the 18 cubic yard dump truck and the production rate of 360 cubic yards per hour shown in Example 5.3, one can calculate how many trucks will be required to ensure that the project will be completed on time.

$$P_r = \frac{100,000\,\text{CY}}{10\,\text{days}} = 10,000\,\text{CY/day}$$

$$N_r = \frac{10,000\,\text{CY/day}}{360\,\text{CY/hr}(8\,\text{hr/day})} = 3.5 = 4\,\text{trucks}$$

One can see from Example 7.1 that no matter what equipment is selected, the crew must have a minimum sustained production rate of at least 10,000 cubic yards per day. Thus, all the crews are assembled and optimized with the required production rate for each in mind.

To use LS, the estimator/scheduler must convert the standard rates of production that are in material units over time, to time over units of space. In Figure 7.8, the unit of space was stations. (By way of review, one station equals 100 linear feet.) Therefore, all the crews must have their production rates converted in the unit days per station to be able to be put into the linear schedule. Again, this can easily be illustrated by Example 7.2.

**Example 7.2:** For the project in Figure 7.8, the subgrade preparation work involves scarifying, compacting, and shaping the existing subgrade to grade. The road will consist of two 12-foot lanes and a 4-foot shoulder on each side. The project is ten stations long. The crew for this feature of work has a sustained production rate of 356 square-yards per day. Determine the production rate in days per station for the activity. The contractor will do one lane plus its shoulder at a time.

$$\text{Roadbed width} = 1\text{ lane}(12\text{ FT}) + 1\text{ shoulder}(4\text{ FT}) = 16\text{ FT}$$

$$\text{Area/station} = \frac{16\text{ FT}(100\text{ FT/STA})}{9\text{ SF/SY}} = 177.8\text{ SY/STA}$$

$$P_s = \frac{177.8\text{ SY/STA}}{356\text{ SY/day}} = 0.5\text{ days/STA/lane}$$

$$\text{Total time} = (0.5\text{ days/STA/lane})(10\text{ STA})(2\text{ lanes}) = 10\text{ days}$$

Looking at the slopes of the lines shown in Figure 7.8, one can see that the crew assigned to installing the stabilized base has a production rate of 0.33 days per station and the paving crew has a production rate of 0.2 days per station.

### 7.3.3 Lines, Bars, and Blocks

Linear schedules consist of the following elements:

- Lines: Lines are used to represent activities that are in continuous movement. The line literally tracks the progress of the crew in time across the project site. The slope of the line represents the crew's production rate. As the line's slope increases, the crew's production rate decreases. Thus, a fairly flat slope indicates a very fast production rate. Figure 7.8 shows three activities that are all represented by lines.
- Bars: A vertical line that indicates a crew working in a single location for a long period of time is called a bar. Often bars are used to represent a series of interrelated activities. In PDM, this would be termed a "hammock activity" [4]. For instance, the construction of a culvert at a particular station would involve a number of separate activities, including preparing the pipe's bed, laying the pipe, construction of headwalls, etc. A single bar covering all the activities involved in the culvert for the entire period of its construction can be used on the linear schedule.
- Blocks: A block is typically a rectangle that literally "blocks" out an area of the project for a specified period. These are used when an activity, like grading, will move back and forth over a specific area rather than through it from one end to the other (a line). Blocks are used if the activity will occupy the space for a relatively long period of time. Thus, other activities shown as lines cannot progress through the block until its duration is completed.

Once again, this process is best explained by talking the reader through an example problem (Example 7.3) for this type of schedule.

**Example 7.3:** A small highway rehabilitation project is awarded and the contractor decides to use linear scheduling to plan the construction sequence and make sure that none of the crews conflict with each other or the other constraints imposed on the project. The project description is as follows:

- Mobilization will occur between STA 9+00 and STA 10+00. It will take 2 days.
- The first task is to demolish 1,900 LF of existing pavement, which has a sustained production rate of 300 LF/day.
- After the pavement is demolished, the next activity is to install cement treated subbase (CTSB). This crew will have a sustained production rate of 267 LF/day.
- On top of the CTSB, asphalt stabilized base (ASB) must be installed, and that crew has a sustained production rate of 200 LF/day.
- Finally, Type A hot-mix asphalt paving will be laid on the completed ASB at a sustained production rate of 400 LF/day.
- A series of small concrete box culverts must be built between STA 9+00 and 11+00. The group will take 8 working days. The specifications restrict putting equipment loads on the new culverts until 3 days after the last culvert is poured. This work is accomplished by a concrete subcontractor and includes excavation, backfill, and final grading for the culverts.
- The last activity in the project is clean-up/demobilization, and it will take 2 days.

The linear schedule will be done using working days and to simplify this example, the assumption will be made that this project will work 7 days per week until it is finished and that the contractor can work the entire width of the roadway without traffic control or detour considerations. Thus, Figure 7.9 shows the first step in assembling the linear schedule for this project. The time is shown on the vertical axes, and the location in stations is shown on the horizontal axis. All the respective crews have had their sustained production rates converted to linear feet per day, which will allow them to be quickly converted to days per station and plotted on the linear schedule as needed. As mobilization will occur in the vicinity of the area between STA 9+00 and 10+00, it is shown as a bar, one station long and two days –, its duration – high.

Next, one must step back and think about the technical sequence of work. Obviously, the first activity will necessarily be the demolition of the existing pavement, which is followed by the CTSB, ASB, and Type A pavement. However, in the STA 9+00 to STA 11+00, the series of box culverts will need to be built BEFORE the new subbase, base, and pavement can be constructed. Therefore, the pavement should be removed in that area first to allow the concrete subcontractor to go to work and get those structures built. Therefore, the demolition of existing pavement activity is scheduled to start at STA 9+00 and proceed to STA 19+00 as shown in Figure 7.9. This allows the concrete subcontractor to get into the area in which it needs to work at the start of day 4, as shown in Figure 7.10.

The "Build Culverts" activity is shown as a bar extending from STA 9+00 to STA 11+00 and 9 days tall. Additionally, the constraint imposed by the contract specifications that no construction wheel loads may be placed on the new culverts until 3 days after the last one is completed is shown in this figure as a block two stations wide and three days high. Once the culvert activity is placed on the graph, the remaining pavement demolition can be added. The pavement demolition at STA 19+00 is completed at the end of day 5. The contractor then moves the crew back to STA 9+00 and completes the pavement demolition from STA 9+00 to STA 0+00. Now that this activity is complete, the CTSB activity can be added to the linear schedule.

Technically, in the areas where there are no culverts, the CTSB can be started as soon as the pavement demolition is completed. However, at STA 9+00 to STA 11+00, the culverts must also be complete. Therefore, there are two options possible for scheduling this activity, and the linear schedule greatly assists in visualizing both. The first option is to start the CTSB crew at STA 19+00 as soon as the demolition crew is clear of that area and work back toward the beginning of the project. When the CTSB crew gets to STA 11+00, it will have to move around the culvert construction area and pick up at STA 9+00. Then, it will need to go back as soon as the load restriction block is over and put the CTSB over and between the culverts.

Looking at Figure 7.11, the lightly dotted line called out as the "Early Start CTSB" schedule shows this option. As the load restriction block does not end until day 14, the early start option forces the CTSB to stand down on day 12 and wait 2.5 days for the load restrictions to be lifted. This is not satisfactory. It would be better to bring the crew on later and have it work continuously until the activity is complete. So the final schedule for the CTSB crew, shown by the heavy dotted line, is to have it start on day 8 rather than day 6, which gives it a half day of float to move back to the culvert area and finish the CTSB between those two stations. Note, when the scheduler converts the linear schedule from working days shown in the figure to actual calendar days, every effort should be made to schedule the work so that the "No Loads" block falls on a weekend or holiday to minimize the loss of production due to this restriction. If this can be done, the scheduler may choose the start the CTSB earlier to take advantage of allowing the culverts to cure over a nonwork period.

This is an excellent example of how the visual format furnished by a linear schedule assists the scheduler/estimator in making good work sequencing decisions that might not be discovered using a PDM network. It is likely if this project had been scheduled using PDM, that the project manager might have brought the CTSB crew onto the job on its early start date (day 6) and then they would have been nonproductive for that 2.5-day period before the load restrictions on the new culverts where lifted. This would have caused the contractor

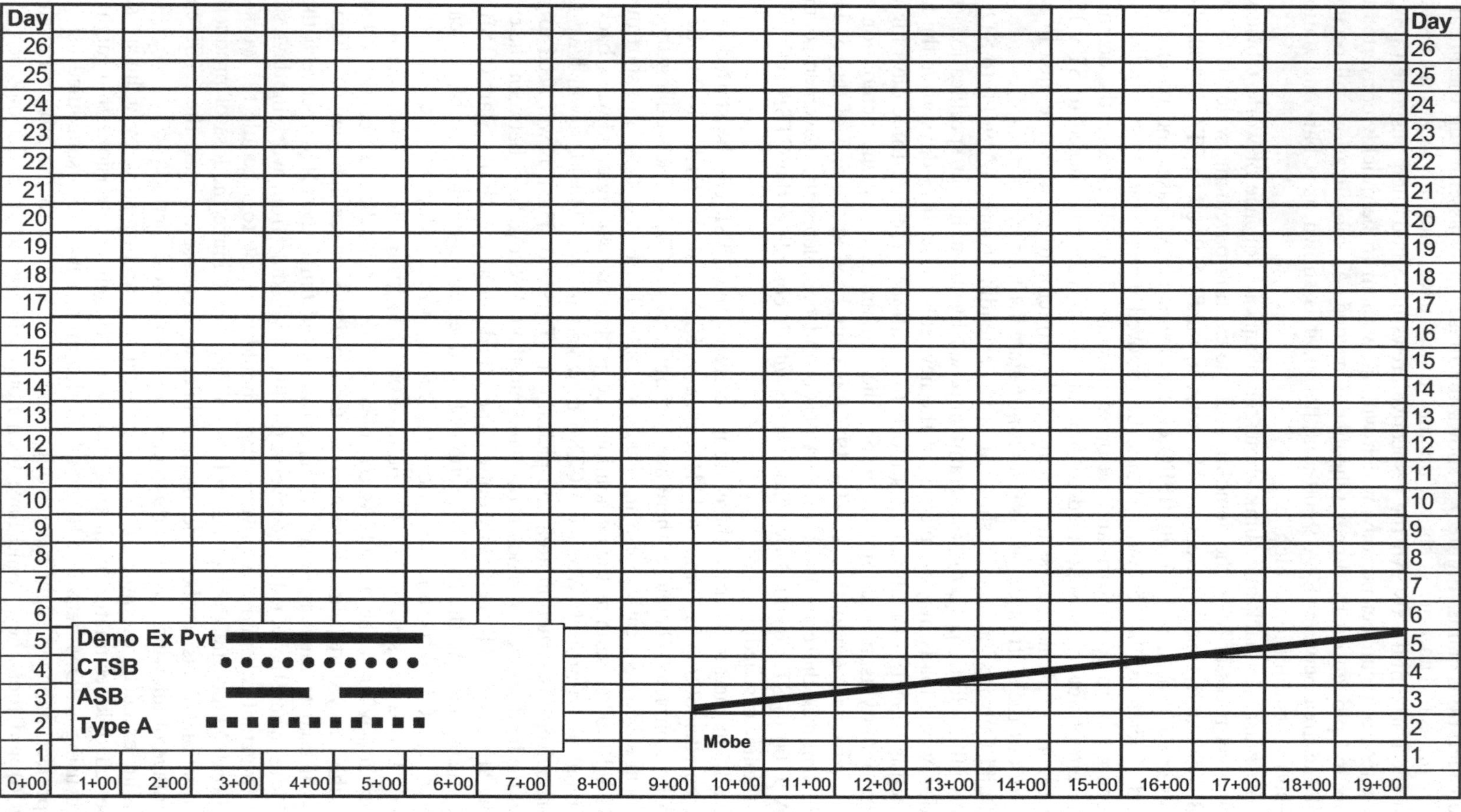


**FIGURE 7.9** Linear Scheduling Example Step 1.

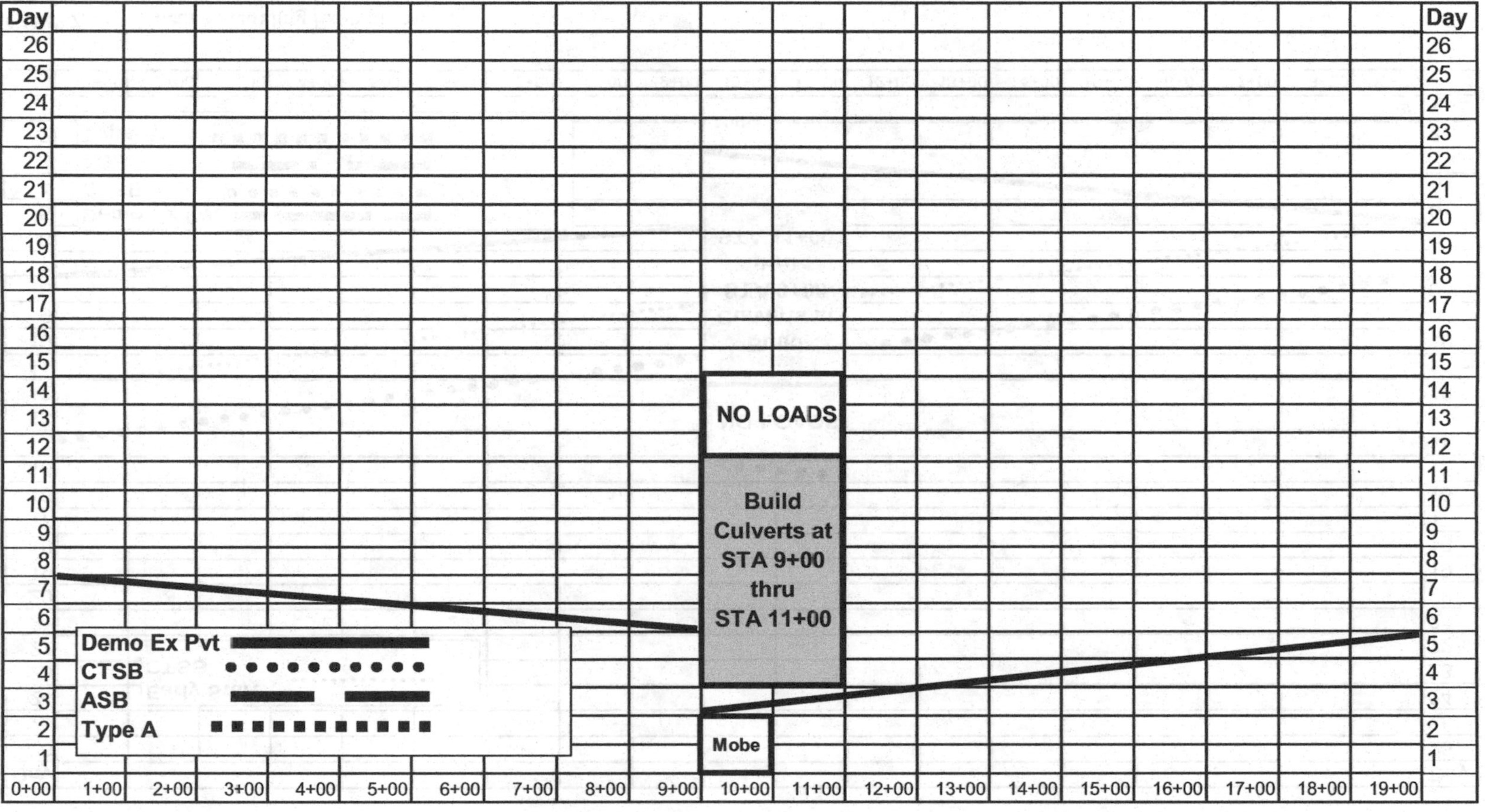


**FIGURE 7.10** Linear Scheduling Example Step 2.

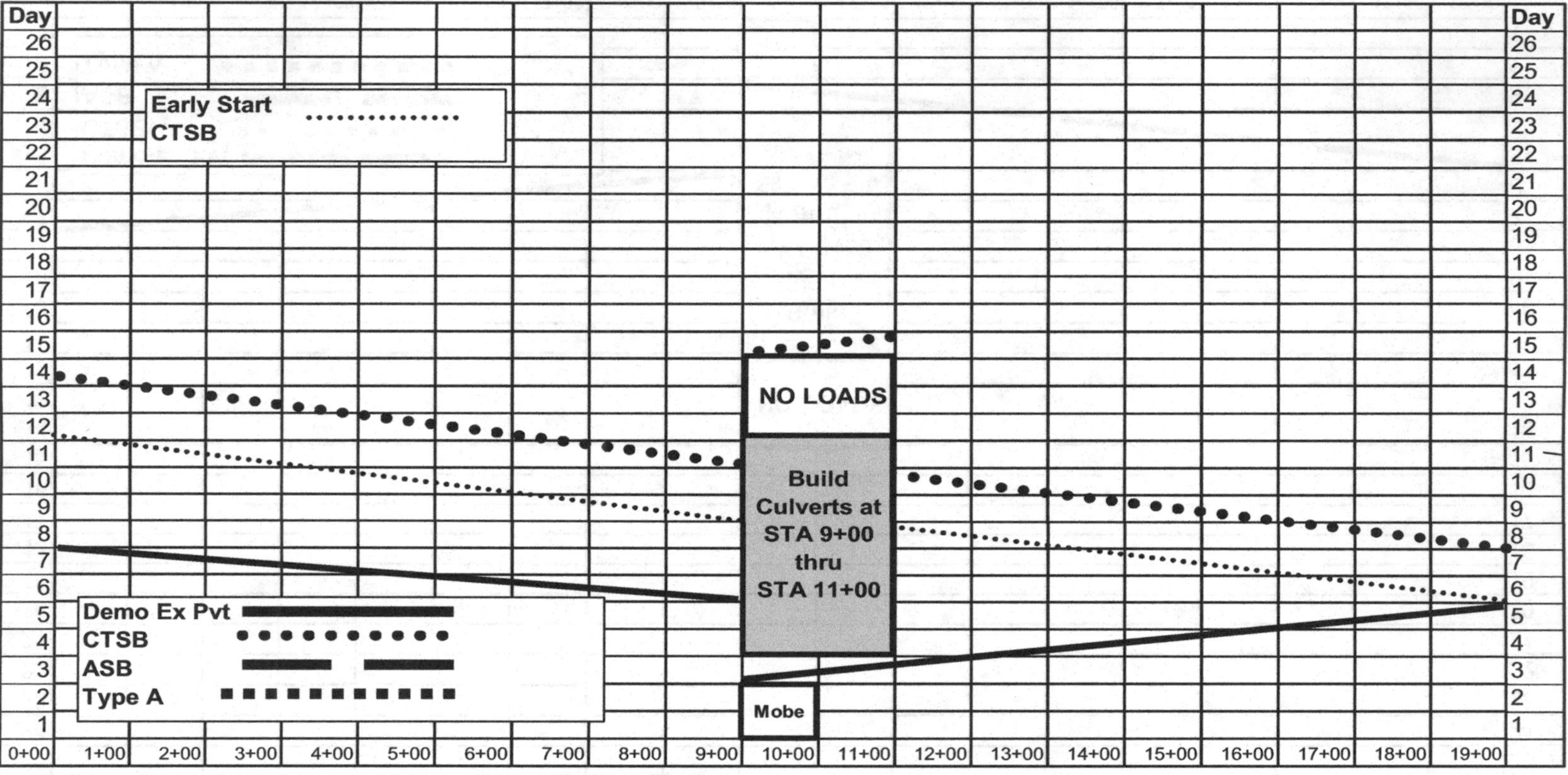


**FIGURE 7.11** Linear Scheduling Example Step 3.

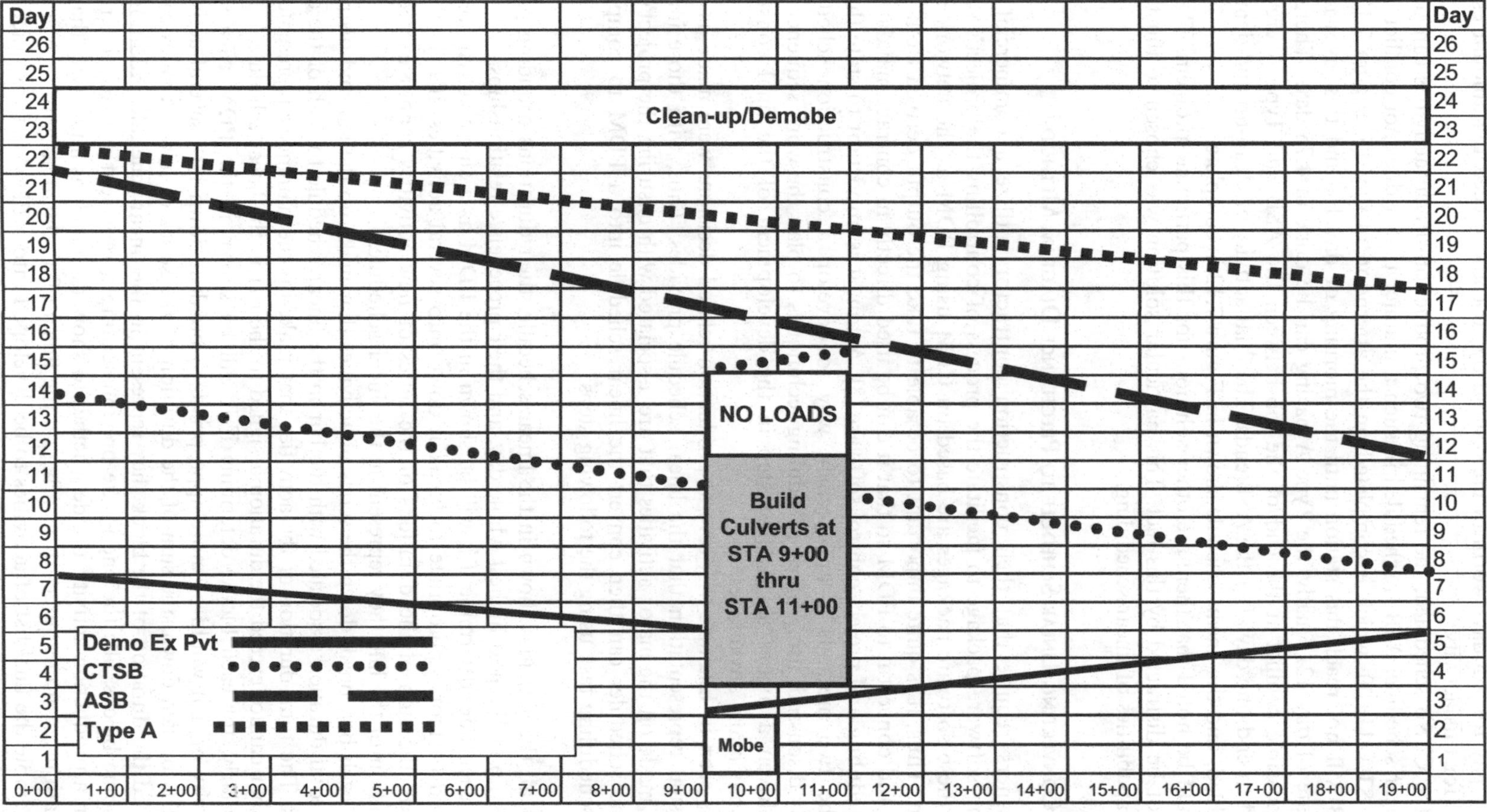


**FIGURE 7.12** Linear Scheduling Example Step 4.

to have to pay for 2.5 days worth of crew stand-by cost and exceed the amount estimated in the bid for this activity.

With the CTSB scheduled, the remaining two production-driven activities can be added to the linear schedule. ASB is scheduled in such a manner that it does not conflict with the CTSB at STA 11+00, which is completed on the afternoon of day 15. Figure 7.12 shows that ASB will not reach this station until the morning of day 16, and this gives the ASB a start date of day 12. Finally, the Type A paving can be done in 4.75 days. Therefore, to avoid a possible conflict at the end of the job between the ASB and Type A crews, it is scheduled to end one day after the ASB ends. The final activity of clean-up and demobilization is added, and the linear schedule shown in Figure 7.12 is complete.

It should be noted that there are other solutions for this particular problem. The reader should not be distracted by this fact. This particular solution was chosen to illustrate the various capabilities of linear scheduling.

### 7.3.4 Converting Linear Schedule to Precedence Diagram Method

Many owners require that their construction contractors utilize a commercial project scheduling software package to facilitate the process of controlling the project's schedule. Most of these software packages are based on CPM using PDM as the network analysis algorithm. Thus, it is quite important to be able to take the output derived from a linear schedule and convert it to PDM so that it can be used directly in contract mandated computer scheduling and project control software [3]. Again, it is important to state that combining the two methods is an excellent way to develop a construction schedule. As previously discussed, using linear scheduling techniques to plan the work sequence of production-driven activities as the first step in the development of the final construction schedule has many advantages.

As was seen in Example 7.3, several production-related decisions were made as a result of the visual representation that the linear schedule provides. Thus, with those important decisions made on the major activities that are associated with ensuring the project's profitability, the scheduler can then convert the linear schedule into a PDM to complete the detailed scheduling task using the following steps:

1. List all the activities shown in the linear schedule, their durations rounding partial days up to the next highest whole day, and their precedence relationships.
2. For each line, determine if it will be shown in the PDM as a single activity or if it would be more appropriate to break it down into several activities. If it is broken down, the sum of the durations of the series of new activities cannot exceed the duration of the line they represent in the linear schedule.
3. For each bar, break down the series of activities that make up the bar and distribute the total duration associated with the bar in the linear schedule to each of the activities. The total duration of the activities that make-up the hammock represented by the bar cannot exceed the duration assigned to the bar in the linear schedule.
4. Similarly, for each block, determine if it will be shown in the PDM as a single activity or if it would be more appropriate to break it down into several activities. It if is broken down, the sum of the duration of the series of new activities cannot exceed the duration of the block they represent in the linear schedule. Additionally, as some blocks, like the one in the previous example, represent a constraint rather than a production activity, decide whether to show the block as an activity that has duration but no resources or as a lag on the finish of a related activity.
5. Assemble the final list of activities to be developed into a PDM network including those minor activities that were not shown on the linear schedule with their

associated durations and precedence relationships and develop the network in the scheduling designated software package.

NOTE: when setting up the initial data project in the software, DO NOT input the contract completion date at this point. Some software will default to making the network start on the designated start date and finish on the designated finish date. Thus, the scheduler may be faced with trying to flush out negative float if the initial network is longer than the period available in which to complete the project, or there will be no critical path (i.e., all activities will have float) if the opposite is true.

6. Check the project completion date computed by the software against the contractual completion date. If the initial schedule is longer than the time allowed, then if possible, adjust the logic used in the network to reduce the overall duration. If this doesn't work, go to those activities where the durations were rounded up and reduce their durations by rounding down.
7. The final product should conform to the contract specifications and fall within the contract period.

To demonstrate how this conversion methodology works, Example 7.3 will be converted to a PDM in Example 7.4.

**Example 7.4:** Figure 7.12 shows that the project can be completed in a total of 24 working days. Table 7.5 is a list of the activities and their respective durations taken directly from the linear schedule in Figure 7.12.

This completes both Step 1, the list of activities from the linear schedule and Step 2, the determination of how the lines on the linear schedule will be broken up. One can see that

**TABLE 7.5**
**PDM Activities and Durations from Figure 7.12**

| Activity Code | Activity | Linear Schedule Type | Duration (Days) | Precedence |
|---|---|---|---|---|
| 010 | Mobilize | Bar | 2 | none |
| 020 | Demolish existing pavement STA 9+00 to 19+00 | Line | 3 | FS 010 |
| 025 | Demolish existing pavement STA 9+00 to 0+00 | Line | 2 | FS 020 |
| 030 | Build culverts STA 9+00 to 11+00 | Bar | 8 | FS 010<br>Consists of 4 individual box culverts including excavation, backfill, formwork, concrete, and final grading |
| 040 | Install CTSB STA 19+00 to 11+00 | Line | 3 | FS 020 with 2 days lag |
| 045 | Install CTSB STA 9+00 to 0+00 | Line | 4 | FS 040 |
| 047 | Install CTSB STA 9+00 to 11+00 | Line | 2 | FS 045, FS 030, & FS 050 |
| 050 | Cure culverts/no loads | Block | 3 | FS 030 |
| 060 | Install ASB STA 19+00 to 0+00 | Line | 10 | SS 040 with 4 days lag<br>FF 047 |
| 070 | Install Type A STA 19+00 to 0+00 | Line | 5 | SS 060 with 6 days lag & FF 060 |
| 080 | Clean-up/demobilize | Bar | 2 | FS 070 |

the scheduler has chosen to break each of the production-driven activities represented by lines into separate activities as they break up on the linear schedule. This allows them to be related to the "Build Culverts" hammock activity and the "No Loads" bar. Additionally, one can see that the work sequencing decisions made on the linear schedule are represented by precedence relationships and lag to ensure that the PDM preserves the logic that went into making those decisions.

The next step will be to breakdown the "Build Culverts" hammock activity into individual activities and replace the single activity shown in Table 7.5. To do this, a list for the activities and durations for a single culvert is generated as follows:

- Excavate and prepare bed 4 hours
- Form box culvert, 8 hours
- Pour and finish concrete, 4 hours
- Backfill, 3 hours
- Final grading, 1 hour

Looking at this list, the scheduler decides to simplify matters a bit by combining the "Backfill" and "Final Grading" for all four culverts into the single activity "Backfill and Final Grading Culverts A-D" and assigns it a duration of 2 days. Next, the excavation and bed preparation activity is broken into two activities, "Excavate and Prepare Bed Culverts A&B" and "Excavate and Prepare Bed Culverts C&D" of 1-day duration each. Similarly, the decision is made to pour the concrete for two culverts per day creating the activities: "Pour Concrete Culverts A&B" and "Pour Concrete Culverts C&D" of 1-day duration each. Finally, each culvert is assigned a separate "Form Culvert" activity with a 1-day duration. The resulting output is shown in Table 7.6 and the fragmentary network (fragnet) for this series of activities is shown in Figure 7.13.

This project has some miscellaneous activities that were not included in the linear schedule. The activities, their durations, and precedence relationships are as follows:

- Guardrail must be built between STA 9+00 and 11+00. It will take 2 days and can begin as soon as the paving is completed in that area.
- Striping of the road will take 2 days and can begin when the pavement is ready.
- Signs at stations 4+00, 7+50, 9+85, and 11+00 can be emplaced after the paving is finished at that location. All the signs can be installed in 1 working day.

**TABLE 7.6**
**Detailed List of "Build Culverts" Activities**

| Activity Code | Activity | Duration (Days) | Precedence |
|---|---|---|---|
| 030 | Excavate and Prepare Bed Culverts A&B | 1 | FS 010 |
| 031 | Excavate and Prepare Bed Culverts C&D | 1 | FS 030 |
| 032 | Form Culvert A | 1 | FS 030 |
| 033 | Form Culvert B | 1 | FS 030 |
| 034 | Form Culvert C | 1 | FS 031 |
| 035 | Form Culvert D | 1 | FS 031 |
| 036 | Pour Concrete Culverts A&B | 1 | FS 033 |
| 037 | Pour Concrete Culverts C&D | 1 | FS 035 |
| 038 | Backfill and Final Grading Culverts A,B,C, & D | 4 | FS 036 & FS 037 |

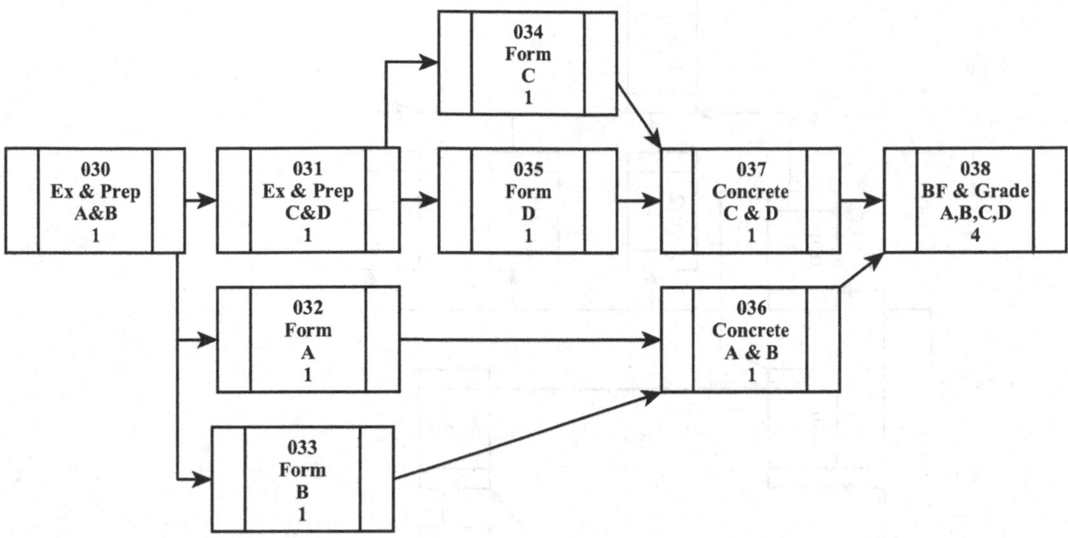


**FIGURE 7.13** PDM Fragmentary Network for "Build Culvert" Work Sequence.

## TABLE 7.7
## Final List of Activities for Converting Figure 7.12 to PDM

| Activity Code | Activity | Duration (Days) | Precedence |
|---|---|---|---|
| 010 | Mobilize | 2 | None |
| 020 | Demolish existing pavement STA 9+00 to 19+00 | 3 | FS 010 |
| 025 | Demolish existing pavement STA 9+00 to 0+00 | 2 | FS 020 |
| 030 | Excavate and prepare bed culverts A&B | 1 | FS 010 |
| 031 | Excavate and prepare bed culverts C&D | 1 | FS 030 |
| 032 | Form culvert A | 1 | FS 030 |
| 033 | Form culvert B | 1 | FS 030 |
| 034 | Form culvert C | 1 | FS 031 |
| 035 | Form culvert D | 1 | FS 031 |
| 036 | Pour concrete culverts A&B | 1 | FS 033 |
| 037 | Pour concrete culverts C&D | 1 | FS 035 |
| 038 | Backfill and final grading culverts A,B,C, & D | 4 | FS 036 & FS 037 |
| 040 | Install CTSB STA 19+00 to 11+00 | 3 | FS 020 with 2 days lag |
| 045 | Install CTSB STA 9+00 to 0+00 | 4 | FS 040 |
| 047 | Install CTSB STA 9+00 to 11+00 | 2 | FS 045 & FS 050 |
| 050 | Cure culverts/no loads | 3 | FS 038 |
| 060 | Install ASB STA 19+00 to 0+00 | 10 | SS 040 with 4 days lag FF 047 |
| 070 | Install type A STA 19+00 to 0+00 | 5 | SS 060 with 6 days lag & FF 060 |
| 072 | Build guardrail STA 9+00 to 11+00 | 2 | SS 070 with 2 days lag |
| 074 | Striping STA 19+00 to 0+00 | 2 | SS 070 with 3 days lag & FF 070 |
| 076 | Install signs | 1 | SS 070 with 4 days lag |
| 080 | Clean-up/demobilize | 2 | FS 070, FS 072, FS 076 |

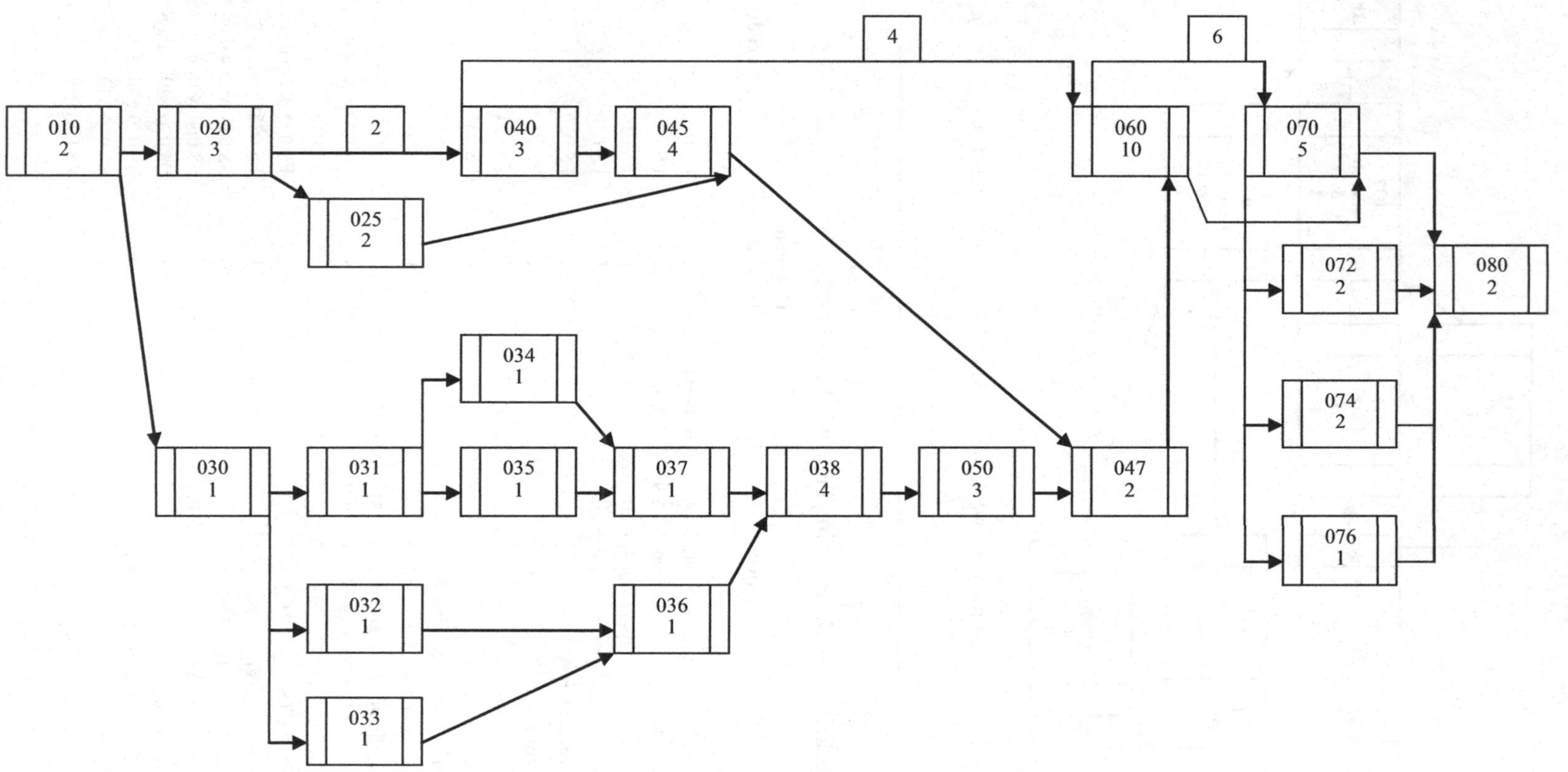


**FIGURE 7.14** Example Linear Schedule Converted to PDM Network.

Adding these activities to the activities in Tables 7.5 and 7.6 completes the activity list for the PDM in Table 7.7 and is shown in Figure 7.14.

Thus, it has been demonstrated that the two scheduling methods, PDM and linear scheduling, can work together to assist the estimator/scheduler/project manager in being able to establish the project's sequence of work in a logical manner that permits the control of the project throughout the construction cycle. To briefly summarize, the following points have been made:

- Linear scheduling furnishes a methodology to focus the scheduling process on those production-driven activities whose successful execution drives an equipment-intensive project's profitability. It does so by reducing the schedule to its most essential portions and graphically manages both time and space on the project site.
- PDM furnishes a methodology for organizing the great amount of detail that attends most construction projects. Its concepts are both well understood and well accepted by both owners and construction contractors, and its use is often required as a part of the construction contract.
- The benefits of both methodologies can be leveraged by using linear scheduling to plan the sequence and timing of work for the major items and then converting that output into PDM. Next, all the remaining items of work can be added to the PDM to produce the final construction schedule for an equipment-intensive project.

## 7.4 DEVELOPING EQUIPMENT RESOURCE PACKAGES (CREWS)

The heart of scheduling equipment-intensive projects is the equipment itself. Previous chapters have shown the reader how to select and then optimize various pieces of equipment to maximize the system's production. The next step is to take the production calculated by the estimator for various sets of equipment and their operators and create a crew. The definition of a crew is as follows:

> A self-contained grouping of machines, operators and other support resources that has been designed to complete a specific item or type of work.

In its simplest form, a crew could consist of a single worker with the appropriate tools. On the other end of the spectrum, the crew can contain a multitude of different pieces of machinery along with operators, laborers, supervisors, support vehicles, and maintenance staff. The makeup of an individual crew is often the prerogative of the estimator who must make certain assumptions to complete the cost estimate for a project. If the project is awarded on a price that was based on the estimator's assumptions, the project manager must ensure that all field personnel knows what those assumptions are and only change them if they are wrong or if there is a better, less costly way to complete the project. To be sure, most estimators are intimately familiar with how their respective companies tend to do business and generally base their crew composition assumptions based on past success.

Example 5.3 illustrates the thought process that goes behind developing a crew designed to haul material a specified distance from one point to another on the project. The final outcome of that problem was that the machinery assigned to the haul crew was a 1.5 cubic yard front-end loader and three 12 cubic yard end dump trucks with sideboards. Once the equipment has been determined, the labor that must go with the equipment must be assigned. In this case, as a minimum, there would be a bucket loader operator and three truck drivers. Depending on the reason for the haul, there might also be a laborer who would act as a spotter for the dump trucks showing then where the next load should be

**TABLE 7.8**
**Equipment Resource Package for Example 5.3**

| | | GRANULAR MATERIAL HAUL CREW | | |
|---|---|---|---|---|
| **Crew # Equipment** | **Number** | **Labor Classification** | **Number** | **Other** |
| 1.5 CY Wheeled front-end Loader | 1 | Loader operator | 1 | |
| 12 CY end dump truck | 3 | Truck driver | 3 | With sideboards |
| ½ Ton pick-up truck | 1 | Supervisor | 1 | Transit |
| NA | | Laborer (dump spotter) | 1 | Hand level and rod |

dumped. There also might be a supervisor who would be assigned a pick-up truck. Thus, the total equipment resource package would look like the one shown in Table 7.8.

### 7.4.1 Rules for Developing Crew Sizes

Thus, the remainder of the crews for a given project would be assembled to permit the estimator to determine the total amount of direct equipment and labor costs that need to be allocated to this particular job. Several rules should be remembered as one develops equipment resource packages into crews for a given project as follows:

1. Develop the crews for the production-driven activities first. These are the items of work that make up the major financial portion of the contract. Therefore, these activities should be given first priority for resources at all times.
2. Never plan to "borrow" a piece of equipment or resource from a crew assigned to a production-driven activity as a short-term savings measure. This could cause that activity to miss its required production and threaten project profitability. It is better to estimate the cost of renting a similar piece of equipment for that limited period than to potentially "rob Peter to pay Paul."
3. Consider the impact on the required production of losing a piece of equipment due to maintenance failure when deciding whether to round up or round down to optimize the size of the equipment spread. If the crew is allocated to a production-driven critical activity, it is better to round up and have an additional piece of equipment in the spread to cover those unavoidable periods of reduced production due to maintenance.
4. For noncritical activities, attempt to develop crews that are sized to be flexible enough to be able to be assigned to more than one specific type of work. In these cases, it is always easier to add workers and equipment to a crew than it is to remove them once they have been used on the jobsite. So, start with the minimum requirement and increase the size of the crew as required.

### 7.4.2 Developing Crew Costs

Once the crews have been developed for a given project, the next step is to calculate the cost of the crew in two different ways: by composite crew hour and by unit cost. The first step in developing a crew cost is to assign the appropriate equipment rental rates and labor rates to each resource in the crew. Table 7.9 shows how this was done for the crew created for Example 5.3.

**TABLE 7.9**
**Crew Cost Table for Example 5.3**

| | | | GRANULAR MATERIAL HAUL CREW | | | |
|---|---|---|---|---|---|---|
| **Crew # Equipment** | **Number** | **Rate $/hr** | **Labor** | **Number** | **Rate $/hr** | **Total $/hr** |
| 1.5 CY wheeled front-end loader | 1 | 32.95 | Loader operator | 1 | 31.20 | 64.15 |
| 12 CY end dump truck | 3 | 58.83 | Truck driver | 3 | 25.00 | 251.49 |
| ½ Ton pick-up truck | 1 | 11.28 | Supervisor | 1 | 37.50 | 48.78 |
| N/A | | | Laborer (dump spotter) | 1 | 18.90 | 18.90 |
| | | | **TOTAL HOURLY CREW COST** | | | 383.32 |
| 2,743 cubic yards per shift | | | | | | |

Example 5.3 calculated that this equipment spread could produce 2,743 cubic yards per 8-hour shift. Thus, the direct unit cost for hauling this material can be calculated as:

$$\text{Crew cost per shift} = \$383.32/\text{hour}(8\ \text{hours/shift}) = \$3,066.56/\text{shift}$$

$$\text{Direct unit cost} = \frac{\$3,066.56/\text{shift}}{2743\ \text{CY/shift}} = \underline{\underline{\$1.12/\text{CY}}}$$

The estimator now has three different cost factors with which to estimate the total cost of this particular activity:

1. Hourly crew cost = $383.32 per hour
2. Daily crew cost = $3,066.56 per shift or per day
3. Direct unit cost of hauling the material = $1.12 per cubic yard

More importantly, these cost and production factors can be translated after contract award to project control metrics that can be used by the project manager to ensure that the project achieves its target profit.

## 7.5 ESTABLISHING PROJECT MANAGEMENT ASSESSMENT PARAMETERS

If the project manager does not fully understand the rationale used by the estimator to assemble the bid, completing that project at a cost that guarantees the margins contained in the bid becomes very difficult. Thus, it is important that the estimator on an equipment-intensive project capture the logic in a manner that is easy for the project manager to translate into project control mechanisms. As such, there are two parameters that are generally tracked to determine if the project is progressing satisfactorily. These are cost performance and time performance. Costs performance metrics assist the project manager in understanding how the project is proceeding on a financially basis. Time performance metrics are used to track the project's actual progress with respect to its schedule. Both are interrelated and equally important. Thus, it is vital that the numbers used in the estimate flow out to the project in the form of project performance metrics.

There are three types of project performance metrics: relative, static, and dynamic [9]. Relative metrics are expressed as a percentage, and as a result, are independent of the size of a project. This allows the project manager to directly compare the performance of small projects with the performance of large projects. The cost and time growth metrics shown below are typical examples of relative metrics.

$$\text{Time Growth} = \frac{\text{Final contract time} - \text{Original contract time}}{\text{Original contract time}} \tag{7.9}$$

Where, Time Growth (percent)
Final contract time (days)
Original contract time (days)

$$\text{Cost Growth} = \frac{\text{Final contract cost} - \text{Original contract cost}}{\text{Original contract cost}} \tag{7.10}$$

Where, Cost Growth (percent)
Original contract Cost ($)
Final contract cost ($)

The second type is static metrics. These metrics are discreet numerical measures that do not change with time. As a result, they are project size dependent, and the project manager can only use them to compare projects that are roughly the same size. Cost per square foot of constructed area and charge days per lane-mile of highway are examples of static metrics. Finally, dynamic metrics are those that vary with time. Dynamic metrics are also project size dependent. These metrics are generally a function of both cost and time. Some include cost, time, and a function of physical size. As a result of their mathematical complexity, project managers must understand the limitations of each and every metric that they choose to use to measure project performance. An example of a typical dynamic metric is construction placement as calculated in Equation 7.11.

$$\text{Construction placement} = \frac{\text{Construction cost}}{\text{Construction time}} \tag{7.11}$$

Where, Construction placement ($/day)
Construction cost ($)
Construction time (days)

The above are project performance metrics that can be used as project control assessment measures on any construction project. On equipment-intensive projects, the focus must remain on production, and, as such, the project management assessment measures should also be developed with a strong production focus.

### 7.5.1 Minimum Required Daily Production

Equation 7.8 for required daily production can be used as a control measure as well as an estimating tool. This is an elegant use of a very simple equation in that the equation calculates a number that is independent of the resources that have been assigned to a given item of work, both the quantity of work and the time allowed in which to accomplish it come from the construction contract. Thus, it can serve as a "floor" requirement that must be met or exceeded each day if the project is going to be completed on time as shown by Example 7.5.

**Example 7.5:** A project requires 100,000 cubic yards of aggregate be moved to a stockpile to support an asphalt batch plant. The contractor has 40 working days in which to build the stockpile. The granular material haul crew shown in Table 7.9 is allocated to this task. Calculate the minimum required daily production and compare it with the crew's sustained production.

$$P_r = \frac{100,000 \text{ CY}}{40 \text{ days}} = 2500 \text{ CY/day}$$

$$\text{Granular material haul crew } P_s = 2743 \text{ CY/day}$$

$$P_s \geq P_r$$

Therefore, this crew will be adequate. However, if the crew loses one of its dump trucks to a maintenance break-down, its production will drop by 33% to 1,860 CY/day. This would cause the actual production to be less than the minimum required daily production. Thus, the project manager might make the decision to regain that daily total by requiring the crew to work roughly 2 hours and 45 minutes of overtime or until the daily total is over 2,500 cubic yards. Thus, it can be seen that having this project assessment parameter can assist the project manager in making day-to-day decisions regarding the control of the project.

### 7.5.2 Expected Daily Production

Expected daily production differs from minimum daily production in that it relates the actual level of resources assigned to a given task to the production that can realistically be achieved. In the above example, the granular material haul crew had a sustained production rate of 2,743 cubic yards per day. Thus, it could be expected to be able to finish the job in 37 days or three days early. Realistically, there will be daily variations both up and down from the expected daily production rate. So the project manager should use it more as an average value rather than an absolute value to gauge the actual production being achieved by a given crew.

### 7.5.3 Allowable Cycle Time Variation

Allowable cycle time variation goes back to the calculation of sustained production itself. This metric furnishes a microscopic measure to assess the ability of a given crew to achieve the production rates around which the estimate was based. As noted in Chapter 5, the estimator does a lot of rounding as the equipment resource allocation is developed, and the production rates for each crew are estimated. For a haul item, the process starts by looking at the round-trip cycle time of the crew. Cycle times are inherently variable in that any number of factors can arise that either increase or decrease the actual cycle time from the cycle time used in the production calculations. Therefore, the project manager needs to know not only the planned cycle time used in the estimate to arrive at the sustained production rate but also the cycle time that is associated with arriving the minimum daily production rate to know what the allowable variation in actual cycle times should be. This can be computed by reorganizing Equation 5.14 as follows:

$$P_s = \frac{60(N)(S_H)(H)}{C_s} \tag{5.14}$$

$C_s$ is the sustained cycle time associated with the sustained production $P_s$. Substituting the minimum required daily production $P_r$, for $P_s$ and maximum allowable cycle time $C_{max}$ for $C_s$, yields Equation 7.12 below.

$$C_{max} = \frac{60(N)(S_H)(H)}{P_r} \tag{7.12}$$

Where $C_{max}$ = maximum allowable cycle time

Thus, using the above example whose $C_s$ = 14 minutes, the maximum allowable cycle time would be calculated as follows:

$$C_{max} = \frac{60(4)(20)(8)}{2500}$$

$$C_{max} = 15.4 \text{ minutes}$$

Therefore, the allowable variation in cycle time would be about 1 minute and 30 seconds. Knowing this, the project manager could ask the foreman for the crew to spot check the cycles time throughout and the day to ensure that the crew would hit its minimum required daily production and its expected daily production each day. This metric furnishes an early warning system that allows the project supervisors to immediately recognize that something is wrong and take action to correct the problem before a substantial amount of time and production are lost.

### 7.5.4 Cost and Unit Targets

Cost and unit targets are project assessment parameters that furnish a historical perspective and lag rather than lead the actual observed progress on a given project. Cost targets seek to measure how closely actual project costs are tracking the assumptions made in the estimate. Unit targets seek to measure how closely actual project production track the production rate assumptions made in the estimate.

The most common and widely used tool for tracking cost target is called earned value. Essentially, earned value assumes that once a unit of work is complete, the contractor has "earned" the value of that unit and can subsequently submit an application for payment. Earned value basically creates a graph of project cash flow versus time. To use this concept as a project control measure, the project manager must strip out all mark-ups and any unbalancing that has been done with the unit prices quoted in the bid documents. A cost target seeks to allow the project manager to control cost without respect to profit. Therefore, to develop a cost target value, the numbers used in the estimate that account for all the costs associated with an item of work must be identified and included in the target value. To arrive at a target cost metric, a specific period of time must be chosen. As many companies do their payroll on a weekly basis, the weekly target cost becomes a convenient project assessment parameter. The target cost value then becomes the value of payroll and equipment ownership costs/rental rates that are associated with a given item of work accomplished using the specified shift length used in the estimate as the basis. Once this value is determined, it can then be compared on a periodic basis with the actual cost value computed using the actual payrolls and equipment hours in the given period.

Using the example shown in Table 7.9, the project manager can compute the weekly cost target value as follows:

$$\text{Weekly target cost} = (5\ \text{days/week})(8\ \text{hours/day})(\$383.22/\text{hour})$$

$$\text{Weekly target cost} = \underline{\underline{\$15{,}328.80}}$$

This value can then be used to compare with the actual total for each week to give the project manager an idea of how well the project is performing against the plan formed by the project cost estimate.

Unit targets are computed in much the same manner, substituting the quantities of work for the costs. Continuing the example, the granular material haul crew had a minimum production of 2,500 cubic yards per day. That can then be extended across a week for a minimum unit target of 12,500 cubic yards per week. It can also be used with the expected production of 2,743 cubic yards per day, to get a second expected unit target value of 13,715 cubic yards per week. Thus, the project manager now has a range of 12,500 to 13, 715 cubic yards a week against which to judge the actual progress and make decisions for future weeks.

It is always valuable to use both cost targets and unit targets together. Equipment management decisions made on the job site are normally centered on recovering lost production and bringing actual production back into the range, as discussed in the previous paragraph. However, in the Example 7.5, the project manager decided to work overtime to recover the production lost by a maintenance problem with one of the dump trucks. The hourly costs of the crew shown in Table 7.9 is based on an 8-hour shift with the labor being computed based on straight time. Once the crew is required to work overtime in this example, the cost of an overtime hour jumps up to $439.62 assuming all the workers will be paid at a rate of 1.5 times their straight time hourly wage. The actual daily cost becomes $4,272.72 instead of $3,065.76, an increase of over $1,200. Thus, while the project manager would have achieved the unit target for the week, the cost target would have been exceeded and the project's actual profit margin would have been reduced from the target profit that was built into the bid.

## 7.6 SUMMARY

The old cliché that "time equals money" was never more appropriate than in the scheduling of equipment-intensive projects. The scale and complexity of these types of construction projects drives the requirement to not only accurately estimate the production rates of various crews but also schedule them in a fashion where the production achieved in critical activities is not interrupted by noncritical activities. This production can only be achieved if those features of work are allocated sufficient equipment and labor to realistically allow those crews to produce at rates that equal or exceed the assumptions used in the estimate. Thus, the understanding of the dynamics of each project is essential to its financial and technical success. Good scheduling practices assist the project manager as well as the estimator in gaining that understanding and ensuring that the bid has not created a "Mission Impossible" situation for the actual execution of the equipment-intensive project.

## REFERENCES


[1] O'Brien, J. and Plotnick, F. (2016). *CPM in Construction Management*. 8th ed. New York: McGraw Hill.

[2] Lopez del Puerto, C. and Gransberg, D.D. (2008). Fundamentals of Linear Scheduling for Cost Engineers. In *2008 Transactions*. Toronto, ON: AACE, International, pp. PS.05.01–PS.05.08.

[3] Gransberg, D.D. (2007). Converting Linear Schedules to Critical Path Method Precedence Diagrams. In *2007 Transactions*. Nashville, TN: AACE, International, pp. PS.05.01–05.08.

[4] Marchman, D.A. (2002). *Construction Scheduling with Primavera Enterprise* ®. Independence, MO: Delmar Publishers, pp. 97–118.
[5] Clough, R.H., Sears, G.A., Sears, S.K., Segner, R.O. and Rounds, J.L. (2015). *Construction Contracting*. 8th ed. Hoboken: John Wiley and Sons, Inc, pp. 384–385.
[6] Jones, C. (2005). *Linear Scheduling. Presentation Slides (April 4, 2005)*. Dallas, TX: Peter Kiewit and Sons, Inc.
[7] Duffy, G.A., Oberlender, G.D. and Jeong, D.H. (2010). Linear Scheduling Model with Varying Production Rates. *Journal of Construction Engineering and Management*, 137(8), pp. 574–582.
[8] Tapia, R.M. and Gransberg, D.D. (2016). Forensic Linear Scheduling for Delay Claim Analysis: Panama Canal Borinquen Dam Case Study. In *Compendium*. 2016 Transportation Research Board Annual Meeting, Paper 16-0557. Washington, DC: National Academies, pp. 1–10.
[9] Gransberg, D.D. and Villarreal, M.E. (2002). Construction Project Performance Metrics. In *2002 Transactions*. Portland, OR: AACE, International, pp. CSC.02.01–CSC.02.05.

## CHAPTER PROBLEMS

1. Discuss the advantages and disadvantages of CPM and LS methods in the context of an equipment-intensive project.
2. Figure 7.15 is a cross-section of a 9,500 lineal foot long Flood By-Pass Channel, and Table 7.10 contains the activities that comprise this project, the production rates associated with each pay item, and the precedence requirements.

   To summarize the sequence of work, since this is a very large-scale project, the contractor will use two full crews, having one start at Station 0+00 and work toward the end, and the other starting at Station 95+00 and work toward the beginning. The reinforced concrete channel walls are constructed on top of the reinforced concrete foundation slab after it has cured for 10 days. Once the walls have cured for 10 days, the excavation can be backfilled. Upon completion of backfill, topsoil can be placed back on the site and remaining spoil material can be hauled away. For academic purposes, assume that the excavation only includes the volume found inside the dotted

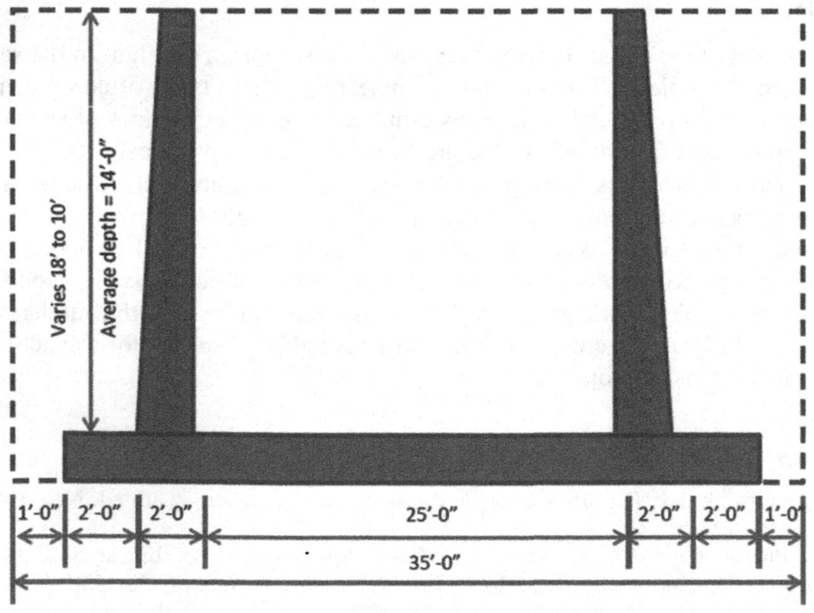


**FIGURE 7.15** Flood By-Pass Channel Cross-section for Problem 2.

**TABLE 7.10**
**Flood By-Pass Channel Activities and Production**

| Activity # | Pay item | Unit | Production Unit/Hour | Precedence |
|---|---|---|---|---|
| A | Mobilization | Workday | 15 | None |
| B | Strip and stockpile topsoil | CY | 150 | FS A |
| C | Excavation–Structural | CY | 80 | FS A with 5 days lag |
| D | Grading | SY | 800 | FS C with 5 days lag |
| E | Erect slab formwork incl. rebar | SF | 60 | FS D with 5 days lag |
| F | Pour and finish slab concrete | CY | 60 | FS E with 5 days lag |
| G | Strip slab concrete formwork | SF | 180 | FS F with 5 days lag |
| H | Erect wall concrete formwork incl. rebar | SF | 100 | FS G with 10 days lag |
| I | Pour and finish wall concrete | CY | 60 | FS H with 5 days lag |
| J | Strip wall concrete formwork | SF | 300 | FS I with 5 days lag |
| K | Backfill–Structural | CY | 120 | FS J with 10 days lag |
| L | Spoil disposal | CY | 250 | FF K |
| M | Spread topsoil | CY | 300 | FF K |
| N | Demobilization including clean-up | Workday | 10 | FF L & M |

line in Figure 7.15 and assuming the project will proceed 7 days a week without regard to weekend or holidays. The contractor will work a 10-hour shift with the loss of 1.5 hours per shift for breaks, etc. Answer the following questions:

a. Develop a CPM based on PDM for the project.
b. How many working days will this project take?
c. Develop a LS for the project including all the activities. The contractor will mobilize in the area from STA 45+00 to 47+00 but the equipment yard will not be in a location that interferes with the production activities.
d. How many working days will this project take?
e. If the crew starting at STA 0+00 encounters material from STA 20+00 to STA 30+00 that reduces its excavation production to 40 CY/hour, what impact will that have on the scheduled project completion days?

# 8 The Buy, Lease, or Rent Decision

## 8.0 INTRODUCTION

In the early 20th Century, new pieces of construction equipment were acquired by an outright purchase. However, as the cost of the machines increased, the need for credit and financing to support heavy equipment purchases became a key feature of the industry. The diversity and enhanced incorporation of emerging technologies has not only created competition in retail prices but also in the areas of financing, terms of use, and methods of payment.

## 8.1 ACQUIRING HEAVY EQUIPMENT

Equipment fleet managers are able to acquire just about any heavy construction machine available on either a temporary or permanent basis through equipment manufacturers, used equipment brokers, and rental companies. Therefore, numerous options must be evaluated when planning heavy equipment acquisition and financing. Traditionally, the equipment purchase process was complete when the buyer selected a specific make and model of machinery at a dealership. The purchaser received financing with a down payment or the trade-in of an older piece of equipment. The new piece of equipment was the loan security. If the buyer defaulted on the loan, the lender could repossess the equipment. This equipment acquisition process typically includes comparing the available financing options and terms that banks, finance companies, leasing agencies, and manufacturers offer. These must be compared to programs offered by major heavy construction equipment manufacturers for leasing, renting, and installment loan products.

The equipment rental industry has leveled the playing field for equipment owners by minimizing the risk associated with equipment purchase and utilization. Being able to rent or lease appropriate equipment greatly increases the type and size of projects that contractors and subcontractors can bid and build. The same applies to public and private entities that use construction machinery to build and maintain their own capital projects.

The used equipment industry also provides a convenient means for equipment to be acquired. Online equipment procurement sites are ubiquitous and provide a virtual marketplace from which contractors can find the necessary data to conduct the necessary analyses to support their equipment acquisition programs. Online equipment advertisements can include operating specifications, work history, operating conditions, and pictures of the actual piece of equipment.

Risk exposure is the primary consideration in the buy, lease, or rent decision and directly associated with financial considerations. Risk evaluation in the equipment acquisition process is necessary to ensure the final selection decision matches corporate cash flow requirements and tax implications. Expected equipment utilization is the most important consideration in achieving the desired return on investment. Future work volume, production needs, project specific or client requirements, long-term company goals, acquisition time, and equipment availability must all be considered when developing a utilization strategy for a piece of equipment.

For this reason, subcontracting equipment-intensive specific tasks minimizes risk and is typically most efficient. Specialty subcontractors that own particular types of equipment and have experience in its proper employment are a cost-effective means to minimize risk for the general contractor. Specialty subcontractors in trades such as earthwork, excavation/demolition and hauling, concrete pumping and soil, and concrete boring are common today.

The Wells Construction Industry Forecast for 2019 found that 96% of the contractors polled still plan to purchase new or used equipment in the coming year [1]. Additionally, 92% of contractors say that they'll maintain or increase the use of leased or rented equipment in the next 12 months. Of that group, 46% cited flexibility as their primary motivation for renting rather than owning heavy equipment. Whether renting, leasing, or owning equipment must pay for itself by earning more money than it costs. The ability to do this is greatly influenced by being able to keep the piece or group of equipment busy. Sustained profitability using standard equipment typically demands continual equipment utilization. The ability to do this is typically market driven and influenced by the overall business strategy of the company. Simply stated, equipment that is not operating is not making money. This places a tremendous responsibility on equipment-intensive civil or road building contractors that own large fleets of equipment to stay busy. To further compound the risk, heavy construction equipment depreciates in value and requires more maintenance with age and use. Finally, the accelerated increase in new technology also creates a limited useful life for most machines due to accelerated obsolescence.

## 8.2 FINANCING METHODS

There are four primary financing methods to support the purchase of construction equipment: conventional financing purchase, the financial lease agreement, the tax lease agreement, and the rental purchase (rent-to-own) agreement. A comprehensive analysis of all available financing options should be done that accounts for company-specific as well as project-specific variables. In doing so, the analyst must make certain assumptions or projections concerning company operating guidelines and desired financial goals. This analysis will typically include, but is not limited to:

- Working capital constraints – actual and desired.
- Desired tax benefits – actual and desired.
- Balance sheet objectives or financial goals – actual and desired.
- Cash flow requirements – actual and desired.
- Equipment obsolescence and ultimate replacement strategy.
- Equipment utilization – actual and desired.
- Opportunity cost.
- Investment strategy goals.

### 8.2.1 Cash Purchases

Putting tax and other considerations aside, an outright cash purchase with funds provided from working capital is the lowest cost method of acquiring needed equipment when funds are available. Service fees, finance charges, and interest expense are eliminated for the purchaser. Customer ownership is immediate, and equipment cost is shown on the balance sheet subject to the depreciation methods used by the customer. Although outright purchase may provide the lowest total cost, other factors should also be considered. The reason most contractors choose to finance their equipment purchases is affected by a number of other considerations, such as the financial impact of increasing the assets on the contractor's balance sheet. A cash purchase converts a liquid asset (cash) into a fixed asset (equipment). Large out-of-pocket purchases can be detrimental to cash flow, thereby reducing working capital and increasing opportunity costs.

### 8.2.2 Conventional Financing Purchase

Many contractors prefer to ultimately own their construction equipment in a manner that permits building equity in it over time. Conventional purchase financing provides the contractor with the capital required to make the purchase through loan arrangements secured by accounts receivable, own equity in the equipment, and use the equipment to collateralize the loan. Often, trade-ins and/or down payments are included in financial agreement. Installment sales contracts allow the contractor to purchase the equipment and pay for it over a fixed period. Installment loans offer many of the same cash flow advantages offered by other financing instruments, as well as the depreciation and tax benefits of ownership. Installment loans are ideal for users desiring immediate equipment ownership. Most heavy equipment dealers will finance new equipment for up to 60 months, with longer terms available under special conditions. Finance terms for used equipment usually run up to 48 months, depending on the equipment and work conditions. Contractor's that prefer paying for equipment using installment loans generally have a more stable predictable workload than those who lease. A general rule-of-thumb for fleets that utilize equipment for 7 or more years is to purchase equipment rather than exercise lease or rental alternatives.

The equipment's anticipated working conditions and the quality of the contractor's maintenance program are other considerations for evaluating the benefits and costs of equipment ownership. Leased or rented equipment must be maintained to certain standards, and that may be difficult for contractors that need to utilize the equipment in harsh working conditions. The contractor may have to pay for extensive overhauls of major components as well as repairs to paint, glass, tires, etc. when the leased or rented equipment is returned. According to Cudworth, the following features make purchasing an attractive alternative [2]:

1. Use and Possession: The owner has absolute control of the use and disposition of the equipment.
2. Flexibility: The owner can sell the equipment, trade it, or use it until it is not economical to repair without having to respond to any creditor. Ownership gives the user complete flexibility regarding servicing, maintaining, and insuring the equipment.
3. Price: The buyer with cash is usually able to get better discounts due to a stronger financial position in the deal.
4. Tax benefits: The owner can take advantage of depreciation and interest tax benefits associated with equipment ownership.
5. Pride of ownership: Ownership can lead to better care and maintenance.

Purchasing a piece of equipment generally becomes more economically attractive if there is a high utilization rate throughout the useful life of the equipment. This point is extremely important because not being able to use the machine enough to cover its cost is the greatest risk and disadvantage of purchasing. Whether the equipment is working or not, the owner's financial obligation to the lender continues. Borrowing large sums of money, sometimes necessary to purchase large heavy equipment, ties up company capital and borrowing power. As a given machine's working capabilities become more specialized, the types of work activities that can be done by the machine become more limited with a direct impact on utilization rates. While this is not necessarily a disadvantage, it is a limitation that must be addressed in the purchasing analysis. Other considerations in determining the cost of owning include licenses, registration, insurance, maintenance, transport, storage, and provision of a qualified operator. These requirements of equipment ownership have to be tracked and controlled. Equipment fleet management and administration costs should be included in the analysis.

### 8.2.3 Leasing

There is an increasing trend toward leasing as a way to finance construction equipment. It is usually easier to gain financial approval for equipment under a lease program than through conventional purchase financing. In its simplest form, an equipment lease is simply a rental agreement. Rent is paid while the equipment is being used. Once the agreement is over, the equipment is returned to the owner. In a true lease, the lease payments are considered an expense of the lessee. The lessee does not own the equipment, and the equipment is not shown as an asset on financial statements. The most significant factor that affects the rent or lease decision is the duration in which the equipment will be required. Leasing is often considered attractive when the equipment is needed for more than 6 months. Most leases run from 18 to 24 months with large expensive equipment leases running as long as 84 months.

Leasing arrangements are a form of finance in which an asset is acquired by a third party, usually a bank, finance company, or dealer, and then leased to the end user for a predetermined period. This arrangement means the leasing party never actually has title to the asset for the term of the lease, although it has the use of the asset during that time. The usual term of the lease will equal the operating life of the asset, and the repayments will be geared to the cost of the asset spread over that time, plus a profit margin for the lessor. Leasing does not have an impact on a company's current or debt-to-worth ratios; thus, it provides a more positive financial condition for bonding purposes. Leasing encourages an orderly equipment replacement strategy, minimizing maintenance costs before they become excessive. Leasing also eliminates the need to dispose of used equipment for the lessee.

Construction equipment leasing allows additional flexibility to deal with cyclical and regional fluctuations in work level. Many leasing companies offer seasonal leases called skip leases that allow the user to schedule payments during busiest months, thus reducing cash flow concerns during seasonal periods when work is slow. Step-up leases start with smaller payments that increase over time, allowing the user to generate revenue while initial payments are lower. Deferred payment leases allow the user to defer initial payments until cash flow is started.

There are two types of common equipment leases offered by most equipment dealers: finance lease and tax lease. The finance lease is a lease that allows the contractor to make lease payments over a specific period and purchase the machine at the end of the lease term. The lessee typically retains the tax benefits. The term for new equipment ranges from 12 to 60 months. For example, at this writing, Caterpillar's finance lease allows the lessee to obtain 100% financing with a fixed finance rate. The lessee has the option of purchasing the piece of equipment for a predetermined amount or returning the piece of equipment at the end of the lease. Payments can be made monthly, quarterly, semi-annually, or annually depending on the qualifications of the lessee. Monthly payments can be setup lower than traditional financing. The tax lease typically offers a lower monthly payment than a finance lease. At the end of the lease term, the contractor has three options: buy the equipment at fair market value, extend the lease for a new fixed term or month-to-month basis, or return the equipment to the dealer.

Lease pricing varies among lessors and is driven by market supply and demand. Pricing is influenced by the lessor's ability to acquire equipment at discounts, financing, calculation of risk factors, assignment of residual value, desired rate of return based on the market, and expenses to carry out the lease transaction and operation. The following factors make leasing an attractive financing solution for the lessee [3,4]:

1. Lower rates and lessee cannot claim tax benefits. When the lessee cannot take advantage of tax benefits associated with equipment ownership, such as depreciation and interest, leasing is more attractive. The lessor can purchase the equipment, claim the tax benefits, and lease the equipment to the lessee. The tax benefits are passed to the lessee in the form of lower lease rates.

2. Cash flow improvement. Compared with a loan, a lease typically gives the lessee a more favorable cash flow, especially during the first years of use.
3. Carried off the balance sheet for financial accounting purposes. The lessor assumes title to the equipment with interest expense capitalized into the lease when the equipment is delivered and accepted by the lessee.
4. Impact on lessee's income. There is usually less effect on income depreciation and interest payments related to the purchase of the equipment during the early years of a properly structured lease.
5. Fixed-rate lease payments. The lessee knows the exact amount of future payments and avoids the risks inherent in fluctuations in the cost of funds, permitting the lessee to predict future financing equipment costs and cash needs more accurately.
6. Faster amortization of the equipment. The lessee under an operating lease may be able to amortize the cost of the equipment faster through tax-deductible rentals than through depreciation and after-tax cash flow.
7. Hedge against inflation. Future lease rentals are paid in inflated currency. The lessor can borrow long to minimize the effect of inflation and pass on this protection to the lessee in the form of long-term level lease payments.
8. Payments coordinated with lessee's cash flow. Payment schedules can be coordinated with earnings generated from the use of the equipment by the lessee. This flexibility may not be available with other financing methods.
9. Long-term financial availability. Leases can be structured for the useful life of the equipment. A lease contract can exceed the period normally available on a term loan. Lessors can offer lease terms due to faster return of capital from cash flow generated by tax benefits.
10. Convenience. For leasing contracts below $5 million, documentation may be simpler and more flexible than other sources of financing.
11. Full financing. Leasing can provide the lessee 100% financing, including shipping and installation charges. A typical equipment loan may require an initial down payment.
12. Earnings from the retained capital. A lease allows the retention of the lessee's capital that can be used elsewhere in the lessee's business.
13. Obsolescence. Leasing avoids equipment obsolescence for the lessee. If equipment becomes obsolete faster than its depreciation life, leasing may be more attractive.
14. Uncertain residual value. As the risk associated with the residual value of the equipment increases, leasing becomes more attractive.

Leasing is an attractive option for acquiring equipment for long-term use and ultimate purchase. Its disadvantages are limited. There are many types of leases, and their conditions will vary between the markets and lessors.

### 8.2.4 Renting

Renting is the final option for acquiring the use of necessary equipment. Equipment rental does not provide an option for accruing equity. Hence, rentals have no impact on the balance sheet. It does, however, impact a contractor's cash position. Rental payments reduce the company's earnings as an operating expense, and since the equipment is not owned, there is no tax benefit from depreciation. Dealer equipment rental programs offer many of the same advantages as leasing programs. The rental contract period provides complete flexibility with contract periods that match the contractor's actual need for the item of equipment.

Nonproductive assets are a risk for any business. Idle construction equipment accrues costs for insurance, maintenance, and depreciation. Most contractors rent approximately 25% of their total equipment requirements [4]. A common rule of thumb recommends an 80/20 balance for equipment purchase and rental. In other words, a contractor should purchase or lease equipment to perform 80% of its forecast work, renting the remaining 20%.

Most rental companies calculate their rates on a monthly basis. Weekly rates usually run about three times daily rates. Similarly, monthly rates run about three times weekly rates. To have some idea of the prevailing rates, there are several sources of rate information.

- Associated Equipment Distributors (AED) *Green Book On-line* [5]. The AED is the trade association of leading distributors and manufacturers of construction equipment. Average rental rates for equipment are listed according to their general characteristics and function.
- United States Army Corps of Engineers. (USACE) *Construction Equipment Ownership and Operating Expense Schedule* [6]. This information is divided into six regions across the nation to account for regional differences. The Region I volume is contained in Appendix A of this book.
- Local agency rental rate schedules: These documents are published by public agencies that routinely contract for equipment-intensive projects, like state departments of transportation, commuter rail agencies, and airports. The Michigan DOT schedule is also contained in Appendix A of the book.
- Rates from local equipment suppliers and rental services.

Renting equipment on an hourly basis is typically the most expensive of the three acquisition alternatives. However, it is generally necessary for work activities not performed on a regular basis, because it minimizes idle time for seldom-used equipment. Renting is the best solution when equipment will be utilized for a short duration. The American Rental Association suggests the following advantages of renting equipment [7]:

1. Minimum equipment for the job. Equipment ownership becomes particularly expensive when the equipment is idle and not utilized. When ownership of basic equipment is combined with rentals as needed, idle time is minimized.
2. Right equipment for the job. Ownership encourages inefficiency through use of wrong size or type of equipment for a given job. Renting can minimize this hidden cost.
3. Warehousing or storage. Warehousing facilities are seldom needed for rental equipment thus reducing overhead.
4. Breakdowns. The renter will typically replace equipment if there is a breakdown, thus minimizing down time due to repairs.
5. Maintenance. Full maintenance is covered on a day-to-day basis. The user needs no repair shop, no spare parts supply, no mechanics, and no maintenance records.
6. Equipment obsolescence. The renter may provide the latest types and models of equipment that are faster and more productive than older models.
7. Disposal cost. Selling used and obsolete equipment is not required.
8. Cost control. Cost is easier to monitor and control with rented equipment. Knowing the true cost of owned equipment is often difficult to determine.
9. Inventory control. Contractors have less inventory loss when equipment is rented. The presence of continuous billing on any rented item tends to establish accountability for that item.
10. Taxes and licenses. Personal property taxes and license costs are eliminated on rented equipment. Leasing cost is 100% deductible.

11. Conservation of capital. The lessees' capital is available for other uses or investment. Contractors should analyze cash requirements and consider renting equipment as a method of conserving working capital.
12. Increase in borrowing capacity. Rented equipment does not result in a liability on the balance sheet. Debt ratios will improve making the lessee firm seem stronger financially.
13. Cost estimating and bid preparation. Renting can increase estimating accuracy because all repair and downtime costs become more predictable.
14. Short-term jobs. Renting is the most economical solution for short-term and specialty jobs.
15. Transportation costs. Renting is the best way to avoid transporting equipment from project to project, thus reducing transportation costs. This is especially beneficial when dealing with heavy equipment requiring special hauling equipment.
16. Equipment testing. Allows use of equipment in the field without purchase, leading to a better understanding of equipment capabilities and suitability for the work.

The greatest disadvantage of renting is the resultant higher unit cost to perform the work. Typically, the hourly rental rate is more than the lease or ownership rate. Higher unit costs will typically result in a less competitive estimate when bidding against equipment owners.

### 8.2.5 Rent to Own (Rental Purchase)

Rent-to-own for projects that last 1 or 2 years with high equipment utilization is a good approach if the user is uncertain about the need for the equipment after the completion of the project. The rent-to-own option can be an attractive alternative to an outright purchase for several reasons. Rent-to-own gives the contractor the opportunity to build a down payment via the application of rental payments toward the purchase price of the equipment. This allows the equipment to begin generating cash toward its purchase. It gives the contractor a trial period to see how the equipment works and a chance to verify its need for future work.

## 8.3 EQUIPMENT FINANCING COMPARISON

Tables 8.1 and 8.2 compare the four means of equipment financing. The tables are adapted from information provided by Caterpillar Financial Services Corporation [8]. Example 8.1 illustrates these concepts.

**Example 8.1:** A contractor needs a Caterpillar D6R bulldozer. The finance amount for the D6 is $290,000. The contractor estimates about 1,500 hours of use each year. The interest rate is 7.85%, and the term is 60 months. Insurance for the D6 is adjusted for the option plan. Table 8.3 shows projected financial amounts for different finance options available through Caterpillar. Based on this situation, the FMV and CVO plans are very close and yield the lowest monthly payment. The primary difference between these two alternatives is the purchase option at the end of the lease term.

## 8.4 RENTAL AND LEASE CONTRACT CONSIDERATIONS

The following discussion is based on typical terms and conditions found in the construction equipment rental industry. These issues and corresponding values will vary according to the terms of the rental agreement, the general characteristics, or clauses that are included in every rental contract, which are [9]:

**TABLE 8.1**
**Comparison of Financial Methods [8]**

| Finance Method | Ownership | Tax Benefits | Equity Accumulation | Payment | Payout at Lease Termination |
|---|---|---|---|---|---|
| Installment sale | Immediate | Contractor | Fastest | Highest | N/A |
| Finance lease | Payout at lease termination | Contractor | Slower | Low | Discounted cost |
| Tax lease | Payout at lease termination | Financier | None | Lower | Premium cost |
| Rent-to-own | Payout at lease termination | Financier | None | Lowest | Fair market value |

**TABLE 8.2**
**Advantages of Financial Methods [8]**

| | |
|---|---|
| **Installment sale** | Contractor owns and depreciates |
| | Contractor builds equity fastest |
| | Contractor has contractual flexibility |
| **Finance lease** | Predetermined purchase amount |
| | Monthly payments lower than purchase financing |
| | Lessee has rights to the depreciation |
| | Slow equity accumulation |
| **Tax lease** | Optional ownership |
| | Low flexible monthly payments |
| | No equity accumulation |
| **Rent-to-own** | Optional ownership |
| | Lowest monthly payments |
| | Lessee treats unit as a rental |

**TABLE 8.3**
**Acquisition Options and Dollar Amounts**

| Acquisition Option | Finance Amount | Rate | Payment | Insurance/ Month | Purchase Option |
|---|---|---|---|---|---|
| Rent | NA | NA | $ 7,500 | NA | NA |
| Installment sale | $290,000 | 7.85% | $ 5,859 | $ 352 | NA |
| Finance lease | $290,000 | 7.85% | $ 4,917 | $ 352 | $68,960 |
| Tax lease–FMV | $290,000 | 7.85% | $ 4,544 | $ 346 | FMV |
| Tax lease–CVO | $290,000 | 7.85% | $ 4,590 | $ 348 | $87,924 |

1. Time basis of the rate. The basis for rental rates is usually a single shift of 8 hours per day, 40 hours per week, or 176 hours per month for a consecutive 30-day period. Many distributors do not rent their equipment by the day or the week, particularly for large equipment. If the equipment will work more than one shift per term, the rate usually increases by 50% for each additional shift.
2. Rental period. The rental period usually begins when the equipment leaves the renter's warehouse and ends when it is returned to the same location. For out-of-town shipments, the renewal period starts on the date of the bill of lading (a list of goods received for transportation) of shipment and ends on the return date of bill of lading.
3. Payment procedures and insurance. Rental payments may be payable in advance or in arrears depending upon the terms and conditions of the rental agreement. The contractor's credit history and bargaining power can have a considerable impact on payment requirements. It is common for the lessee to provide all insurance coverage.
4. Normal wear and tear. Normal wear and tear are defined as the deterioration resulting from the use of equipment under normal circumstances, provided that the equipment is properly maintained and serviced. This clause considers who is responsible for the repair and maintenance of the equipment and payment for those services. This issue should be addressed at the time of the rental in order to prevent misunderstanding and problems.
5. Fuel and lubricants. The lessee usually provides all wearable supplies, such as fuel and lubricants.
6. Operators. Operators and field specialists are generally not included in the rental rate. Their salaries and expenses are typically extra. If provided by the renter, the operator should be trained and certified to operate the equipment if necessary.
7. Freight charges. The common practice is that the lessee pays for the freight charges from the shipping point to the destination and return. The lessee generally pays the additional charge when loading, unloading, dismantling, or assembling is required. Many suppliers add a surcharge based on the current price of fuel.
8. Condition of equipment. The equipment is to be returned in the same condition as it was in at the time of delivery except for normal wear and tear. The lessee usually is responsible for paying clean-up charges for excessively dirty equipment.
9. Cancellation and extension of contracts. These clauses outline the user's privilege to cancel the contract, term extensions, and the result of late payment.

The basic agreement between lessor and lessee may include many different conditions and provisions. The variation of these provisions depends on the capacity of the parties and the potential implications for tax treatment. The following is a detailed checklist of items that should be addressed in a rental or lease agreement [9].

1. Identification of both lessee and lessor.
2. Identification of equipment, attachments, or accessories by make, model, serial number, and hour meter reading if appropriate.
3. Lease or rental period, including the start and end dates, duration, use location, and any rental or lease period extension provisions.
4. Payment or rate including the actual dollar rate per unit of time, date of initial and periodic payments, payment location, deferred payment, deposit(s), nonpayment periods, payment for partial periods, and penalties or interest.
5. Loading, unloading, and transportation responsibilities, inspections, compliance, and special provisions.

6. Assumption of operation, repair, and maintenance responsibilities and payment, including notification of the lessor and field inspections for major repairs or component replacement.
7. Proof of required permits, insurance, and bonds for transport, assembly, and use.
8. Indemnity and "hold harmless" clauses and provisions.
9. Specification of ownership or title to the equipment or terms for transfer.
10. Subletting, use by other parties, or movement of equipment from the authorized location without the express permission of the lessor.
11. Damage or misuse definitions, repair provisions, "normal wear and tear" limitations, and return provisions and expectations for equipment, tires and tracks, attachments, or accessories.
12. Fuel, oil, grease, and filter replacement.
13. Purchase option provisions such as time, deadlines, amount or rate, options, discounts, required paperwork, warranties, and other conditions if applicable.
14. Arbitration rights and responsibilities.
15. Termination of agreement conditions and financial obligations.
16. Miscellaneous legal provisions or conditions regarding the contract.

## 8.5 THE BUY, LEASE, OR RENT DECISION

Table 8.4 is adapted from Coomb's and Palmer's book, *Construction Accounting and Financial Management* [10]. The table suggests the optimal approach for equipment acquisition based on customer need or criteria.

As stated before, the buy, lease, or rent decision is most influenced by how long the equipment is needed. A short period of utilization favors renting and a longer period favors leasing or purchase. Along with discussed financial analysis and comparison, there are many nonquantitative areas to be considered prior to the decision.

**TABLE 8.4**
**Customer Criteria for Equipment Acquisition [10]**

| Customer Criteria | Cash Purchase | Finance Purchase | Lease | Rent |
|---|---|---|---|---|
| Wants ownership | x | x | | |
| Optional ownership | | | x | |
| Use and return only | | | | x |
| Off-balance sheet accounting | | | x | x |
| 100% financing | | x | x | |
| Trade-in value | x | x | | |
| Expense 100% of payments | | | x | x |
| Need tax write-off | x | x | | |
| Lowest total cost (ownership) | x | | | |
| Lowest monthly payment (use) | | | x | |
| Uncertain future work | | | x | x |
| Avoid debt | | | x | x |
| Try out equipment | | | x | x |
| Improve cash flow | | | x | |
| Plan equipment replacement | | x | x | x |
| Minimize equipment disposal concerns | | | x | x |

- Work volume.
- Nature and types of construction projects.
- Client requirements and expectations.
- Reputation and company perception to potential clients.
- Funding capabilities.
- Long-term financial goals.
- Relationship with equipment supplier.
- Company ownership policy.

A study of the acquisition and finance alternatives comes after identification of the need for a piece of equipment. This evaluation is a key component in the financial planning of the construction firm.

## REFERENCES

[1] Fargo, W. (2019). Construction Industry Forecast 2019, Wells Fargo Bank, N.A. www.wellsfargomedia.com/wholesale/2019/cif/2019_Construction_Industry_Forecast.pdf. (Accessed November 23, 2019).
[2] Cudworth, E.F. (1989). *Equipment Leasing Partnerships*. Chicago: Probus Publishing Company.
[3] Fetridge, K.R. (1991). Buy, Lease or Rent? *Concrete Construction*. The Aberdeen Group. www.concreteconstruction.net/author/keith-r-fetridge. (Accessed November 23, 2019).
[4] MacManamy, R. (1991). Contractors Take the Rental Option (Equipment). 3rd Quarter Cost Report. *Engineering News Record*, 227(13), p. 41.
[5] Associated Equipment Distributors (AED). (2019). *Green Book On-line*. www.equipmentwatch.com. (Accessed November 23, 2019).
[6] US Army Corps of Engineers (USACE). (2016). *Construction Equipment Ownership and Operating Expense Schedule*, Region I. Document EP 1110- 1-8(Vol. 1). Washington, DC: USACE.
[7] American Rental Association. (2019). *Rental Management Sourcebook*. www.ararental.org/. (Accessed November 23, 2019).
[8] Caterpillar Financial Services Corporation. (2019). CAT Financial: For the Life of Your Business. www.CatFinancial.com. (Accessed November 23, 2019).
[9] Park, W.R. and Jackson, D.E. (1984). *Cost Engineering Analysis*. 2nd ed. New York: John Wiley and Sons.
[10] Coombs, W.E. and Palmer, W.J. (1994). *Construction Accounting and Financial Management*. 5th ed. New York: McGraw-Hill Book Company.

## CHAPTER PROBLEMS

1. List and explain the three most important advantages found in purchasing a piece of construction equipment outright in your opinion.
2. List and explain the three most important advantages found in leasing a piece of construction equipment outright in your opinion.
3. List and explain the three most important advantages found in renting a piece of construction equipment outright in your opinion.

# 9 Sustainability in Equipment Fleet Management

## 9.0 INTRODUCTION

Construction, by definition, is disruptive to the environment as contractors change the landscape, replacing it with infrastructure that is meant to promote economic growth and increased well-being for the society. As with most economic activities, this development comes with an environmental cost. Past research estimates that 20% to 30% of global GHG emissions can be attributed to the construction industry, making construction a prime contributor to the global carbon footprint. These estimates consider direct and indirect activities/processes associated with preconstruction, construction, and postconstruction activities related to the operation and maintenance of the constructed product.

This chapter focuses on sustainability and environmental issues associated with the construction phase and its major source of emissions: diesel-powered off-road construction equipment such as bulldozers, loaders, backhoes, rollers, dump trucks, and others. The chapter devotes most of its effort explaining the issues related to fuel combustion emissions (also called exhaust emissions), their potential health and environmental impacts, and relevant regulatory efforts aimed toward mitigating those impacts. It will present a straightforward methodology to quantify different types of exhaust emissions for a given piece of equipment or fleet that can be used to evaluate options for alternative equipment choices based on the level of pollutants they produce. The chapter concludes with a discussion of how the current low bid award system constitutes a major barrier to enhancing the sustainability of construction project execution. The chapter proposes a potential solution to adapt alternative procurement methods to surmount that issue, encouraging construction contractors to deploy the state-of-the-art in sustainable equipment.

## 9.1 SUSTAINABILITY IN CONSTRUCTION EQUIPMENT MANUFACTURING AND MANAGEMENT

Sustainability is commonly conceptualized as a reasonable balance between three pillars: economic, social, and environmental. This chapter adopts a sustainability model based on the premise that sustainability initiatives are motivated by the need or desire to achieve given environmental goals. Thus, any financial costs associated with enhanced sustainability are justified by the social benefits accrued by the environment. This means that environmental risks are first assessed, and strategies to address each risk are then formulated based on the results of that assessment. Ideally, the intended strategies should be both economically feasible and socially acceptable. Preferably, those strategies should offer tangible socio-economic benefits. This holistic sustainability model can be framed around an environmental Life-Cycle Assessment (LCA), as shown in Figure 9.1.

LCA is a standardized method to assess potential environmental impacts across the lifecycle of a product, from the acquisition of raw materials to manufacturing, use, and final disposal [1]. In the context of construction equipment fleet management, the range of environmental risks analyzed at each LCA stage are classified into the following five major categories:

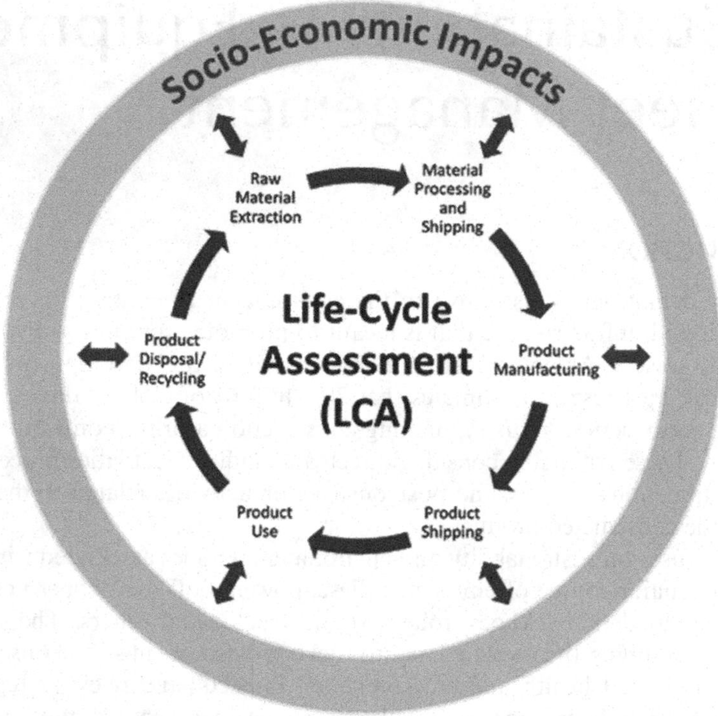


**FIGURE 9.1** Sustainability Model [1]

- Risk of air pollution
- Risk of land pollution
- Risk of water pollution
- Risk to flora and fauna
- Risk for noise and vibration

### 9.1.1 Influence of Equipment Manufactures on Sustainability Outcomes

Among all the different parties involved in the lifecycle of a product, the manufacturer is usually the one in the best position to mitigate environmental impacts. Changes in product design and manufacturing processes decrease the use of virgin raw materials, eliminate practices that lead to pollution, and reduce waste products. Likewise, products can be designed to minimize disruptions to the natural environment during use by their intended final users (Product Use stage in Figure 9.1), as well as to maximize recycling and reuse at the end of their useful life. Construction equipment manufacturers have recognized and accepted this responsibility. During the last couple of decades, they have heavily invested in a continuous effort toward improving sustainability around their products, working along with their partners and suppliers. The following are a few examples of those efforts:

- Engine and general equipment improvements to maximize fuel efficiency, reducing fuel burn and emissions. Examples of these improvements include:

- Functionality enhancements to reduce the amount of energy or equipment hours required to complete a task. For instance, the use of advanced compaction technologies to reduce the number of passes of a compactor.
- Enhanced fuel and combustion systems with less moving parts, reducing emissions and operating costs.
- Some fuel-powered internal elements and functions have been replaced by electric alternatives reducing fuel use.
- Systems developed to keep engine speed low during idling time.
- The development of autonomous equipment systems allowing for less idling time and greater productivity since the operation of the equipment can better align with the optimal performance recommendations provided by the manufacturer.

- An increasing amount of parts and components are designed to be reused or recycled, while at the same time, equipment units are being built with fewer parts – creating less to dispose of and replace.
- Single pieces of equipment have been developed or modified to carry out tasks traditionally performed by two or more machines, mitigating the environmental implications of larger equipment fleets. For example, there is a commercially available machine called a "spray paver" that combines the function of a paver and a tack truck by incorporating a tack spraying system into the paver tractor.
- Long-lasting components and fluids, along with extending maintenance intervals, help in reducing maintenance- and repair-related waste.
- Oil renewal systems to extend oil change intervals.
- Engines that are compatible with biodegradable oil options, reducing the use of fossil fuels.
- Environmentally friendly coating and painting products.
- Ecological fluid draining systems to reduce the risk of spillage.

### 9.1.2 Influence of Equipment Fleet Managers on Sustainability

Once a piece of equipment leaves the assembly line and is placed in service, the equipment fleet manager becomes the leading influencer of its environmental performance. It occurs during the Product Use LCA stage (see Figure 9.1), which is the most intense LCA stage in terms of exhaust emissions, with approximately 90% of all lifecycle carbon dioxide ($CO_2$) emissions occurring during that stage. Management efforts to optimize equipment performance and fuel efficiency during equipment operations bring the manufacturers' sustainability initiatives to fruition. As a result, the decisions made by fleet managers in terms of a capital purchase, maintenance, repair, and replacement of machinery combined with the decisions regarding the employment of the construction equipment in the field have a greater potential to increase construction site sustainability than incremental improvements to the systems that generate the emissions.

Research has found that GHG emissions from construction equipment could be reduced as much as 50% via effective fleet management [2]. Effective management practices include matching loading equipment to hauling equipment, the optimization of cycle times with the optimum number of units, and the reduction of idling, all of which are covered in Chapters 4 and 5. Additionally, manufacturers recommend equipment fleet managers to closely monitor the performance of their equipment in order to implement corrective actions promptly when performance issues are identified. Equipment manufacturers offer a number of products to monitor machines or fleets, including the use of sensors and wireless technologies. For example, Caterpillar offers an application called Project Link™ to track equipment location, operating hours, productivity, and idle time from mobile devices [3]. There are many other applications aimed to facilitate equipment management efforts, providing economic benefits, and maximizing equipment performance. In general, improvements in equipment performance are expected to provide indirect environmental benefits. It has been demonstrated that equipment performance is linked to equipment-related emissions, with fuel consumption decreasing as equipment performance increases.

## 9.2 FUEL COMBUSTION AND AIR QUALITY

The combustion of diesel fuel produces the energy required to operate most pieces of construction equipment. The combustion process releases a range of emissions to the air, causing air quality deterioration and adverse health effects. Combustion emissions are more commonly referred to as exhaust emissions. The factors that affect the amount and composition of exhaust emissions include fuel consumption rates, engine age and load, type of fuel, and environmental conditions such as altitude and temperature.

Exhaust emissions can be classified into three groups: 1) GHG emissions, 2) pollutant emissions, and 3) nontoxic emissions. GHG emissions are known for their contribution to global warming. Pollutant emissions, on the other hand, have been identified as the cause of serious health problems, particularly respiratory diseases. Nontoxic emissions are considered harmless and usually ignored in environmental assessments; hence, they are not further addressed in this chapter. Table 9.1 lists the most relevant emissions under each exhaust emission group, as described in more detail in the following sections.

### 9.2.1 Greenhouse Emissions and the Carbon Footprint

GHG is a term used to refer to a small group of atmospheric gases with the capacity to trap solar heat, warming up the Earth's surface. As the levels of GHGs in the atmosphere increase, more heat is retained, making the temperature at the surface level higher. This phenomenon is known as the greenhouse effect and is thought to be the dominant cause of global warming.

The three GHGs listed in Table 9.1 are the top three GHG emissions produced through fuel combustion. They are also the three largest contributors to the global greenhouse effect. There are other less relevant GHGs not associated with exhaust emissions. Therefore, all references to GHG emissions in the rest of this chapter are limited to carbon dioxide ($CO_2$), methane ($CH_4$), and nitrous oxide ($N_2O$) emissions.

Carbon footprint is a term commonly associated with GHG emissions. It refers to the total amount of GHG emissions released by an individual, organization, event, or economic activity. The carbon footprint is estimated in three steps. First, emissions for each GHG are estimated using a unit of mass (i.e., grams, kilograms, or tons). The unit of mass is commonly given by the scale of the assessment. Depending on the context, they could also be accompanied by a unit of time. For example, industrial emissions at the national or global

**TABLE 9.1**
**Classification of Exhaust Emissions**

| Greenhouse Emissions | Pollutant Emissions | Non-Toxic Emissions |
|---|---|---|
| Carbon dioxide ($CO_2$) | Carbon monoxide (CO) | Water vapor ($H_2O$) |
| Methane ($CH_4$) | Hydrocarbons (HC) | Nitrogen ($N_2$) |
| Nitrous oxide ($N_2O$) | Nitrogen oxides (NOx) | |
| | Particulate matter (PM) | |

level are typically calculated in terms of tons per year, while the GHG emissions produced by a single excavator could be estimated in grams or kilograms per hour.

In order to be able to aggregate GHG emissions from different gases, it is first necessary to express all of them in the same terms. In the second step, all gases are converted into their equivalent amounts of $CO_2$, representing the amount of $CO_2$ emissions that would cause the same level of global warming. Those amounts are known as carbon dioxide equivalent ($CO_2e$) emissions. This conversion is performed by multiplying the amount of each gas by its respective global warming potential (GWP) factor provided by the Intergovernmental Panel on Climate Change (IPCC) [4]. According to IPCC's 2014 assessment, the GWP factor for $CH_4$ is 28, and the one for $N_2O$ is 265, which means that one gram of $CH_4$ or $N_2O$ is about 28 or 265 times more potent and harmful, respectively, than a single gram of $CO_2$. However, $CO_2$ is still considered the most significant GHG due to large amounts of this gas in the atmosphere. About 75% of the GHG atmospheric content corresponds to $CO_2$.

The third and final step in the estimation of the carbon footprint is simply aggregating all $CO_2e$ values. The aggregation requires all $CO_2e$ values to be expressed in the same units of mass and time. Otherwise, additional unit conversions should be performed as needed to meet this requirement. A more detailed description of the carbon footprint calculation process and some estimation examples are provided later in Section 9.3.1 of this chapter.

### 9.2.2 Pollutant Emissions

Exhaust emissions are considered one of the primary sources of pollutants, with carbon monoxide (CO), hydrocarbons (HC), nitrogen oxides (NOx), and particulate matter (PM) posing the greatest threat to human health. All of them have been linked to respiratory issues at different degrees of severity. CO and PM have also been associated with heart and vascular diseases, and there is a proven direct link between HC emissions exposure and cancer.

Increasingly stringent emission regulations, such as those discussed in the next section, have achieved a significant reduction in pollutant emissions during the last 20 years. The amount of pollutants in a modern diesel engine is less than 0.4% of the total exhaust emissions, while GHGs account for over 12%. It does not mean that the negative health effects of pollutants can now be disregarded.

### 9.2.3 Exhaust Emission Regulations

While the environmental impacts of GHGs are usually seen as a mid- to long-term issue, the severity of the health problems attributed to pollutant emissions are seen as a more immediate issue by health experts and policymakers around the world. For this reason, pollutant emissions have been the main concern of exhaust emission regulations. In the case of the US, emission regulations have been introduced through the Clean Air Act (CAA) [5], which were initially enacted in 1955, and revised in 1970, 1977, and 1990. In general terms, the CCA assigns the US Environmental Protection Agency (EPA) the responsibility for developing and enforcing emission standards for a wide range of air pollution sources.

Off-road equipment, such as agricultural and construction equipment, were not regulated by the EPA until the 1990 CAA revision that ordered the EPA to develop a separate set of emission standards for this type of equipment. It should be noted that EPA off-road equipment emission standards are only directed to equipment and engine manufacturers. There are no current federal regulations in the US to control emissions from the actual use of equipment in construction projects.

The first set of off-road emission standards, called Tier 1, was adopted in 1994 with the first Tier 1 engines produced in 1996. Three increasingly stringent versions of these standards have been adopted since then: Tiers 2, 3, and 4. All engines produced before the Tier 1 standards are referred to as Tier 0 engines. Rather than imposing emission standards on all types of engines at the same time, standards implementation has been phased along a few years and according to eight horsepower groups, as shown in Table 9.2. This table also shows an interim or transitional phase for the implementation of Tier 4 standards ($T4_i$). EPA emission standards for off-road diesel engines are published in the US Code of Federal Regulation, Title 40, Part 89. These standards align to a certain degree with European off-road emission standards.

A comparison between Tier 1 and Tier 4 standards shows that emission regulatory efforts have achieved a significant reduction of pollutant emissions during the last couple of decades. In the case of NOx and PM, Tier 4 standards require emission levels over 95% lower than those originally established for Tier 1 engines. Likewise, Tier 4 maximum emissions for CO and HC are 85% and 69%, respectively, lower than those stated in Tier 1 standards [6].

Unlike pollutant emissions, GHG emissions have not yet been heavily regulated in the construction equipment industry. However, GHG reduction and control regulations are expected to appear in the near future as global warming effects become prevalent. Likewise, the general trend of increasingly rigorous environmental protection regulations allows one to reasonably predict upcoming emission standards to regulate the use of off-road equipment on the construction site. To be prepared for these future equipment management challenges, equipment fleet managers must learn how to assess and forecast exhaust emissions, as well as how to optimize their fleets to meet emission standards without affecting other project goals such as

**TABLE 9.2**
**Progressive Phasing of EPA Diesel Emission Standards [7]**

| | Horsepower Groups (hp) | | | | | | | |
|---|---|---|---|---|---|---|---|---|
| **Year** | **25–50** | **51–75** | **76–100** | **101–175** | **176–300** | **301–600** | **601–750** | **750+** |
| **1995** | T0 | T0 | T0 | T0 | T0 | T0 | T0 | T0 |
| **1996** | T0 | T0 | T0 | T0 | T1 | T1 | T1 | T0 |
| **1997** | T0 | T0 | T0 | T1 | T1 | T1 | T1 | T0 |
| **1998** | T0 | T1 | T1 | T1 | T1 | T1 | T1 | T0 |
| **1999** | T1 | T1 | T1 | T1 | T1 | T1 | T1 | T0 |
| **2000** | T1 | T1 | T1 | T1 | T1 | T1 | T1 | T1 |
| **2001** | T1 | T1 | T1 | T1 | T1 | T2 | T1 | T1 |
| **2002** | T1 | T1 | T1 | T1 | T1 | T2 | T2 | T1 |
| **2003** | T1 | T1 | T1 | T2 | T2 | T2 | T2 | T1 |
| **2004** | T2 | T2 | T2 | T2 | T2 | T2 | T2 | T1 |
| **2005** | T2 | T2 | T2 | T2 | T2 | T2 | T2 | T1 |
| **2006** | T2 | T2 | T2 | T2 | T3 | T3 | T3 | T2 |
| **2007** | T2 | T2 | T2 | T3 | T3 | T3 | T3 | T2 |
| **2008** | $T4_i$ | $T4_i$ | T3 | T3 | T3 | T3 | T3 | T2 |
| **2009** | $T4_i$ | $T4_i$ | T3 | T3 | T3 | T3 | T3 | T2 |
| **2010** | $T4_i$ | $T4_i$ | T3 | T3 | T3 | T3 | T3 | T2 |
| **2011** | $T4_i$ | $T4_i$ | T3 | T3 | $T4_i$ | $T4_i$ | $T4_i$ | $T4_i$ |
| **2012** | $T4_i$ | $T4_i$ | $T4_i$ | $T4_i$ | $T4_i$ | $T4_i$ | $T4_i$ | $T4_i$ |
| **2013** | T4 | T4 | $T4_i$ | $T4_i$ | $T4_i$ | $T4_i$ | $T4_i$ | $T4_i$ |
| **2014** | T4 | T4 | $T4_i$ | $T4_i$ | T4 | T4 | T4 | $T4_i$ |
| **2015+** | T4 | T4 | T4 | T4 | T4 | T4 | T4 | T4 |

*Note:* hp = horsepower; $T4_t$ = Tier 4 interim.
*Source:* California Air Resources Board. https://ww3.arb.ca.gov/msprog/ordiesel/faq/emissionfactorsfaq.pdf.

cost and schedule objectives. The following sections of this chapter are intended to provide equipment fleet managers with the knowledge and tools required to address these challenges.

## 9.3 ESTIMATION OF EXHAUST EMISSIONS

Ideally, all sources of exhaust emissions, including construction equipment, should be equipped with portable emissions measurement systems (PEMS), providing real-time and reliable emissions data for each piece of equipment. These systems would allow the development of historical emissions databases to facilitate emission management and assessment efforts. There are no existing regulatory requirements to monitor emissions levels after a piece of equipment leaves the manufacture's plant. Thus, PEMS are rarely found as permanent elements in construction equipment units. In the future, PEMS will furnish a means to track emissions once emission regulations start addressing exhaust emission during construction.

Given the absence of PEMS as standard components of construction equipment units, the current alternatives to estimate exhaust emissions are the dynamometer measures taken during the testing and certification of new engines. A dynamometer by itself is just a device used to simulate the use of an engine under different conditions. However, when paired with an emission measurement system, it provides exhaust emission readings at different engine loads. Engine load is the instantaneous power delivered by an engine relative to its maximum design load or capacity. An engine operating at a 100% load factor is running at its maximum horsepower and fuel-burning rate. Such operating conditions rarely occur during extended periods. Most of the time, an engine is only using a fraction of its maximum capacity, burning less fuel and producing fewer emissions.

The EPA has used dynamometer measures to produce emission factors for different GHG and pollutants. These factors represent the average amount of emissions emitted by a given piece of equipment per unit of fuel burned (e.g., grams of $CO_2$ per gallon of diesel fuel [g/gallon]) or per unit of power used (e.g., grams of NOx per horsepower-hour [g/hp-hr]). Likewise, the EPA provides separate sets of guidelines for the estimation of GHG and pollutant emissions using their respective emission factors. EPA's GHG and pollutant emission estimation guidelines are explained below in Sections 9.3.1 and 9.3.2, respectively.

There are several factors influencing fuel consumption and exhaust emission rates. Regardless of the complexity and robustness of the method used to calculate exhaust emissions, there is always some unavoidable uncertainty affecting the accuracy of those calculations. The EPA's guidelines are assumed to produce reasonably accurate emission estimates using only a few relevant emission influencing factors, as explained below.

### 9.3.1 GHG EMISSIONS AND CARBON FOOTPRINT CALCULATION

As explained in Section 9.2.1, the carbon footprint from exhaust emissions for a piece of equipment is equal to the amount of $CO_2$ emitted plus the $CO_2e$ emissions of $CH_4$ and $N_2O$. Taking into consideration the GWP factors for $CH_4$ and $N_2O$, the calculation process of the carbon footprint for a single equipment unit can be synthesized into Equation 9.1. Likewise, the carbon footprint for an entire fleet is just the aggregation of the carbon footprints of all its equipment units, as shown in Equation 9.2.

$$CF_i = E_{CO_2i} + E_{CH_4i} \times 28 + E_{N_2Oi} \times 265 \tag{9.1}$$

$$FCF = \sum_{i}^{n} CF_i \tag{9.2}$$

Where $CF_i$ is the carbon footprint due to exhaust emissions; $E_{CO_2i}$, $E_{CH_4i}$, and $E_{N_2Oi}$ are the amounts of $CO_2$, $CH_4$, and $N_2O$ emissions for equipment unit $i$; and $FCF$ is the total fleet carbon footprint that results after aggregating the carbon footprints of all the $n$ pieces of equipment forming the fleet. The carbon footprint units in Equations 9.1 and 9.2 are the same as the units of the GHGs in Equation 9.1.

#### 9.3.1.1 GHG Emission Factors

In order to use Equations 9.1 and 9.2, an equipment fleet manager should first be able to estimate the expected amounts of $CO_2$, $CH_4$, and $N_2O$ to be emitted by the equipment unit or fleet under consideration. Here is where the EPA's emission factors and guidelines play their part. EPA's emission factors for GHGs are provided in grams per gallon of fuel burned (g/gal). Thus, the total amount of emissions (in grams) for a given GHG is equal to the total amount of diesel fuel consumed by the equipment (in gallons) multiplied by the corresponding emission factor. This calculation is generalized in Equation 9.3. The emission factors for $CO_2$, $CH_4$, and $N_2O$ are listed in Table 9.3.

$$E_j = FC \times EF_j \tag{9.3}$$

Where $E_j$ is the expected emission amount of GHG $j$, in grams, emitted during a given period of time; $j$ is either $CO_2$, $CH_4$, and $N_2O$, as applicable; $FC$ is the expected/recorded fuel consumption, in gallons, during the same period of time; and $EF_j$ is the emission factor for GHG $j$ (see Table 9.3).

As shown in Table 9.3, $CH_4$ and $N_2O$ exhaust emissions are considerably small in comparison with $CO_2$ emissions. Hence, $CH_4$ and $N_2O$ are usually excluded from emission analyses. However, the EPA recommends the inclusion of these two gases in the assessment of GHG emissions since their actual relevance would only be known after finishing the assessment. It should also be noted that any expected amounts of $CH_4$ and $N_2O$ are still to be converted into $CO_2e$, increasing those amounts 28 and 265 times, respectively.

#### 9.3.1.2 Fuel Consumption and Operating Hours

When having the GHG emission factors, the only additional input required for the calculation of the carbon footprint from off-road exhaust emissions is the amount of fuel burned. If the calculation of the carbon footprint is part of a postconstruction analysis, actual amounts of fuel consumed can be obtained from the project records. However, that type of data would not be available if the analysis is conducted during the planning phase in an attempt to predict exhaust emission during construction. If that is the case, either historical fuel consumption rates recorded from previous projects or average rates provided by the manufacturer can be used in Equation 9.3.

**TABLE 9.3**
**EPA's GHG Emission Factors [6]**

| Greenhouse Gas | $CO_2$ | $CH_4$ | $N_2O$ |
|---|---|---|---|
| Emission factor (g/gal) | 10210.00 | 0.57 | 0.26 |

*Source:* EPA Greenhouse Gas Inventory Guidance – Direct Emissions from Mobile Combustion Sources (2016), Tables A-1 and B-8.

*Note:* EPA's $CO_2$ emission factor is applicable to all types of diesel-powered off-road equipment, while emission factors for $CH_4$ and $N_2O$ are specific to diesel-powered construction and mining equipment.

Some manufacturers provide average fuel consumption rates for different engine load factors. For example, Caterpillar uses three engine load factor ranges to provide fuel consumption rates for its equipment: low, medium, and high load factors. Table 9.4 is an example of how Caterpillar provides fuel consumption information for three equipment units: a track-type tractor (model: D10), a scraper (model: 657G), and an asphalt paver (model: AP1000F).

Even though the tractor and the scraper in Table 9.4 have the same load factor ranges, in one hour of operation, the scraper could easily consume more than twice the amount of fuel consumed by the tractor during the same time; which, in turn, would be reflected in a greater amount of exhaust emissions from the scraper. In comparison with the tractor and the scraper, the asphalt paver would produce the lowest amount of emissions since it presents the lowest hourly consumption rates. This comparison also shows that the asphalt paver usually works under lower engine loads, with its highest load factor ranging between 40% and 50%, while the highest ranges in the other two pieces of equipment are between 65% and 80%. The Typical Application column in Table 9.4 is intended to assist equipment fleet managers in the selection of the load factor range that best matches the intended application of the equipment under consideration.

The application of the GHG exhaust emissions assessment guidelines provided in this section is better illustrated through Example 9.1:

**TABLE 9.4**
**Caterpillar Equipment Fuel Consumption Rates – Example [3]**

| Equipment | Load Factor | Typical Application | Fuel Consumption (Gal/Hour) |
|---|---|---|---|
| Track-type tractor Model: D10 | Low: 35%–50% | Pulling scrapers, most agricultural drawbar, stockpile, coal pile, and finish grade applications. No impact. Intermittent full throttle operation. | 11.4–16.3 |
| | Medium: 50%–65% | Production dozing in clays, sands, gravels. Push loading scrapers, borrow pit ripping, most land clearing applications. Medium-impact conditions. Production landfill work. | 16.3–21.1 |
| | High: 65%–80% | Heavy rock ripping. Push loading and dozing in hard rock. Working on rock surfaces. Continuous high impact conditions. | 21.1–26.0 |
| Scraper model: 657G | Low: 35%–50% | Level or favorable grades on good haul roads and low rolling resistance. Easy-loading materials, partial loads. No impact. Average use, but with considerable idling. | 17.6–26.1 |
| | Medium: 50%–65% | Adverse and favorable grades with varying loading and haul road conditions. Long and short hauls, near full. Some impact. Typical road building use. | 26.1–34.6 |
| | High: 65%–80% | Rough haul roads. Loading heavy clay, continuous high total resistance conditions with steady cycling. Overloading. High impact conditions, such as loading ripped rock. | 34.6–43.1 |
| Asphalt paver Model: AP1000F | Low: 20%–30% | Narrow width paving, low production. | 2.5–4.0 |
| | Medium: 30%–40% | 3–4 m (10–12 feet) width, 50–75 mm (2–3 inch) lift. | 4.0–5.0 |
| | High: 40%–50% | Wide width, deep lift paving. | 5.0–6.5 |

*Source:* *Caterpillar Performance Handbook*, Edition 48, (2018). Chapter 25.

**Example 9.1:** An equipment fleet manager has decided to use a CAT D10 track-type tractor to push-load a CAT 657G scraper across a heavy clay load. Under this configuration, both the tractor and the scraper are expected to work at a medium engine load factor. What would be the expected total carbon footprint from exhaust emissions for these two equipment units if they work together for 60 hours? The midpoints of the hourly fuel consumption ranges in Table 9.4 can be used for this calculation ([16.3 gal/hour + 21.1 gal/hour]/2 = 18.70 gal/hour for the dozer; and [26.1 gal/hour + 34.6 gal/hour]/2 = 30.35 gal/hour for the scraper).

The first step to solve this problem is to determine the expected amount of fuel to be consumed by each equipment unit according to the average hourly fuel consumption (*FC*) rates provided in Table 9.4.

$$FC_{\text{Tractor}} = 18.70\ ^{\text{gal}}/_{\text{hr}} \times 60\ \text{hr} = 1,122\ \text{gallons}$$

$$FC_{\text{Scraper}} = 30.35\ ^{\text{gal}}/_{\text{hr}} \times 60\ \text{hr} = 1,821\ \text{gallons}$$

Equation 9.3 and the emission factors from Table 9.4 can now be used to estimate the amounts of $CO_2$, $CH_4$, and $N_2O$ to be emitted by each equipment unit.

$$E_{\text{CO}_2\text{\_Tractor}} = 10,210\ ^{\text{g}}/_{\text{gal}} \times 1,122\ \text{gal} = 11,455,620\ \text{g}$$

$$E_{\text{CH}_4\text{\_Tractor}} = 0.57\ ^{\text{g}}/_{\text{gal}} \times 1,122\ \text{gal} = 639.54\ \text{g}$$

$$E_{\text{N}_2\text{O}_{\text{Tractor}}} = 0.26\ ^{\text{g}}/_{\text{gal}} \times 1,122\ \text{gal} = 291.72\ \text{g}$$

$$E_{\text{CO}_2\text{\_Scraper}} = 10,210\ ^{\text{g}}/_{\text{gal}} \times 1,821\ \text{gal} = 18,592,410\ \text{g}$$

$$E_{CH_4\_Scraper} = 0.57\ ^{\text{g}}/_{\text{gal}} \times 1,821\ \text{gal} = 1,037.97\ \text{g}$$

$$E_{\text{N}_2\text{O\_Scraper}} = 0.26\ ^{\text{g}}/_{\text{gal}} \times 1,821\ \text{gal} = 473.46\ \text{g}$$

The next step is to calculate the carbon footprint for each equipment unit using Equation 9.1.

$$\text{CF}_{\text{Tractor}} = 11,455,620\text{g} + 639.54\text{g} \times 28 + 291.72\text{g} \times 265 = 11,550,832.92\ \text{g\_CO}_2e$$

$$CF_{\text{Scraper}} = 18,592,410\text{g} + 1,037.97\text{g} \times 28 + 473.46\text{g} \times 265 = 18,746,940.06\ \text{g\_CO}_2e$$

Finally, the total carbon footprint is calculated by combining the carbon footprints from both equipment units using Equation 9.2, yielding a carbon footprint of about 30.4 tons of $CO_2e$.

$$\text{FCF} = 11,550,832.92\text{g} + 18,746,940.06\text{g} = 30,397,772.98\text{g}_{\text{CO}_2e} \approx 30.4\ \text{ton\_CO}_2e$$

### 9.3.2 Pollutant Emissions Calculation

Given that exhaust emission standards and testing procedures are mainly intended for pollutants, the EPA has access to more detailed testing data on pollutant emissions. This data has allowed the EPA to develop pollutant emission estimating procedures that offer better accuracy than those proposed for GHGs. Pollutant emission procedures are slightly more complex, but they are still very straightforward and easy to follow. There are three main differences between the pollutant and GHG emission estimating procedures:

1. Pollutant emission factors are given in grams per horsepower-hour (g/hp-hr) instead of grams per gallon of fuel consumed;
2. Emission factors are adjusted to account for transient operations and engine deterioration;
3. Emissions from different pollutants cannot be aggregated into a single emission measurement.

Once the adjusted emission factor for a given pollutant and equipment unit has been calculated, the emitted amount for that pollutant is estimated, as shown in Equation 9.4, while pollutant emissions at the fleet level are calculated using Equation 9.5. The amount for each pollutant is reported by separate since there are no equivalent units to combine them into a single emission measurement.

$$E_{ji} = AEF_{ji} \times HP_i \times OH_i \tag{9.4}$$

$$FE_j = \sum_{i}^{n} E_{ji} \tag{9.5}$$

Where $E_{ji}$ is the amount of emissions of pollutant $j$ from equipment unit $i$; $AEF_{ji}$ is the adjusted emission factor for pollutant $j$ and equipment unit $i$; $HP_i$ is the rated horsepower of equipment unit $i$; $OH_i$ is the expected/recorded number of operating hours for equipment unit $i$; and $FE_j$ are the total amount of emissions of pollutant $j$ produce by all the $n$ pieces of equipment forming the fleet.

### 9.3.2.1 Pollutant Emissions Factors

The adjusted emission factor for a given pollutant on a given equipment unit is calculated as follows:

$$AEF_{ji} = EF_{ji} \times TAF_{ji} \times DF_{ji} \tag{9.6}$$

Where $AEF_{ji}$ is the adjusted emission factor for pollutant $j$ and equipment unit $i$; $EF_{ji}$ is the unadjusted factor for pollutant $j$ and equipment unit $i$; $TAF_{ji}$ is the transient adjustment factor for pollutant $j$ and equipment unit $i$; and $DF_{ji}$ is the deterioration factor for pollutant $j$ and equipment unit $i$. The adjusted and unadjusted emission factors are expressed in grams per horsepower-hour (g/hp-hr), while the two adjustment factors are unitless. Unadjusted emission factors are referred to as "zero-hour, steady-state emission factors" by the EPA, meaning they are calculated from data obtained from steady-state tests conducted on new engines at the plant. However, exhaust emission levels change over time, and the steady-state operation of the engine during testing does not represent the typical operating conditions of the equipment, which is the reason for the use of the transient and deterioration adjustment factors.

Unadjusted emission factors are mainly a function of the rated horsepower provided by the manufacturer and the engine technology of the equipment under consideration. The type of technology of an engine is given by the emission standards applicable to that engine at the moment of its fabrication. In other words, it refers to the Tier of the engine, which can be determined from Table 9.2 with the engine power (horsepower) and the model year.

EPA has further subdivided interim and final Tier 4 technologies for the calculation of emission factors. This subdivision is based on the after-treatment configuration of the engines. After-treatment systems are intended to reduce the amount of pollutants in fuel combustion emissions. There are two after-treatment systems commonly used in Tier 4 engines: diesel particular filter (DPF) and selective catalytic reduction (SCR).

As shown in Table 9.5, a Tier 4 engine could include both after-treatment systems, only one, or none of them. The manufactures provide information about after-treatment technologies on specific equipment units. The last two columns of Table 9.5 show the labels used by the EPA to report emission factors for Tier 4 engines in Table 9.6. This table provides the emission factor for each pollutant based on the engine power and technology of the equipment unit under consideration. EPA's pollutant emission factors are provided across the same seven horsepower groups shown in Table 9.2.

#### 9.3.2.2 Transient Adjustment Factor

Transient adjustments on emission factors are intended to account for changes in exhaust emissions due to the actual in-use conditions of the equipment. These conditions may vary significantly among different types of equipment, depending on the type of work that they perform. This is another way to account for the typical engine load ranges previously discussed for GHG emissions.

The EPA provides transient adjustment factors for 26 different types of off-road construction equipment, as shown in Table 9.7. Tier 4 engines are required to be tested and certified under transient conditions. Therefore, the transient adjustment for those engines is not. Adjustment factors for Tier 4 engines for all pollutants are assumed to be one.

#### 9.3.2.3 Deterioration Adjustment Factor

The deterioration adjustment factor is intended to represent the influence of equipment aging on exhaust emissions. Emission levels increase with the age of the engine as the fuel combustion system, and the equipment, in general, becomes progressively less effective. Deterioration factors are calculated as a function of the age of the engine, technology type (Tier), level of usage of the equipment unit along with its operating life, and the pollutant under consideration. These are direct and indirect inputs of the deterioration factor equation (Equations 9.7 and 9.8).

$$DF_{ji} = 1 + A_{ji} \times FLE_i \tag{9.7}$$

$$FLE_i = \begin{cases} \frac{COH_i \times LF_i}{ML_i}; \frac{COH_i \times LF_i}{ML_i} \leq 1 \\ 1; \frac{COH_i \times LF_i}{ML_i} > 1 \end{cases} \tag{9.8}$$

**TABLE 9.5**
**Tier 4 Technology Type Subdivision [8]**

| | | Tier 4 – Technology Subdivision | |
|---|---|---|---|
| **DPF** | **SCR** | **Tier 4 Interim (I)** | **Tier 4 Final (F)** |
| No | No | Tier 4IA | Tier 4FA |
| No | Yes | Tier 4IB | Tier 4FB |
| Yes | No | Tier 4IC | Tier 4FC |
| Yes | Yes | Tier 4ID | Tier 4FD |

*Source:* US Environmental Protection Agency. Exhaust and Crankcase Emission Factors for Off-road Compression-Ignition Engines in MOVES2014b (2018).

**TABLE 9.6**
**Pollutant Emission Factors [7]**

| | | Emission Factors (G/Hp-Hr) | | | |
|---|---|---|---|---|---|
| **Engine Power (Hp)** | **Technology Type** | **CO** | **HC** | **$NO_x$** | **PM** |
| 25 to 50 | Tier 0 | 5.000 | 1.800 | 6.900 | 0.800 |
| | Tier 1 | 1.5323 | 0.2789 | 4.7279 | 0.3389 |
| | Tier 2 | 1.5323 | 0.2789 | 4.7279 | 0.3389 |
| | Tier 4A | 1.373 | 0.420 | 3.905 | 0.161 |
| | Tier 4FA | 0.408 | 0.136 | 2.762 | 0.027 |
| | Tier 4FC | 0.047 | 0.018 | 2.184 | 0.001 |
| 51 to 75 | Tier 0 | 3.490 | 0.990 | 6.900 | 0.722 |
| | Tier 1 | 2.366 | 0.521 | 5.599 | 0.473 |
| | Tier 2 | 2.366 | 0.367 | 4.700 | 0.240 |
| | Tier 4A | 0.978 | 0.177 | 3.030 | 0.149 |
| | Tier 4FA | 0.267 | 0.074 | 2.787 | 0.024 |
| | Tier 4FC | 0.055 | 0.018 | 2.218 | 0.001 |
| 76 to 100 | Tier 0 | 3.490 | 0.990 | 6.900 | 0.722 |
| | Tier 1 | 2.366 | 0.521 | 5.599 | 0.473 |
| | Tier 2 | 2.366 | 0.367 | 4.700 | 0.240 |
| | Tier 3B | 2.366 | 0.184 | 3.000 | 0.200 |
| | Tier 4IA | 0.392 | 0.087 | 2.521 | 0.115 |
| | Tier 4IC | 0.027 | 0.007 | 1.955 | 0.007 |
| | Tier 4FA | 1.342 | 0.075 | 3.274 | 0.186 |
| | Tier 4FB | 0.101 | 0.012 | 0.136 | 0.015 |
| | Tier 4FC | 0.000 | 0.000 | 2.064 | 0.007 |
| | Tier 4FD | 0.000 | 0.008 | 0.091 | 0.000 |
| 101 to 175 | Tier 0 | 2.700 | 0.680 | 8.380 | 0.402 |
| | Tier 1 | 0.867 | 0.338 | 5.652 | 0.280 |
| | Tier 2 | 0.867 | 0.338 | 4.100 | 0.180 |
| | Tier 3 | 0.867 | 0.184 | 2.500 | 0.220 |
| | Tier 4IA | 0.200 | 0.035 | 1.976 | 0.033 |
| | Tier 4IB | 0.262 | 0.017 | 2.029 | 0.011 |
| | Tier 4IC | 0.009 | 0.003 | 1.888 | 0.004 |
| | Tier 4ID | 0.075 | 0.007 | 0.224 | 0.000 |
| | Tier 4FB | 0.052 | 0.007 | 0.144 | 0.011 |
| | Tier 4FC | 0.004 | 0.003 | 1.828 | 0.002 |
| | Tier 4FD | 0.023 | 0.010 | 0.096 | 0.001 |
| 176 to 300 | Tier 0 | 2.700 | 0.680 | 8.380 | 0.402 |
| | Tier 1 | 0.748 | 0.309 | 5.577 | 0.252 |
| | Tier 2 | 0.748 | 0.309 | 4.000 | 0.132 |
| | Tier 3 | 0.748 | 0.184 | 2.500 | 0.150 |
| | Tier 4IA | 1.642 | 0.114 | 2.467 | 0.108 |
| | Tier 4IB | 0.247 | 0.008 | 1.100 | 0.011 |
| | Tier 4IC | 0.052 | 0.011 | 1.116 | 0.001 |
| | Tier 4ID | 0.000 | 0.028 | 0.110 | 0.007 |
| | Tier 4FB | 0.020 | 0.008 | 0.148 | 0.009 |
| | Tier 4FC | 0.214 | 0.011 | 1.153 | 0.000 |

*(Continued)*

**TABLE 9.6 (Cont.)**

| | | Emission Factors (G/Hp-Hr) | | | |
|---|---|---|---|---|---|
| **Engine Power (Hp)** | **Technology Type** | **CO** | **HC** | **$NO_x$** | **PM** |
| | Tier 4FD | 0.015 | 0.010 | 0.079 | 0.002 |
| 301 to 600 | Tier 0 | 2.700 | 0.680 | 8.380 | 0.402 |
| | Tier 1 | 1.306 | 0.203 | 6.015 | 0.201 |
| | Tier 2 | 0.843 | 0.167 | 4.335 | 0.132 |
| | Tier 3 | 0.843 | 0.167 | 2.500 | 0.150 |
| | Tier 4IA | 1.642 | 0.114 | 2.467 | 0.108 |
| | Tier 4IB | 0.247 | 0.008 | 1.100 | 0.011 |
| | Tier 4IC | 0.053 | 0.012 | 1.115 | 0.001 |
| | Tier 4ID | 0.000 | 0.028 | 0.110 | 0.007 |
| | Tier 4FB | 0.020 | 0.008 | 0.148 | 0.009 |
| | Tier 4FC | 0.214 | 0.011 | 1.153 | 0.000 |
| | Tier 4FD | 0.015 | 0.010 | 0.079 | 0.002 |
| 601 to 750 | Tier 0 | 2.700 | 0.680 | 8.380 | 0.402 |
| | Tier 1 | 1.327 | 0.147 | 5.822 | 0.220 |
| | Tier 2 | 1.327 | 0.167 | 4.100 | 0.132 |
| | Tier 3 | 1.327 | 0.167 | 2.500 | 0.150 |
| | Tier 4IA | 1.642 | 0.114 | 2.467 | 0.108 |
| | Tier 4IB | 0.247 | 0.008 | 1.100 | 0.011 |
| | Tier 4IC | 0.053 | 0.012 | 1.115 | 0.001 |
| | Tier 4ID | 0.000 | 0.028 | 0.110 | 0.007 |
| | Tier 4FB | 0.020 | 0.008 | 0.148 | 0.009 |
| | Tier 4FC | 0.214 | 0.011 | 1.153 | 0.000 |
| | Tier 4FD | 0.015 | 0.010 | 0.079 | 0.002 |

*Source:* US Environmental Protection Agency. Exhaust and Crankcase Emission Factors for Off-road Compression-Ignition Engines in MOVES2014b (2018).

Where $DF_{ji}$ is the deterioration factor for pollutant $j$ and equipment unit $i$; $A_{ji}$ is the relative deterioration factor for pollutant $j$ and equipment unit $i$; $FLE_i$ is the fraction of the median life expended by the equipment $i$; and $COH_i$, $LF_i$, and $ML_i$ are the cumulative engine operating hours, the average load factor, and the median life of the engine at full load (in hours), respectively, for equipment unit $i$.

The EPA has found that engine deterioration has a different impact on different types of engine technologies and pollutants. The relationship between engine deterioration, engine technologies, and specific pollutants is quantified by the EPA in the form of relative deterioration factors ($A$ in Equation 9.7). Table 9.8 shows the relative deterioration factors suggested by the EPA.

The fraction of the median life expended ($FLE$) is calculated by dividing the equipment life expended ($COH \times LF$) by the median life of the engine at full load ($ML$). The life expended by an old engine may exceed its median life, giving an $FLE$ value greater than one. If that is the case, $FLE$ is set as one under the assumption that, at that point, the engine has reached a level of deterioration that can be offset by maintenance.

## TABLE 9.7
## EPA's Transient Adjustment Factors for Off-Road Construction Equipment [7]

| Type of Equipment | Transient Adjustment Factor | | | | | |
|---|---|---|---|---|---|---|
| | CO | HC | NOx | | PM | |
| | T0 – T3 | T0 – T3 | T0 – T2 | T3 | T0 – T2 | T3 |
| Pavers | 1.53 | 1.05 | 0.95 | 1.04 | 1.23 | 1.47 |
| Tampers/Rammers | 1.00 | 1.00 | 1.00 | 1.00 | 1.00 | 1.00 |
| Plate compactors | 1.00 | 1.00 | 1.00 | 1.00 | 1.00 | 1.00 |
| Rollers | 1.53 | 1.05 | 0.95 | 1.04 | 1.23 | 1.47 |
| Scrapers | 1.53 | 1.05 | 0.95 | 1.04 | 1.23 | 1.47 |
| Paving equipment | 1.53 | 1.05 | 0.95 | 1.04 | 1.23 | 1.47 |
| Surfacing equipment | 1.53 | 1.05 | 0.95 | 1.04 | 1.23 | 1.47 |
| Signal boards | 1.00 | 1.00 | 1.00 | 1.00 | 1.00 | 1.00 |
| Trenchers | 1.53 | 1.05 | 0.95 | 1.04 | 1.23 | 1.47 |
| Bore/Drill rigs | 1.00 | 1.00 | 1.00 | 1.00 | 1.00 | 1.00 |
| Excavators | 1.53 | 1.05 | 0.95 | 1.04 | 1.23 | 1.47 |
| Concrete/Industrial saws | 1.53 | 1.05 | 0.95 | 1.04 | 1.23 | 1.47 |
| Cement/Mortar mixers | 1.00 | 1.00 | 1.00 | 1.00 | 1.00 | 1.00 |
| Cranes | 1.00 | 1.00 | 1.00 | 1.00 | 1.00 | 1.00 |
| Grades | 1.53 | 1.05 | 0.95 | 1.04 | 1.23 | 1.47 |
| Off-highway trucks | 1.53 | 1.05 | 0.95 | 1.04 | 1.23 | 1.47 |
| Crushing/Proc. equipment | 1.00 | 1.00 | 1.00 | 1.00 | 1.00 | 1.00 |
| Rough terrain forklifts | 1.53 | 1.05 | 0.95 | 1.04 | 1.23 | 1.47 |
| Rubber tire loaders | 1.53 | 1.05 | 0.95 | 1.04 | 1.23 | 1.47 |
| Rubber tire dozers | 1.53 | 1.05 | 0.95 | 1.04 | 1.23 | 1.47 |
| Tractors/Loaders/Backhoes | 2.57 | 2.29 | 1.10 | 1.21 | 1.97 | 2.37 |
| Crawler dozers | 1.53 | 1.05 | 0.95 | 1.04 | 1.23 | 1.47 |
| Skid steer loaders | 2.57 | 2.29 | 1.10 | 1.21 | 1.97 | 2.37 |
| Off-highway tractors | 1.53 | 1.05 | 0.95 | 1.04 | 1.23 | 1.47 |
| Dumpers/Tenders | 2.57 | 2.29 | 1.10 | 1.21 | 1.97 | 2.37 |
| Other construction eqpt. | 1.53 | 1.05 | 0.95 | 1.04 | 1.23 | 1.47 |

*Source:* US Environmental Protection Agency. Exhaust and Crankcase Emission Factors for Off-road Compression-Ignition Engines in MOVES2014b (2018). Note: T0 = Tier 0; T2 = Tier 2; T3 = Tier 3.

## TABLE 9.8
## EPA's Relative Deterioration Factors [7]

| Pollutant | Relative Deterioration Factor ($A$) | | | |
|---|---|---|---|---|
| | Tier 0 | Tier 1 | Tier 2 | Tier 3 + |
| CO | 0.185 | 0.101 | 0.101 | 0.151 |
| HC | 0.047 | 0.036 | 0.034 | 0.027 |
| NOx | 0.024 | 0.024 | 0.009 | 0.008 |
| PM | 0.473 | 0.473 | 0.473 | 0.473 |

*Source:* US Environmental Protection Agency. Exhaust and Crankcase Emission Factors for Off-road Compression-Ignition Engines in MOVES2014b (2018).

The total number of operating hours accumulated by a piece of equipment (*COH*) can usually be obtained from historical equipment records. If such records are not available, the equipment fleet manager's professional judgment would be sufficient for this parameter. Using a rough estimate for cumulative operating hours is better than ignoring the impact of engine deterioration.

Something similar happens with the determination of average load factors (*LF*) and median life estimates (*ML*) on Equation 9.8. Those input values usually involve a considerable degree of uncertainty. However, as part of the data processing and analysis efforts required for the development of pollution emission assessment procedures, the EPA has estimated average load factors for 27 different types of construction equipment, as well as engine median life values (in hours) for three different horsepower ranges. EPA's average load factors and median life values are shown in Tables 9.9 and 9.10. It should be noted that values in Table 9.10 are calculated for an engine continuously running at its maximum capacity, which is an unrealistic scenario during actual equipment use; nonetheless, this is

**TABLE 9.9**
**EPA's Average Engine Load Factors [9]**

| Equipment Description | Load Factor |
|---|---|
| Pavers | 0.59 |
| Tampers/Rammers (unused) | 0.43 |
| Plate compactors | 0.43 |
| Concrete pavers (unused) | 0.59 |
| Rollers | 0.59 |
| Scrapers | 0.59 |
| Paving equipment | 0.59 |
| Surfacing equipment | 0.59 |
| Signal boards | 0.43 |
| Trenchers | 0.59 |
| Bore/Drill rigs | 0.43 |
| Excavators | 0.59 |
| Concrete/Industrial saws | 0.59 |
| Cement & mortar mixers | 0.43 |
| Cranes | 0.43 |
| Graders | 0.59 |
| Off-highway trucks | 0.59 |
| Crushing/Proc. equipment | 0.43 |
| Rough terrain forklifts | 0.59 |
| Rubber tire loaders | 0.59 |
| Rubber tire dozers | 0.59 |
| Tractors/Loaders/Backhoes | 0.21 |
| Crawler tractors | 0.59 |
| Skid steer loaders | 0.21 |
| Off-highway tractors | 0.59 |
| Dumpers/Tenders | 0.21 |
| Other construction equipment | 0.59 |

*Source:* US Environmental Protection Agency. Median Life, Annual Activity, and Load Factor Values for Off-road Engine Emissions Modeling (2010).

**TABLE 9.10**
**EPA's Median Engine Life in Hours at Full Load [9]**

| Engine Power (Hp) | Median Engine Life (Hours) |
|---|---|
| 16 to 50 | 2,500 |
| 51 to 300 | 4,667 |
| 300 + | 7,000 |

*Source:* US Environmental Protection Agency. Median Life, Annual Activity, and Load Factor Values for Off-road Engine Emissions Modeling (2010).

the input for Equation 9.8. The actual expected life of an engine under normal operating conditions should not be used in this equation. Alternatively, average engine load factors could be used in Equation 9.8, such as the load factor ranges in Table 9.4.

This section closes with Example 9.2, illustrating the complete pollutant emission estimating procedure described in this section.

**Example 9.2:** The same equipment fleet manager from Example 9.1 wants to estimate the amount of pollutant emissions associated with the same tractor-scraper configuration (CAT D10 track-type tractor and CAT 657G scraper), but this time using the average engine load factor, as well as the other input parameters suggested by the EPA. What would be the expected total amount of CO, HC, NOx, and PM exhaust emissions for these two equipment units if they work together for 60 hours? Use the information provided in Table 9.11 to answer this question.

The equipment fleet manager begins by calculating the adjusted emission factor (*AEF*) for each pollutant on each piece of equipment using Equation 9.6. However, it is first necessary to determine the engine technology for each equipment unit using Table 9.2. Both the tractor and the scraper are within the 301 to 600 horsepower range and based on their model years; it is possible to infer that they have Tier 2 and Tier 3 engines, respectively.

With that information and the other specifications provided in this problem, the equipment fleet manager can obtain the adjustment emission factors, transient adjustment rates, and relative deterioration factors from Tables 9.6 to 9.8. Likewise, Table 9.9 provides EPA's average load factors for tractors and scrapers, and Table 9.10 provides their median engine lives if operated at fuel capacity. Table 9.12 summarizes the information obtained from those tables.

Next, the cumulative operating hours given with the problem, and the load factors and median engine lives shown in Table 9.12 are used along with Equation 9.8 to estimate the

**TABLE 9.11**
**Example 9.2 Equipment Data**

| Equipment | Model Year | Engine Power (Hp) | Cumulative Operating Hours |
|---|---|---|---|
| Track-type tractor (CAT D10) | 2002 | 570 | 17,000 |
| Scraper (CAT 657G) | 2010 | 600 | 6,800 |

**TABLE 9.12**
**Example Summary Information**

| Equipment | Pollutant | Unadjusted Emission Factor ($EF$) (G/Hp-Hr) | Transient Adjustment Factor ($TAF$) | Relative Deterioration Factor ($A$) | Load Factor ($LF$) | Median Engine Life ($ML$) (Hours) |
|---|---|---|---|---|---|---|
| CAT D10 | CO | 0.843 | 1.53 | 0.101 | 0.59 | 7,000 |
| | HC | 0.167 | 1.05 | 0.034 | | |
| | NOx | 4.335 | 0.95 | 0.009 | | |
| | PM | 0.132 | 1.23 | 0.473 | | |
| CAT 657G | CO | 0.843 | 1.53 | 0.151 | 0.59 | 7,000 |
| | HC | 0.167 | 1.05 | 0.027 | | |
| | NOx | 2.500 | 1.04 | 0.008 | | |
| | PM | 0.150 | 1.47 | 0.473 | | |

fraction of the median life expended by each equipment unit. The outputs from those equations are then used on Equation 9.7 to estimate the deterioration factors for each pollutant on each piece of equipment with their respective relative deterioration factors as shown below.

$$FLE_{\text{Tractor}} = \frac{17,000\,\text{hrs} \times 0.59}{7,000\,\text{hrs}} = 1.433 > 1 \rightarrow \text{FLE}_{\text{Tractor}} = 1$$

$$DF_{\text{CO, Tractor}} = 1 + 0.101 \times 1 = 1.101$$

$$DF_{\text{HC, Tractor}} = 1 + 0.034 \times 1 = 1.034$$

$$DF_{\text{NOx, Tractor}} = 1 + 0.009 \times 1 = 1.009$$

$$DF_{\text{PM, Tractor}} = 1 + 0.473 \times 1 = 1.473$$

$$FLE_{\text{Scraper}} = \frac{6,800\,\text{hrs} \times 0.59}{7,000\,\text{hrs}} = 0.573 < 1 \rightarrow FLE_{\text{Scraper}} = 0.573$$

$$DF_{\text{CO, Scraper}} = 1 + 0.151 \times 0.573 = 1.087$$

$$DF_{\text{HC, Scraper}} = 1 + 0.027 \times 0.573 = 1.015$$

$$DF_{\text{NOx, Scraper}} = 1 + 0.008 \times 0.573 = 1.005$$

$$DF_{\text{PM, Scraper}} = 1 + 0.473 \times 0.573 = 1.271$$

The equipment fleet manager can now use Equation 9.6 to calculate the adjusted emission factor for each pollutant and equipment unit with their respective unadjusted emission and transient adjustment factors, and with the deterioration factors calculated above.

$$AEF_{\text{CO, Tractor}} = 0.843\,\text{g/hp} - \text{hr} \times 1.53 \times 1.101 = 1.420\,\text{g/hp} - \text{hr}$$

$$AEF_{\text{HC, Tractor}} = 0.167\,\text{g/hp} - \text{hr} \times 1.05 \times 1.034 = 0.181\,\text{g/hp} - \text{hr}$$

$$AEF_{NOx,\ Tractor} = 4.335\,g/hp - hr \times 0.95 \times 1.009 = 4.155\,g/hp - hr$$

$$AEF_{PM,\ Tractor} = 0.132\,g/hp - hr \times 1.23 \times 1.473 = 0.239\,g/hp - hr$$

$$AEF_{CO,\ Scraper} = 0.843\,g/hp - hr \times 1.53 \times 1.087 = 1.401\,g/hp - hr$$

$$AEF_{HC,\ Scraper} = 0.167\,g/hp - hr \times 1.05 \times 1.015 = 0.178\,g/hp - hr$$

$$AEF_{NOx,\ Scraper} = 2.500\,g/hp - hr \times 1.04 \times 1.005 = 2.612\,g/hp - hr$$

$$AEF_{PM,\ Scraper} = 0.150\,g/hp - hr \times 1.47 \times 1.271 = 0.280\,g/hp - hr$$

Finally, the equipment fleet manager can use the adjusted emission factors calculated above, the engine power for each piece of equipment, the operating period under consideration (60 hours), and Equations 9.4 and 9.5 to calculate the total amount of pollutants emitted by the tractor and the scraper.

$$E_{CO,\ Tractor} = 1.420\,g/hp - hr \times 570\,hp \times 60\,hrs = 48,566\,g$$

$$E_{HC,\ Tractor} = 0.181\,g/hp - hr \times 570\,hp \times 60\,hrs = 6,201\,g$$

$$E_{NOx,\ Tractor} = 4.155\,g/hp - hr \times 570\,hp \times 60\,hrs = 142,112\,g$$

$$E_{PM,\ Tractor} = 0.239\,g/hp - hr \times 570\,hp \times 60\,hrs = 8,179\,g$$

$$E_{CO,\ Scraper} = 1.401\,g/hp - hr \times 600\,hp \times 60\,hrs = 50,451\,g$$

$$E_{HC,\ Scraper} = 0.178\,g/hp - hr \times 600\,hp \times 60\,hrs = 6,410\,g$$

$$E_{NOx,\ Scraper} = 2.612\,g/hp - hr \times 600\,hp \times 60\,hrs = 94,029\,g$$

$$E_{PM,\ Scraper} = 0.280\,g/hp - hr \times 600\,hp \times 60\,hrs = 10,090\,g$$

$$FE_{CO} = 48,566\,g + 50,451\,g = 99,019\,g \approx 99\,kg\ of\ CO$$

$$FE_{HC} = 6,201\,g + 6,410\,g = 12,611\,g \approx 13\,kg\ of\ HC$$

$$FE_{NOx} = 142,112\,g + 94,029\,g = 236,141\,g \approx 236\,kg\ of\ NOx$$

$$FE_{PM} = 8,179\,g + 10,090\,g = 18,269\,g \approx 18\,kg\ of\ PM$$

A comparison between the pollutant emissions estimated above and the GHG emissions calculated in Example 9.1 for the same two equipment units shows the considerable difference in the amounts of exhaust emissions between these two types of gases. The total amount of emissions for all pollutants together is about 1.2% of the total amount of $CO_2$ produced by the same equipment. Again, this difference does not mean that the impact of pollutant emissions can be disregarded. Both GHG and pollutant emissions impact the economic, social, and natural environment. As such, both of them should be equally considered on exhaust emission reduction and control initiatives implemented by equipment fleet managers.

## 9.4 PUTTING CONSTRUCTION SUSTAINABILITY IN PERSPECTIVE

Enhancing sustainability on the construction project itself requires a broader context than currently exists. Traditionally, enhancing sustainability has focused on reducing the impact of projects after they are built during the Product Use LCA stage shown in Figure 9.1. Current green rating systems, such as the U.S. Green Building Council Leadership in Energy and Environmental Design (LEED ®) Guidelines and Greenroads ® Guidelines, are focused on design. These programs typically include some construction and maintenance factors, such as the recycling of construction materials. The underlying assumption is that the contractor will merely execute the sustainable design requirements contained in the construction documents to achieve the project's sustainable design objectives. The missing component is metrics that assess the relative sustainability of different construction contractors' preferred means and methods. For instance, a paving project using a portable batch plant will include minimum air quality standards in the contract. If one competitor owns a batch plant that minimally meets the standard and another owns the latest technology which exceeds the standard, there are no contractual means to differentiate the relative value of a bid that is based on a more sustainable set of means and methods. The new technology's ownership cost may force the contractor to submit a higher price, making it impossible for it to win [10]. Hence, the owner is unable to enhance overall project sustainability beyond the minimums described in the construction documents even if it maintains a sincere commitment to minimizing environmental impact. The bottom-line is explicit. The low bid procurement system creates a financial conundrum that stifles the ability to increase the sustainability of the means and methods used to build an equipment-intensive project. Regardless of whether the owner or the contractor possesses a strong desire to reduce the environmental impact of the construction process, the project will be built by the low bidder using equipment with the lowest ownership costs that meet the minimum standards found in the environmental regulations.

### 9.4.1 Incentivizing Sustainability in the Construction Procurement Process

Design-bid-build (DBB) project delivery is the dominant method for procuring construction in the US. However, there have been variations on the method that do not rely on a pure low bid award that can be adapted to surmounting the barrier to enhancing sustainability described in the previous section. The US Department of Transportation's Federal Highway Administration (FHWA) authorized "cost plus time bidding" (called A+B bidding) and incentive/disincentive (I/D) contracting on federal-aid funded projects through its Special Experimental Program 14 in 1995. Both the owners and contractors have embraced incentivizing construction contracts as a fair and equitable method for the owner to accelerate the construction and enhance quality after the design is complete. Research has shown that incentivizing highway construction not only improves quality but also produces time and cost benefits. The following quotations from previous studies on A+B and I/D projects show the effectiveness of both procurement tools:

- "Combined incentives can be used to achieve cost and schedule effectiveness in a project. The contractor is always interested in increased profit . . ." [11].
- "The benefit/cost ratio in I/D contracts averaged 2.46 where the benefit is defined as the cost savings to the public in terms of road-user delay costs and the cost is defined as the money paid as an incentive to the contractor" [11].
- "A+B projects that applied the I/D contracting strategy demonstrated the power of I/D clauses: many of these types of projects achieved or surpassed the agency's

goal of early project completion. In conclusion, it is recommended that A+ B contracting be used with an I/D provision to motivate contractors to meet a scheduled completion date" [12].

Incentivizing construction contracts has been proven to be an effective tool to achieve the owner's project goals. Both A+B bidding and I/D contracting could be adapted to enhance construction sustainability by incentivizing construction contractors. It would allow them to propose means and methods that exceed statutory environmental minimums for air and water quality, minimized noise pollution, and decreased energy consumption during project construction.

A+B contracts establish a value for time based on road-user delay cost by setting a value for a construction working day. The contractor's bid price is the "A" part. The "B" component is the product of the number of days proposed to complete the project by each contractor and the value of a working day. The two components are added together, and the contract is awarded to the lowest A+B bid. Hence, the owner can justify awarding to a contractor whose bid price is higher than the lowest bidder but whose schedule is faster. I/D contracts establish a time or quality-based value and a standard for the given project. At project completion, the contractor is paid a bonus (incentive) if it exceeds the desired standard, and an equal amount is deducted (disincentive) if the contractor fails to achieve the established standard.

The two approaches have also been combined to yield A+B+I/D contracting. A+B+I/D has been successfully employed to accomplish the following:

- Accelerate schedule: I/D were used to enhance the quality of the finished pavement using the International Roughness Index as the standard.
- Enhance overall quality: I/D were based on a percentage within limits for a series of quality control test results as the standard.
- Minimize project impact to the traveling public: I/D were based on hours of lane closures as a standard.

The A+B and A+B+I/D alternatives are classified as "best value" award methods. Best value is based on the premise that the contract award is made on the basis of something other than price alone [10]. A+B is price plus time. In best value awards, the owner seeks to justify *not awarding* to the lowest, minimally qualified bidder based on increased value offered by a higher competing bid. Thus, the value of employing a more sustainable fleet of equipment could be combined with the price to provide an objective best value decision criterion. Additionally, since A+B bidding, I/D, and A+B+I/D contracting have been successfully implemented on a variety of projects for over a decade, it provides an opportunity to extend the concept into a framework from which to incentivize construction sustainability. The next section will illustrate possible alternatives for accomplishing that objective.

### 9.4.2 Bidding the Value of Sustainable Equipment

Owners have used best value approaches to include a full range of desired value-adding factors from the qualifications of key contractor personnel to the proposed design itself into construction project award determinations. An objective assessment of sustainability can also be included in the current system. The assessment can be accomplished by merely requiring competing contractors to furnish the necessary information to objectively evaluate competing bidders' means and methods using commonly accepted sustainability standards, like CO2 emissions, carbon footprint, energy efficiency, recycling, and the like. Example 9.3 illustrates a potential approach to use A+B, I/D, and A+B+I/D awards as the contractual technique to achieve the owner's project sustainability goals.

**Example 9.3:** An owner wants to encourage competing contractors to reduce energy consumption during the construction of an equipment-intensive project. It states in its invitation for bids that the value of a gallon of fuel for sustainability purposes is $8.00. The contractors are required to include their diesel fuel consumption in their bids. The owner's field personnel will check and document fuel consumption during project execution daily. The below example will demonstrate the application of A+B, I/D, and A+B+I/D on this particular scenario.

Table 9.13 shows the results of the bid opening. Contractor B is the second low bidder and has selected means and methods that minimize fuel consumption (90,000 gallons versus the 100,000 gallons by the lowest bidder). Therefore based on the A+B award, Contractor B would be award the contract as shown in the following calculation.

$$A + B\ award\ determination:\ \$2,050,000 + (90,000\ gal.x\$8.00/gal) = \$2,770,000$$

If the contract was advertised as purely low bid with I/D provisions, the award would have gone to Contractor A with the lowest bid price. Table 9.14 shows that Contractor A would have earned a $16,000 incentive payment for consuming less fuel than the amount in its bid. The table also illustrates that if the project was awarded to Contractor B on an A+B+I/D basis and it subsequently used of 1,000 gallons more than the amount of fuel shown in its bid, an $8,000 disincentive would have been assessed.

Example 9.3 is both simple and straightforward. Minimizing fuel consumption also minimizes exhaust emissions, which addresses the GHG and pollution issues discussed earlier in this chapter. It demonstrates the potential to build sustainability into the contract award decision-making process without needing to depend on complicated sustainability rating systems discussed in Section 9.1. Sustainability values could be created to address other construction environmental issues like air quality (perhaps using a percent below/within limits approach for particulate samples), water quality, noise pollution, carbon footprint, etc. based on existing socio-economic concepts like sustainability return on investment.

## 9.5 SUSTAINABILITY IN EQUIPMENT FLEET MANAGEMENT SUMMARY

To achieve environmental and sustainable construction goals requires a pragmatic approach to making sure that construction contractors can achieve those goals while continuing to make the necessary profits to keep their companies thriving. The major change is to shift

**TABLE 9.13**
**Example A+B Bidding Results with B as the Sustainability Metric**

| Contractor | Bid Price (A) | Gals (1,000) | $/Gal | Fuel Factor (B) | Total A+B |
|---|---|---|---|---|---|
| A | $2,000,000 | 100 | $8.00 | $800,000 | $2,800,000 |
| B | $2,050,000 | 90 | $8.00 | $720,000 | $2,770,000 |
| C | $2,050,500 | 98 | $8.00 | $784,000 | $2,834,500 |

**TABLE 9.14**
**Example I/D Final Payment Results**

| Contractor | Bid Price (A) | Gals (1,000) | Gals Used (1,000) | $/Gal | Total Final Payment |
|---|---|---|---|---|---|
| A | $2,000,000 | 100 | 98 | $8.00 | $2,016,000 |
| B | $2,050,000 | 90 | 91 | $8.00 | $2,042,000 |
| C | $2,050,500 | 98 | 97 | $8.00 | $2,058,500 |

the process. Change it from being "design-centric," where all sustainability is achieved by producing the greenest design possible within the project's constraints, to make it "construction-centric," where contractual mechanisms are put in place that will not penalize the industry for keeping up with the state-of-the-art in environmentally friendly construction machinery of all types. Owners must carefully analyze sustainability goals for the construction phase and ensure that the project is sustainable throughout its entire lifecycle. Implementing a strategy to enhance project sustainability must necessarily include incentivizing sustainable means and methods; in doing so, it requires a shift in the traditional project procurement process.

## REFERENCES


[1] US Environmental Protection Agency (EPA). (2019). *Design for the Environments Life-Cycle Assessments.* Washington, DC: EPA. www.epa.gov/saferchoice/design-environment-life-cycle-assessments=. (Accessed November 22, 2019).

[2] Szamocki, N., Kim, M. K., Ahn, C. R. and Brilakis, I. (2019). Reducing Greenhouse Gas Emission of Construction Equipment at Construction Sites: Field Study Approach. *Journal of Construction Engineering and Management*, 145(9), 10.1061/(ASCE)CO.1943-7862.0001690.

[3] Caterpillar, Inc. (2018). *Caterpillar Performance Handbook*. 48th ed. Peoria: Caterpillar Inc.

[4] Intergovernmental Panel on Climate Change (IPCC). (2014). *Climate Change 2014 – Synthesis Report: Contribution of Working Groups I, II and III to the Fifth Assessment Report of the IPCC*. R. K. Pachauri and L.A. Meyer (eds.). IPCC, Geneva, Switzerland, p. 151.

[5] US Environmental Protection Agency (EPA). (2007). *The Plain English Guide to the Clean Air Act. Office of Air Quality Planning and Standards.* Publication No. EPA-456/K-07-001. Washington, DC: EPA.

[6] Dallmann, T. and Menon, A. (2016). *Technology Pathways for Diesel Engines Used in Non-Road Vehicles and Equipment*. Washington, DC: International Council on Clean Transportation (ICCT).

[7] US Environmental Protection Agency. (2016). *EPA Greenhouse Gas Inventory Guidance – Direct Emissions from Mobile Combustion Sources.* Washington, DC: EPA.

[8] US Environmental Protection Agency. (2018). *Exhaust and Crankcase Emission Factors for Off-Road Compression-Ignition Engines in MOVES2014b.* Washington, DC: EPA.

[9] US Environmental Protection Agency. (2010). *Median Life, Annual Activity, and Load Factor Values for Off-Road Engine Emissions Modeling*. Washington, DC: EPA.

[10] Gransberg, D.D. and Lopez del Puerto, C. (2017). Incentivizing Construction Contracts to Enhance Sustainability in Construction Projects. 2017 AACE International Annual Meeting, Orlando, FL, pp. OWN-2472.1–OWN-2472.9.

[11] Arditi, D. and Yasamis, F. (1998). Incentive/Disincentive Contracts: Perceptions of Owners and Contractors. *Journal of Construction Engineering and Management*, 124(5), pp. 361–373.

[12] Choi, K., Kwak, Y.K., Pyeon, J. and Son, K. (2012). Schedule Effectiveness of Alternative Contracting Strategies for Transportation Infrastructure Improvement Projects. *Journal of Construction Engineering and Management*, 138(3), pp. 323–330.

## CHAPTER PROBLEMS

1. Calculate the expected daily total carbon footprint from exhaust emissions for the asphalt paver shown in Table 9.4 under medium load factors? The planned shift is 10 hours and the crew will take 30 minutes to get the machine operating in the morning and 15 minutes to shut it down for the night.
2. Using the average engine load factor, as well as the other input parameters suggested by the EPA, calculate the expected total amount of CO, HC, NOx, and PM exhaust emissions for the paver in problem 1 if it operates for 4 weeks 6 days per week?
3. Propose a sustainability metric that could be used in an A+B award to incentivize contractors to address another environmental issue. Demonstrate its use by applying it to a hypothetical project in the same manner shown in Table 9.13.

# 10 Construction Equipment Automated Machine Guidance Systems

## 10.0 INTRODUCTION

Automated machine guidance (AMG) is a technology used on equipment-intensive construction projects to enhance location referencing and increase equipment fleet production. AMG-enabled construction equipment has onboard computers connected to sensors mounted on key elements of the machine. AMG provides real-time vertical and horizontal guidance/control to the equipment operator with a combination of 3-Dimensional (3D) modeled data correlated with global positioning system (GPS) technology. AMG has been found to facilitate the completion of projects accruing not only cost and time benefits but also enhancing quality and project safety. In most cases, AMG-enabled equipment are able to achieve designed grades on the first pass, reducing the time and cost found in traditional grade staking by a survey crew. Federal Highway Administration (FHWA) research found that AMG "can increase productivity by up to 50 percent on some operations and cut survey costs by as much as 75 percent [1]." When fully implemented with 3D design and global information systems (GIS) input, AMG facilitates "seamless electronic data transfer – from the initial surveying, to the development of digital terrain models (DTMs), through design and construction, to final inspection and verification[2]."

The integration of AMG on equipment-intensive projects requires complex workflow processes. The fleet manager must select appropriate surveying methods and technologies, as well as software design and engineering analytic tools. These must be married to selected machinery via sensor technologies, data interoperability, and transfer mechanisms. Lastly, human–machine interaction during construction must be considered and training provided for the operators of AMG-enabled equipment. The result of the process is construction equipment whose machinery operations are directed with a high level of precision, improved speed, and enhanced accuracy of the constructed product. Implementing AMG technology will potentially improve the efficiency, quality, and safety of equipment-intensive construction projects.

The rest of the chapter will discuss the various components of AMG and how it is related to the equipment fleet management process. It will include details on the role of GPS as the basis for making accurate adjustments to machine operations in real-time. 3D digital terrain models will be briefly discussed as the foundation for defining project vertical and horizontal measurements. Finally, the linkage between AMG and GIS platforms will be covered to provide information on applying as-built AMG data to a public agency's asset management program.

## 10.1 GLOBAL NAVIGATION SATELLITE SYSTEM

Global navigation satellite system (GNSS) is a term used to refer to any set of satellites orbiting the Earth with the purpose of determining the position of GNSS receptors on the Earth's surface. The most popular GNSS, and the one preferred by construction equipment manufacturers, is the GPS, but this is just one of the four systems currently orbiting over the Earth. GPS is owned by the US and operated by the US Air Force. The other systems are operated by Russia (GLONASS), China (BeiDou), and the European Union (Galileo).

### 10.1.1 Fundamentals of GPS

The first GPS satellites were launched by the US Department of Defense in 1978, but the system did not become fully functional until 1993. It was initially intended for national defense and military purposes, but the US government decided to make it available for civilian and commercial use even before it became fully functional. Hence, GPS satellites can nowadays be accessed by anyone from virtually any computer or mobile device.

Currently, there are 28 satellites in orbit with plans to launch more. Figure 10.1 illustrates the concept of GPS constellation. In most areas of the US, there are seven satellites visible to GPS base stations at any given time. GPS receptors are mounted on construction equipment to allow onboard computer systems to measure the 3D position of the equipment (latitude, longitude, and altitude). Most commercial GPS receptors do not provide the accuracy and reliability needed to ensure the required level of quality. Therefore, those used by construction equipment manufacturers are more advanced and require the installation of a local stationary GPS receiver (base station), as shown in Figure 10.2, for which location is precisely known. The stationary receiver compares the position provided by the satellites against its own known position and calculates a position correction factor that is then sent to mobile GPS receivers on the equipment units to correct their own positions.

**FIGURE 10.1** GPS Constellation.

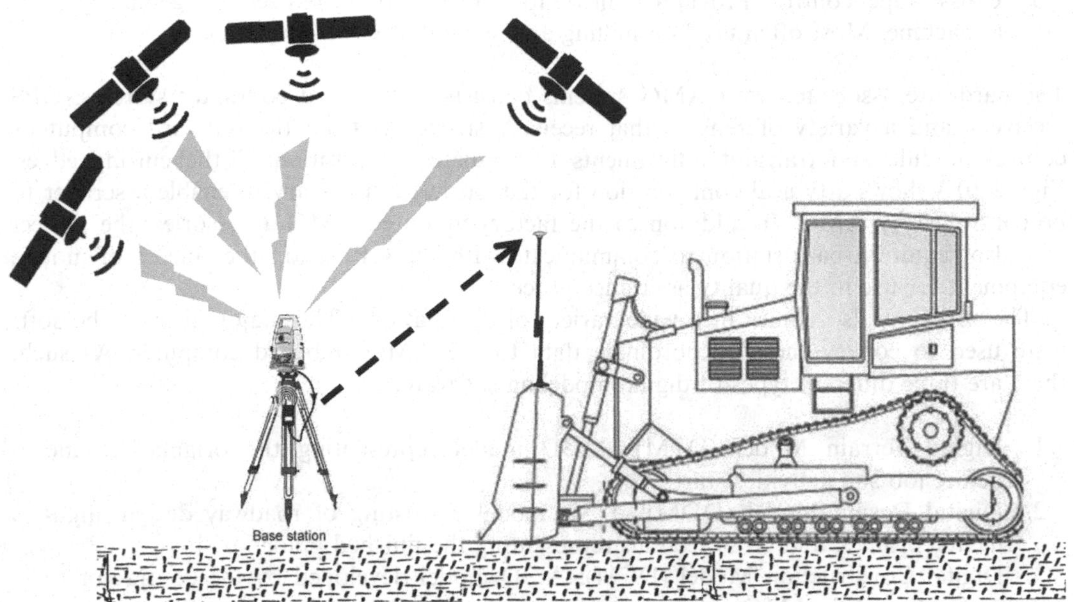


**FIGURE 10.2** Automated Machine Guidance and Global Positioning System Interface.

The machine's onboard computer uses AMG references to position the cutting edges or pavement molds using GPS satellites, robotic total stations, lasers, or combinations of these methods. The finished-grade for a given location is computed by the 3D model of the proposed constructed facility that has been uploaded to the onboard computer. The cutting edges or pavement molds are subsequently adjusted automatically for small differences in elevation. The system can also provide the cut or fill quantities through the computer interface to the machine operator for large differences in elevation. Typically, the equipment operator will manually grade the surface until it is close to design elevations; the operator then turns over control to the AMG system to complete the finish grading.

### 10.1.2 Construction Equipment AMG Hardware and Software

The technology required to implement AMG in the field requires both hardware and software. Fortunately, this is a growth industry with three major manufacturers of surveying and positioning equipment who have heavily invested in AMG-related devices. Those companies are Leica Geosystems, Topcon Corp., and Trimble Navigation Limited. The products developed by these and other firms can be grouped into the following five categories of control technologies:

1. GPS control: Provides 3D grade control and is suitable for bulk earthwork accomplished by bulldozers, motor graders, excavators, and scrapers.
2. Laser control: Provides 2D grade control and is primarily used for finish grading on bulldozers and motor graders. It also can measure lift thickness to tight tolerances of 3 to 6 millimeters.
3. Robotic total station control: Provides 3D control for a single machine.
4. Sonic control: Provides vertical control via an ultrasonic sensor. Requires an external reference such as a string line or a curb. Accuracy is good to 1.0 millimeters (0.04 inch).

5. Cross slope control: Provides control to maintain the cross-section geometry of a machine. Most often used on milling and string-less concrete paving.

The hardware associated with AMG systems consists of onboard computer systems, GPS receivers, and a variety of sensors that receive instructions from the onboard computer's control module and transmit adjustments to regulate the position of the cutting edges. Figure 10.3 shows a typical configuration for the hardware necessary to enable a scraper to be controlled by AMG. In addition to the factory-mounted AMG accessories, the project will also require a base station to communicate with the GPS and other pieces of mobile equipment for use in the quality assurance process.

The software also comes in a wide variety of applications. The major piece is the software used to convey the 3D coordinate data to the AMG onboard computer. As such, there are three different types of digital modeling software:

1. Digital Terrain Model (DTM): A 3D model representing the original ground before job site activities start.
2. Digital Design Model (DDM): A 3D model consisting of roadway design alignments, profiles, and cross sections representing the finished grade.
3. Digital Construction Model (DCM): A 3D model developed by the contractor to use with specific AMG equipment.

One can see that the aforementioned three models follow the project development and delivery process. The DTM corresponds to the acquisition of the project's initial topography, which provides the baseline for the DDM where the engineers overlay the geospatial design information. The combination of the two creates the basis for the development of quantities of work that are articulated in the project's solicitation and upon which competing contractors base their bids. Finally, the winning contractor will take the DDM and produce the DCM for each piece of AMG-enabled equipment to complete the construction.

In addition to the terrain models, each control system listed above will have its own set of software requirements that link the various components of the 2D or 3D controllers and feed the AMG-enabled construction machinery. The final software consideration is driven by the owner's need to capture as-built construction data for use after construction is complete in its asset management program. The following are the types of software that are commonly required to inform asset management decisions:

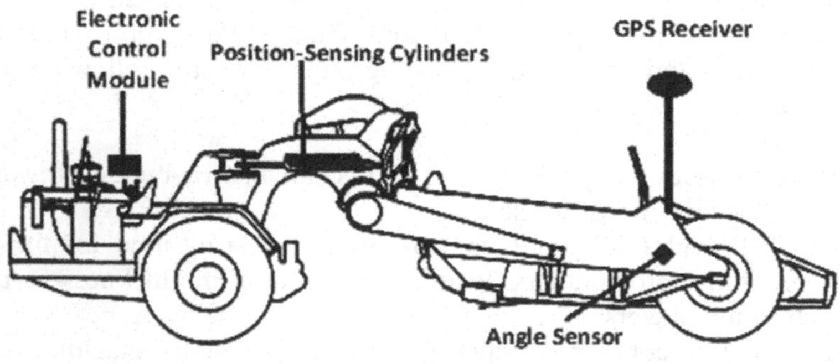


**FIGURE 10.3** AMG-Enabled Scraper.

- Digital imaging packages that permit the measurement of geometric changes, such as settlement, to be tracked over time.
- Software that is compatible with sensor systems used by the owner to monitor maintenance data collection for items like pavement condition, surface friction, etc.
- Software that provides an interface between digital as-built models and administrative submittals.

### 10.1.3 AMG Workflow Processes

The implementation of AMG creates a need for the equipment fleet manager to develop a new set of management skills. Since AMG is a combination of electronic, mechanical, and information technology, the management process becomes one of developing and controlling the various workflow processes inherent to AMG. Table 10.1 details the three major workflow processes that an equipment fleet manager must address to successfully implement AMG on the project site.

Explaining the details is beyond the scope of the book; however, the reader is referred to NCHRP Web-only Document 250: *Use of Automated Machine Guidance within the Transportation Industry* [2] for an informative set of detailed workflow diagrams and narratives on the processes and technologies found in AMG. AMG is not applicable to all projects. It is best used on projects with large quantities of earthwork or paving and/or new alignments. Overlays with new profiles or cross slope construction also benefit from AMG. That being said, the Table 10.1 workflow processes should be managed in light of the following limitations [1]:

- "Widening with narrow strip additions 10
- Designs, such as overlays, that are not based on an existing DTM.

**TABLE 10.1**
**AMG Workflow Processes**

| Workflow Process | Task |
|---|---|
| Project design Concept/Surveying | • Project initial topographic survey; Control point establishment<br>• Geometric design of features of work<br>• Environmental studies<br>• Public outreach<br>• Right of way identification and acquisition<br>• Final project design<br>• Procurement and construction contract award |
| Contractor AMG design | • 3D data preparation<br>• Conversion of contract document data to specific application format<br>• Verify 3D model<br>• Revise input data as required |
| Field AMG implementation | • Synchronize AMG equipment with field controls<br>• Customize digital terrain model for each operator<br>• Upload adapted models into AMG equipment<br>• Construct according to model-driven data<br>• Conduct field quality assurance verification |

- Designs that do not exist in a 3D digital environment
- Structures
- Projects that are under a tree canopy, in narrow canyons, or next to tall buildings that interfere with GNSS signals (note that robotic total stations or traditional methods are viable solutions)
- Design difficulties that would prevent the creation of an accurate and complete DTM
- Lack of training on AMG
- Construction specifications that do not allow the use of AMG
- Lack of equipment for [owner] construction management staff to perform quality assurance [1]."

The major reason for employing AMG is to increase equipment productivity without sacrificing quality. In this application, quality is measured in terms of geospatial accuracy. Table 10.2 comes from the Michigan DOT [3] and describes its expectations for accuracy on contractor's AMG-enabled equipment. Table 10.3 contains the accuracy requirements for the field tools used to check AMG-produced grades for quality assurance verification.

## 10.2 GEOGRAPHIC INFORMATION SYSTEMS

A geographic information system (GIS) is a database designed to store and analyze spatially organized, object-oriented data in a manner that permits the user to visualize the information contained in the database. GIS facilitates all forms of spatial analysis and typically is displayed as a map of the designated information that can be used by project teams to better understand the project and its environment. The primary benefit found in GIS is that it promotes geospatial data collaboration, which is the delivery and exchange of

**TABLE 10.2**
**AMG Accuracy Requirements [3]**

| | Required Accuracy | |
|---|---|---|
| **AMG Operation** | **Horizontal** | **Vertical** |
| GPS machine guidance | 0.04 foot | 0.07 foot |
| Total station machine guidance | 0.02 foot | 0.02 foot |
| Laser augmented GPS (LAGPS) machine guidance | 0.04 foot | 0.02 foot |
| Total station string-less paving | 0.02 foot | 0.02 foot |

**TABLE 10.3**
**Quality Assurance Field Tool Accuracy Requirements [3]**

| | Required Accuracy | |
|---|---|---|
| **Field Tool** | **Horizontal** | **Vertical** |
| GPS rover | 0.04 foot | 0.07 foot |
| Total station | 0.02 foot | 0.02 foot |
| Laser augmented GPS | 0.04 foot | 0.02 foot |

geospatial data and information during project delivery. When used in conjunction with AMG, GIS acts as the archive for the information generated by the AMG equipment.

### 10.2.1 GIS Applications for Equipment Management

GIS is an extremely robust technology that has found a myriad of uses in a variety of applications. In the construction equipment management arena, GIS furnishes the means to make equipment management decisions based on data-driven information that is displayed spatially. On a construction project, GIS becomes the mechanism where construction progress can be mapped directly from data acquisition systems found on the project site. That map can be updated in real-time as new data is acquired, and the progression of GIS maps can then function as a historical record on what was built, where it was built, and when it started and finished.

Layers can be added to the GIS map that displays an unlimited amount of information of interest to the fleet manager. Typically, a fleet manager would want to at least track weather, quality, schedule, material, and safety data. Figure 10.4 is a framework that illustrates the flow of data and its conversion into actionable information on an intelligent construction site

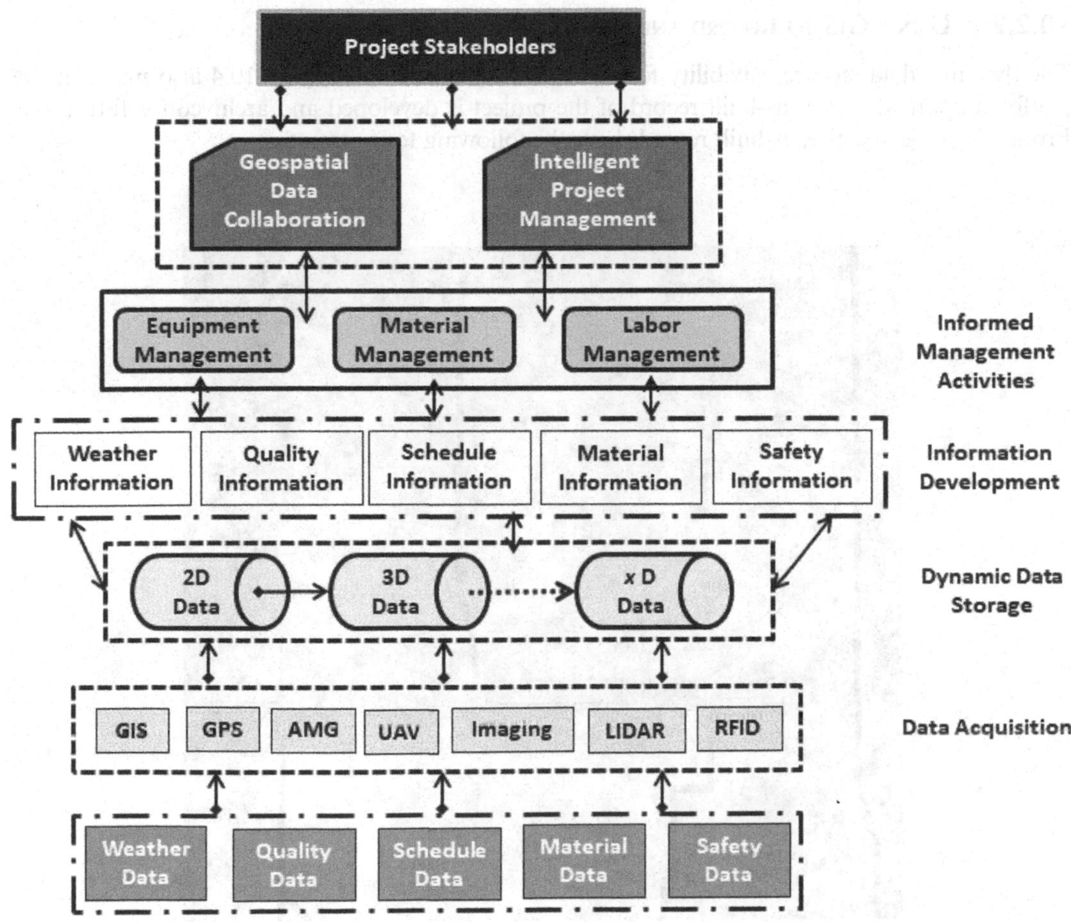


**FIGURE 10.4** Data-Driven, Intelligent Construction Management Framework after Zhou et al. [4].

that is equipped with a variety of data acquisition systems, including GIS and AMG. It shows a one-way flow of data to the database and then two-way communication from there upwards. This two-way feature makes the database dynamic, allowing its users to continually refresh the information required to make equipment, material, and labor-management decisions. It also shows the capacity for geospatial collaboration among and between the various stakeholders on the particular project.

Figure 10.5 is an example of a GIS record of construction progress. The figure shows various numbered and color-coded tabs. Each of these represents a specific information set that can be clicked on the computer screen to gain immediate access to the information. In this instance, the tabs are contained in the quality data layer, and the color codes represent quality control test results taken at each location. Green is a satisfactory test result. Red indicates a nonconformance requiring further action. The project manager can click on each tab, and it will pop-up a document with the information contained on that particular test. The teardrop icons indicate locations where materials were tested for contamination and since they have not yet been assigned a number, the viewer knows that the test results are not yet complete. Thus, this powerful system is able to visually communicate an immense amount of information at a single glance with additional details residing in the dynamic database being only a mouse-click away on the interactive screen.

### 10.2.2 Using GIS to Record and Archive Progress

The dynamic data storage capability found in GIS and shown in Figure 10.4 also makes it the platform upon which the as-built record of the project is developed and archived for future use. From a legal perspective, as-built records have the following four purposes:

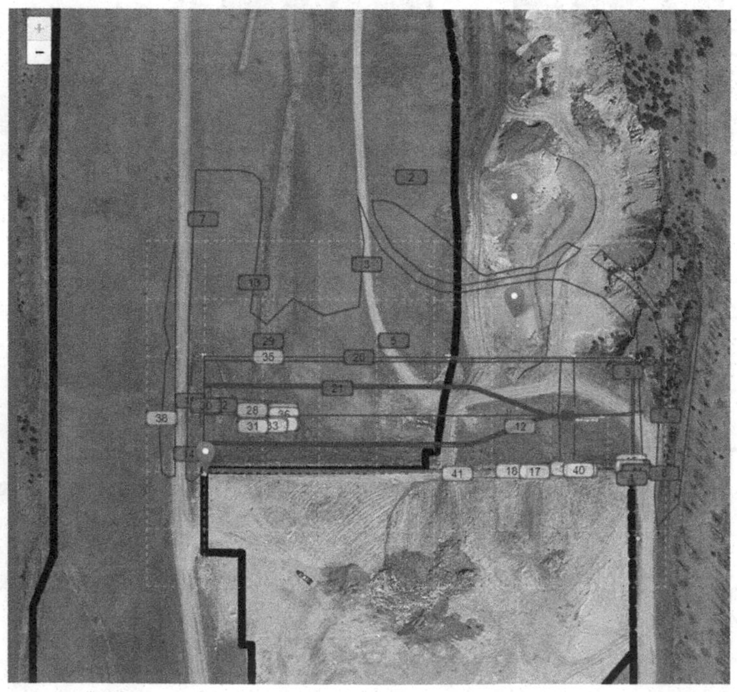

**FIGURE 10.5** GIS Construction Progress Example.

1. To serve as a repository for all directed changes that occurred during project construction.
2. To serve as the contractor's certified record attesting to exactly what was built.
3. To serve as the design basis for future improvements to the project during its life cycle.
4. To serve as land-use history, documenting the materials and other matter that has been installed or emplaced to facilitate future environmental assessments.

In a traditional construction project that does not use 3D digital design and construction documents, the as-built record is viewed as an onerous task that must be constantly monitored. Changes to the approved design have to be recorded either on paper copies or added to computer-aided design files. In many cases, the owner receives the as-builts some time after the construction is complete and the facility has been opened. Experience has shown that the ultimate quality of as-builts is generally poor and that inaccuracies introduced into plans become legal issues if those documents are used to portray the existing site condition on a future project.

The dynamic database feature found in GIS virtually eliminates this issue because it automatically collects as-built measurements from the AMG-enabled equipment and updates the TDM to show actual grades and elevations, not to mention it computes actual quantities of work for payment purposes. The as-built TDM can then become a valuable resource for the owner during the life cycle of the completed project. For example, an urban highway project's as-built TDM will contain 3D coordinates for all the underground utility systems installed or relocated during the course of the project, making it easy for the owner to plan other projects around the actual locations of the utilities. The potential applications for GIS continue to grow as the technology matures, and its inherent ability to present information in a visual manner that can be manipulated to the desire of the user is its greatest selling point.

## 10.3 AMG SUMMARY

The "intelligent construction site" is currently a work in progress. However, all the components are available, and full implementation of the concept will only be generated by owners who desire to harness the computing power on hand today to improve their decision-making processes. The fleet manager of the past used a lot of data to make the equipment management decisions discussed in this book. However, that data has to be deliberately collected, reduced, and filed manually. The times have changed and now the machinery itself can collect the data and its onboard computers can transmit that data to other computers that can manipulate it and turn it into actionable information in a fraction of the time. AMG and GIS are the proverbial "tip of the iceberg." Chapter 11 will discuss the future challenges to be faced in construction equipment management.

## REFERENCES

[1] Federal Highway Administration. (2016). EDC 2012 Initiatives – Three-Dimensional Modeling. www.fhwa.dot.gov/everydaycounts/edctwo/2012/3d.cfm. (Accessed December 19, 2019).

[2] White, D.J., Jahren, C.T., Vennapusa, P., Westort, C. and Alhasan, A.A. (2018). *Use of Automated Machine Guidance within the Transportation Industry.* NCHRP Web-only Document 250. The National Academies Press. https://doi.org/10.17226/25084. (Accessed December 19, 2019).

[3] Michigan Department of Transportation (MDOT). (2016). *Automated Machine Guidance*, CMU 16-001 AMG Attachment. http://mdotwiki.state.mi.us/construction/index.php/File:CMU16-001_AMG_Attachment_Final.pdf. (Accessed December 18, 2019).

[4] Zhou, H., Wang, H. and Zeng, W. (2018). Smart Construction Site in Mega Construction Projects: A Case Study on Island Tunneling Project of Hong Kong-Zhuhai-Macao Bridge. *Frontiers of Engineering Management*, 5(1), pp. 78–87.

## CHAPTER PROBLEMS

1. Do some internet research and determine the benefits and limitations of string-less paving.
2. The Intelligent construction site framework shown in Figure 10.4 has five categories of information that fleet managers would like to collect. List at least four different types of data that could be collected for each of those five areas.
3. Develop a list of at least three uses for GIS on the intelligent construction site beyond construction equipment management.

# 11 The Future of Construction Equipment Management

## 11.0 INTRODUCTION

In Chapter 1, Figure 1.2 chronicled the evolution of construction equipment from the 18th Century to the present, documenting the transition from hand tools to construction machinery. It also shows the advent of digital technology replacing mechanical technology to perform a variety of tasks that allow the machine to run more efficiently and achieve higher levels of cost-effective production. The short discussion of the future in Chapter 1 concluded that new models for production estimation, equipment selection, and a different skillset for the equipment "operators" to accommodate the future technologies will be available for use on equipment-intensive construction projects.

This chapter will briefly discuss three emerging technologies that the authors believe will establish the foundation for the future of equipment management. The intent is not to delve into these three areas in depth but rather to introduce and make the reader aware of the current state-of-the-practice as a means to set the stage for absorbing future developments as they occur and enter the industry.

## 11.1 CIVIL INTEGRATED MANAGEMENT SYSTEMS

The growth of construction technologies has been both rapid and continuous. While each individual development brings its own costs and benefits, there is no denying that the impact of technology on construction projects will be felt for decades in the future. In the equipment fleet management arena, there are a number of emerging technologies that have already been implemented, and equipment users are already reaping the benefits of enhanced production, better machine availability, and more efficient systems. Two of these, intelligent compaction and autonomous equipment, will be discussed later in the chapter. Taken individually, each emerging technology provides a single advance in the evolution of equipment management, but taken together and optimized for the specific project; emerging technology has the power to revolutionize the industry.

Most of the emerging technology provides a means to convert data into information that can be used to inform equipment management decisions, regardless of the specific decision. Technology also provides a means to more knowledgeably address construction cost, schedule, and quality risk. As a result, it is important to have a framework from which to build a program that can assimilate advancements as they emerge. Civil integrated management (CIM) is that framework for the equipment fleet management sector. Developed by the Federal Highway Administration, it seeks to assemble all promising technologies in a manner that associates each with those phases of the design and construction process where they can be best exploited. The Federal Highway Administration defines CIM in the following manner:

> Civil Integrated Management (CIM) is the technology-enabled collection, organization, managed accessibility, and the use of accurate data and information throughout the life cycle of a transportation asset. The concept may be used by all affected parties for a wide range of purposes, including planning, environmental assessment, surveying, construction, maintenance, asset management, and risk assessment.
>
> [1]

The reader is referred to NCHRP Report 831: *Civil Integrated Management (CIM) for Departments of Transportation* for a detailed discussion on the topic. Table 11.1 is a summary of the CIM tools that the NCHRP report details, and it shows those that can be directly or indirectly applied to the construction equipment management process. The following sections will discuss those features of CIM that specifically relate to the construction equipment management process.

### 11.1.1 3-Dimensional Construction Documents

The building construction industry has used 3-dimensional (3D) building information modeling (BIM) for quite some time, and BIM has proven its effectiveness as a means to not only identify and eliminate clashes between various trades in complex building designs, but more importantly, construction contractors have found it can be used to construct buildings in virtual space, allowing them to optimize the construction sequence of work. BIM software packages come with the ability to extend the process into the 4th and 5th dimensions, which are schedule and cost, respectively. Hence, the utility of multidimensional construction documentation is proven and can easily be transferred to construction sectors outside the building field.

As discussed in Chapter 10, AMG systems are reliant on 3D digital terrain models that conceptualize the final construction in geospatial format, using an x-y-z coordinate system. Once those models are uploaded to the AMG processors, they permit the computer to adjust a piece of equipment's hydraulic cutting edges hydraulically to match the design grades, slopes, alignments, etc. Figure 11.1 illustrates the conversion from traditional 2D design to 3D design. It should be noted that there is no longer a requirement to design in 2D, and current commercial software packages allow the engineer to directly design in 3D, eliminating the need for the conversion shown in the Figure 11.1.

Once a 3D design for the project is available, a whole host of applications beyond machine guidance can be brought to bear on the project. The following are just a few examples of the application of CIM tools that are currently in use in equipment management.

- 4D Modeling: The 4th dimension is schedule. The elements of the 3D model can be associated with the sequence of work that is planned for the project including the durations corresponding to each equipment-related construction activity to provide the ability to validate the construction schedule before it is finalized. This also allows the contractor to "build" the project in virtual space to optimize the final sequence of work. When combined with the linear scheduling method detailed in Chapter 7, the simulation models discussed in Chapter 6 can be used in conjunction with the optimized sequence of work to quantify both cost and schedule risk on equipment-intensive projects.
- Geographic Information Systems (GIS): Taking the project design and associating it geospatially using GIS allows the locational tracking of equipment production, quality control outcomes, and document the progress of equipment-related tasks.
- Radio Frequency Identification (RFID): Installing RFID tags on pieces of equipment permits the fleet manager to track equipment locations in real-time. This is particularly effective for projects where a substantial amount of hauling over public roads is required. If a truck breaks down, the RFID tag allows the fleet manager to pinpoint its location so a repair team can be dispatched with no time lost finding the broken machinery.

**TABLE 11.1**
**Civil Integrated Management (CIM) Tools [Adapted from 1]**

| Tool Type | CIM Tool | Description | Equipment Management Application | |
|---|---|---|---|---|
| | | | Direct | Indirect |
| | 2D digital design tools | Use for design data exchange and organization | X | |
| Modeling | 3D, 4D, and nD Modeling tools | Resolving spatial conflicts among design entities (e.g. utility conflicts) and construction activities (temporal conflicts) | X | |
| | Traffic modeling and simulation tools | Impact studies (e.g. traffic delays) and aids public information when combined with design visualization | | X |
| | Project information management systems | Manages & allows sharing of documents, databases, model-based data during project-delivery processes | | X |
| | Asset information management systems | Used for archiving asset data after construction and supports inventory asset management during O&M lifecycle | | X |
| Data management | Geographical information systems (GIS) | Associates databases with geospatial positioning information; offers benefits for planning and programming, environmental assessment, surveying, and asset management | X | |
| | Digital signatures | Eliminates need to print, sign, and scan and allow continuous flow of digital documents | | X |
| | Mobile digital devices | Allows field access of digital documents | X | |
| | LiDAR | Accurate and dense point cloud data for design, quantity estimates, and 3D models | X | |
| | Aerial imagery | Design and computation of earthwork, mapping, photogrammetry | X | |
| | Global positioning system (GPS) | Mapping, surveying, automated machine guidance (AMG) | X | |
| | Robotic total stations (RTS) | Remote-controllable total stations from observation point (one-operator) measurements, used for AMG (final earthwork, paving, etc.) | X | |
| | Ground penetrating radar (GPR) | Locate underground utilities, groundwater, tunnels, and other objects | X | |
| Sensing | Radio frequency identification (RFID) | Track materials, equipment, utilities, etc. | X | |
| | Real time network (RTN) | Continually operating reference stations (CORS) for real-time positioning for surveying, AMG, QA/QC checks | X | |
| | Integrated measurement system (IMS) | Feedback control system with sensors and GPS for temperature control for intelligent compaction (IC) | X | |
| | Unmanned aerial vehicles (UAV) | UAVs collect geo-referenced images and point clouds for surveying and quality control; rapid data collection, high precision, image resolution | X | |

**FIGURE 11.1** 3-Dimensional Conversion from 2-Dimensional Plans.

- Unmanned Aerial Vehicles (UAV): UAVs have a myriad of potential uses on equipment-intensive projects. On the Panama Canal Expansion project, which was completed in 2016, they were used to provide the imagery required to calculate daily cut and fill production. They have also been to verify equipment cycle times using digital video feeds that are then timed. Differences from planned cycle times are evaluated, and resource allocation decisions can then be made while the project is in progress.

### 11.1.2 Digital As-built Plans and Files

A perennial problem in the entire construction industry has been the development and submission of final as-built record documents. Before 2D computer design, as-builts were developed from a set of "red-line" documents that were kept up to date on the project site, documenting any variations from the original contract plans and specifications. Additionally, paper files of construction submittals, shop drawings, samples, QC test results, and other input were maintained to document the quality assurance process and extensions to the contract design. The manual entry nature of the process made the final project as-builts documents highly susceptible to error. Errors in as-built drawings become expensive when they are used as the basis of design for the next project, and those errors are unintentionally induced in the new design. This is especially true when the errors are in underground features of work like utility lines.

CIM provides the capability to have a paperless construction administration system if so desired. The 3D design model can be updated in virtual space, and the need to redraw the record as-builts from the red-line documents goes away. Construction submittals can be digitally associated with a component of the 3D design model and easily accessed after construction completion using typical computer database queries. Figure 11.2 is an example of a 3D as-built utility drawing. One can easily see how a designer could take this and directly reference it geospatially to any new construction that must be done after the project in the figure was complete.

### 11.1.3 Postconstruction Equipment-Related Project Operations and Maintenance

The asset information management system CIM tool will be valuable to asset owners after construction is completed. All projects require operations and maintenance. Systems need servicing, and most heavy civil assets require periodic upgrades and rehabilitation to permit them to function as intended for their service lives. Having an accurate inventory of all the various components of a physical asset like a bridge or a power plant is mandatory to being able to manage its continued functionality.

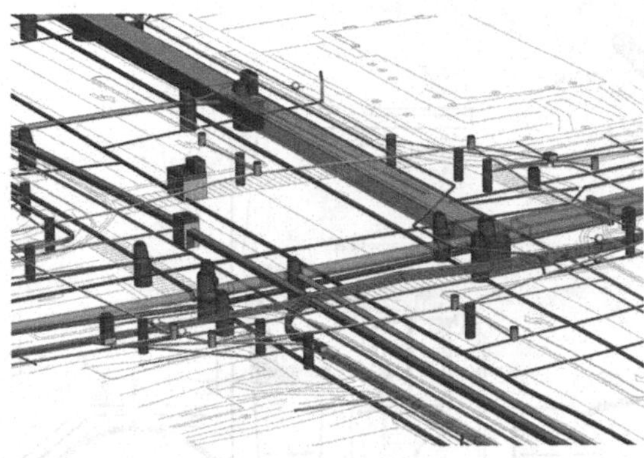

**FIGURE 11.2** 3-Dimensional Utility As-Builts.

## 11.2 INTELLIGENT COMPACTION

The key quality control indicator on projects with structural fills, backfilled trenches, asphalt pavements, and building foundation pads is compaction. To achieve the required density a variety of compaction equipment is used, and it covers the entire area needing compaction. Whereas, compaction testing theory is based on a representative sample of point measurement tests collected across the breadth and length of the compacted area in each lift. According to one study [1],

> conventional test methods for roadway compaction cover less than one percent of roadway; whereas, intelligent compaction (IC) offers a method to measure 100 percent of a roadway. IC offers the ability to increase compaction uniformity of soils and asphalt pavements, which leads to decreased maintenance costs and an extended service life.

Hence, marrying the roller and the technology to measure compaction while it is rolling is an elegant solution that promises potentially huge long-term benefits to the constructed project's owner.

Equipment that furnishes this capability are divided into two categories:

- Roller-integrated continuous compaction control (CCC) systems. Equipment of this type have an integrated GPS system to develop a geographic information system (GIS) record of the worksite and the ability to continuously assess mechanistic soil properties, such as stiffness and modulus via monitoring roller vibrations.
- Intelligent soil compaction (ISC) systems. These systems provide the two capabilities found in CCC systems and adds an automatic feedback control to modify vibration amplitude and frequency while moving.

For this chapter, the term IC will be used to discuss both types. IC uses onboard technologies to furnish measurement values (MV) of soil stiffness and modulus from the lift under compaction (usually 6 to 12 inches), as well as the underlying layers up to a depth of 5 feet. Correlating the MVs with GPS, spatial references provide real-time compaction data to be gathered and spatially assess the level of resultant compaction in both soils and pavements. Figure 11.3 is a schematic of how the IC system is configured. The information then is used to identify areas where the required compaction has not been achieved, and to prevent unnecessary over-compaction [2]. Figure 11.3 illustrates how the data is processed and visibly identifies the

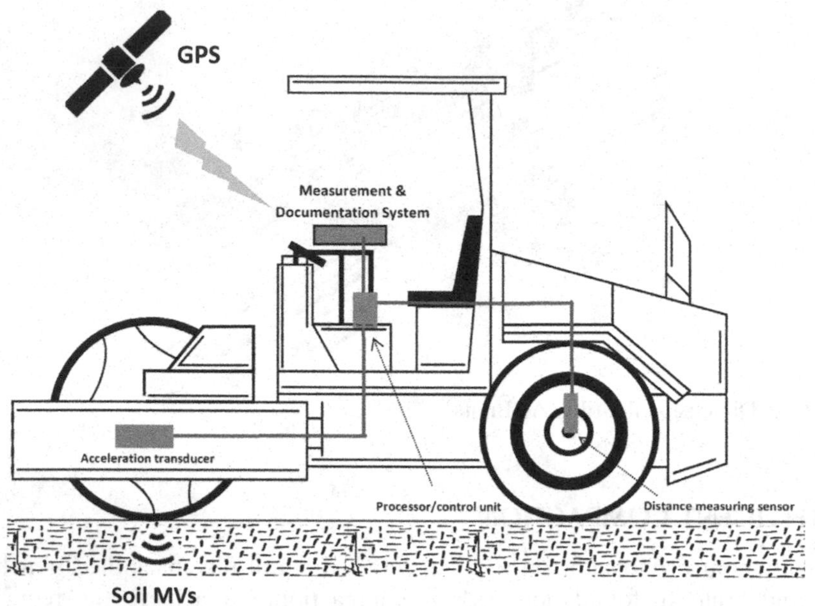


FIGURE 11.3 Intelligent Compaction Schematic.

resultant compaction and allows the operator to recompact areas with lower than desired densities. Figure 11.4 illustrates the output from the IC process and graphically designates those areas where the current densities are below standard. Thus, the operator can immediately return to those patches without the need to wait for AC test results.

### 11.2.1 Intelligent Compaction Applications

Vibratory rollers with IC capability are generally available and are in use throughout the nation. Table 11.2 is a listing of the manufacturers and the type of measurements their

FIGURE 11.4 Intelligent Compaction Output.

**TABLE 11.2**
**Intelligent Compaction Roller Manufacturers and Measurement Type [2]**

| Measurement Value | Manufacturer |
|---|---|
| Compaction meter value (CMV) | Caterpillar, Dynapac, Hamm, Volvo |
| Compaction control value (CCV) | Sakai |
| Stiffness ($k_s$) | Ammann, Case |
| Vibration modulus ($E_{vib}$) | Bomag |

equipment is capable of taking. There are four different methodologies currently available for making compaction measurements. They are as follows [2]:

- Compaction Meter Value (CMV): This approach measures the ratio of vertical drum acceleration amplitudes at fundamental (operating) vibration frequency and its first harmonic in the frequency domain.
- Compaction Control Value (CCV): This approach measures the algebraic relationship of multiple vertical drum vibration amplitudes, including fundamental frequency, and multiple harmonics and subharmonics in the frequency domain.
- Stiffness ($k_s$): This approach measures the vertical drum displacement and drum-soil contact force.
- Vibration Modulus ($E_{vib}$): This approach also measures the vertical drum displacement and drum-soil contact force.

As of 2016, 18 state departments of transportation have drafted IC specifications. In most cases, these require the contractor to supply conventional testing with techniques like the nuclear density gauge (NDG) and core sampling to validate the IC output [3]. IC is used to compact both soils and hot mix asphalt pavement. The quality control standards are based on a percentage difference between MVs roller passes, which are correlated with in-situ point measurements to set target IC MVs. For soil compaction, a number of specifications retain the construction of a test strip from the classic roller compaction specification. Specifications for pavements set a target number of roller passes based on calculating percentage difference in MVs and then follow that by establishing target values for the MVs correlated with the NDG. At this writing, IC specifications and quality control methods are a work in progress, and the reader should be careful to check the details on a project by project basis.

### 11.2.2 Intelligent Compaction Benefits

The fundamental reason for adopting IC is to enhance production rates on structural fills and pavement projects. Currently, the production rate is capped by the ability to complete the necessary quality control testing to verify that compaction specifications have been met. When the tests fail, the roller team must recompact the failed lifts. In many cases, contractors are required to scarify the failed areas before recompacting, and this greatly impacts the project's sustained production rates. This is particularly critical in flood control and dam projects where earthen hydraulic structures like levees included an impermeable clay core. If the core is not properly compacted, it will not be impermeable. This condition can lead to dam or levee failure and catastrophic flooding. IC permits the effectiveness of the compaction effort to be measured without the need to stop the roller team production. The ability to spatially correlate compaction MVs automatically also provides the inspection force with a mechanism to assess the effectiveness of the contractor's crews.

Two research studies have quantified the benefits of IC. The first looked at the life cycle cost of the (NDG) as compared to other available options to measure compaction [4]. The NDG is the fastest compaction testing method, and that aspect makes it the test method of choice in most projects because it minimizes the disruption of the contractor's production to take tests. However, because of the use of radioactive material in the device, the NDG entails significant ownership costs including special storage requirements, training, and certification of technicians, and significant hazardous waste disposal costs at the end of its useful life. The study was conducted by the Missouri DOT and found that the NDG's equivalent annual cost was four times greater than the dynamic cone penetrometer and ten times greater than the sand cone method. Hence, the speed and convenience provided by the NDG come with significant life cycle costs, making IC a very attractive option.

The second study was conducted for the Wyoming DOT and compared the costs of measuring compaction using the conventional methods to the cost of IC [3]. That study found the following results:

> IC, when used effectively, can produce a 37 percent decrease in construction costs based on a hypothetical thick asphalt overlay on one lane mile long roadway and a 54 percent decrease in costs on a new roadway section; and improved pavement performance based on compaction uniformity using IC can yield approximately $15,000 per lane mile per year in cost savings.
>
> [3]

These studies lead one to conclude that the future of IC is extremely promising. The intersection of high technology, common construction equipment, and the ability to relate the constructed product's quality to the location using GPS provides a capability to not only improve the service life of compacted structures but also to achieve the improvement at a reduced cost. As such, IC provides a single current example of what can possibly be achieved with other types of construction equipment.

## 11.3 AUTONOMOUS CONSTRUCTION EQUIPMENT

As the ability to manage a variety of tasks using computers has become more robust, the advent of fully autonomous construction machinery became a reality. This construction equipment does not require a human operator on the machine. The advances brought by GPS, AMG, and CIM tools have been combined to produce construction robots, which not only increase production but also reduce construction site safety hazards to humans by removing them from the site. Construction equipment that is remotely controlled is also available with the operator working at an off-site work station and no longer susceptible to the physical challenges involved with riding a machine all day long.

The Rio Tinto Mining Company uses 69 Komatsu autonomous dump trucks on its Pilbara iron ore mine in Western Australia. A report on this initiative found that "… driverless vehicles deliver their loads more efficiently, minimizing delays and fuel use, and are controlled remotely by operators who exert more control over their environment and ensure greater operational safety" [5] Suncor Energy, a Canadian oil company, uses 400-ton autonomous haulers and has observed that the driverless vehicles can make 22 trips per day whereas the same unit with a driver makes 20. While the use of autonomous and semiautonomous construction equipment is still limited, the early demonstrated benefits clearly show that the potential will only be limited to the constraints of technology to address the broad set of challenges faced on equipment-intensive projects. Those benefits are found in the following three areas:

- Efficiency: One major benefit is associated with the telematics (remote diagnostics) that use sensors connected to the internet or cell-phone network to continuously monitor equipment functions and ensure that it runs at its peak performance. This extends the economic life of the machine. For example, the Australian iron mine haulers' tires lasted 50% longer than tires on haulers with operators. The telematics kept the running horsepower at the minimum required for the rolling resistance and saved fuel over operated haulers [5].
- Productivity: The production of autonomous equipment is not reduced by operator fatigue issues and other human factors, which in turn reduces cycle time variability and higher sustained production rates. Semiautonomous equipment operators control the equipment from within a much more controlled, comfortable environment, not bouncing down a hot and dusty haul road. As a consequence, the operator fatigue increases at a much lower rate, which enhances sustained production. These factors pay a dividend in the bidding process by reducing the uncertainty inherent to equipment production assumptions, allowing the contractor to submit a more competitive price.
- Safety: Removing the operator from the machine also removes the operator from the danger zone. The enhanced consistency achieved by computerization of the machines operation also reduces the potential for machine on machine conflicts and reduces accidents.

Thus, with autonomous machinery, the construction industry now has the means to enhance its potential maximum production by eliminating the weakest link in the production chain: the human equipment operators. Humans get weary and slow down. Robots don't. Humans are subject to constraints on their ability to produce imposed by statutory work rules and union contracts. Robots don't have to join unions, get paid overtime, health benefits, or take vacations.

## 11.4 SUMMARY

The future of construction equipment management is going to be driven advances in the available technology. CIM contains a growing set of tools that will be available to future fleet managers. The ability to produce virtual construction documents that can digitally communicate with other systems that have the capability to translate geospatial coordinate data puts the equipment manager in a position where the fleet can be employed in a highly efficient manner – from estimating its costs during the project's procurement through increased production during construction to a more comprehensive data-driven system to ensure the fleet is maintained in a manner that keeps it operating at its most efficient level of performance.

As previously stated in Chapter 1, construction equipment management has become a computer and data-driven activity. The fleet managers of the 21st Century must be computer literate, technology savvy, and masters of analytics. Advancements in equipment technology will continue to occur, and fleet managers must be prepared to exploit them as they arrive.

## REFERENCES

[1] O'Brien, W.J., Sankaran, B., Leite, F.L., Khwaja, N., De Sande Palma, I., Goodrum, P. and Johnson, J. (2016). *Civil Integrated Management (CIM) for Departments of Transportation*, NCHRP Report 831, Volume 1 (No. Project 10-96).

[2] Mooney, M.A., Rinehart, R.V., White, D.J., Vennapusa, P., Facas, N. and Musimbi, O. (2010). *Intelligent Soil Compaction Systems*, National Cooperative Highway Research Program, Report 676. Washington, DC: Transportation Research Board.
[3] Savan, C.M., Ng, K.W. and Ksaibati, K. (2016). Benefit-Cost Analysis and Application of Intelligent Compaction for Transportation. *Transportation Geotechnics*, 9, pp. 57–68.
[4] McLain, K.W. and Gransberg, D.D. (2016). Life Cycle Cost Evaluation of Alternatives to the Nuclear Density Gauge for Compaction Testing on Design-Build Projects. *Journal of Structural Integrity and Maintenance*, 1(4), pp. 197–203.
[5] Alderton, M. (2018). The Robots Are Coming! Driverless Dozers and the Dawn of Autonomous Vehicle Technology in Construction, *Construction*. www.autodesk.com/redshift/autonomous-vehicle-technology-in-construction/. (Accessed December 10, 2019).

## CHAPTER PROBLEMS

1. Do some internet research and find one use of autonomous equipment and answer the following questions:
   a. What specific safety hazards have been avoided by this particular application?
   b. How was productivity improved?
   c. What other type of project would this equipment work well on and why?
2. Do some internet research on IC and determine if the benefits cited in Section 11.2 are still the same. If not, what did you find?
3. Do some internet research and identify a potential CIM tool that is not listed in Table 11.1. Describe its use and speculate on how you would integrate it on an equipment-intensive construction project.

# Appendix A

## *Corps of Engineers Construction Equipment Ownership and Operating Expense Schedule*

For infortmation see https://www.crcpress.com/Construction-Equipment-Management-for-Engineers-Estimators-and-Owners/Gransberg-Benavides/p/book/9781498788489

# Appendix B

## *Heavy Equipment Product Guides*

This appendix contains a sample of standard equipment manufacturer's equipment guides for use in solving the types of problems described in Chapter 4.

Grove RT600E-Product Guide
Model 777-Product Guide
MD 485B-M20

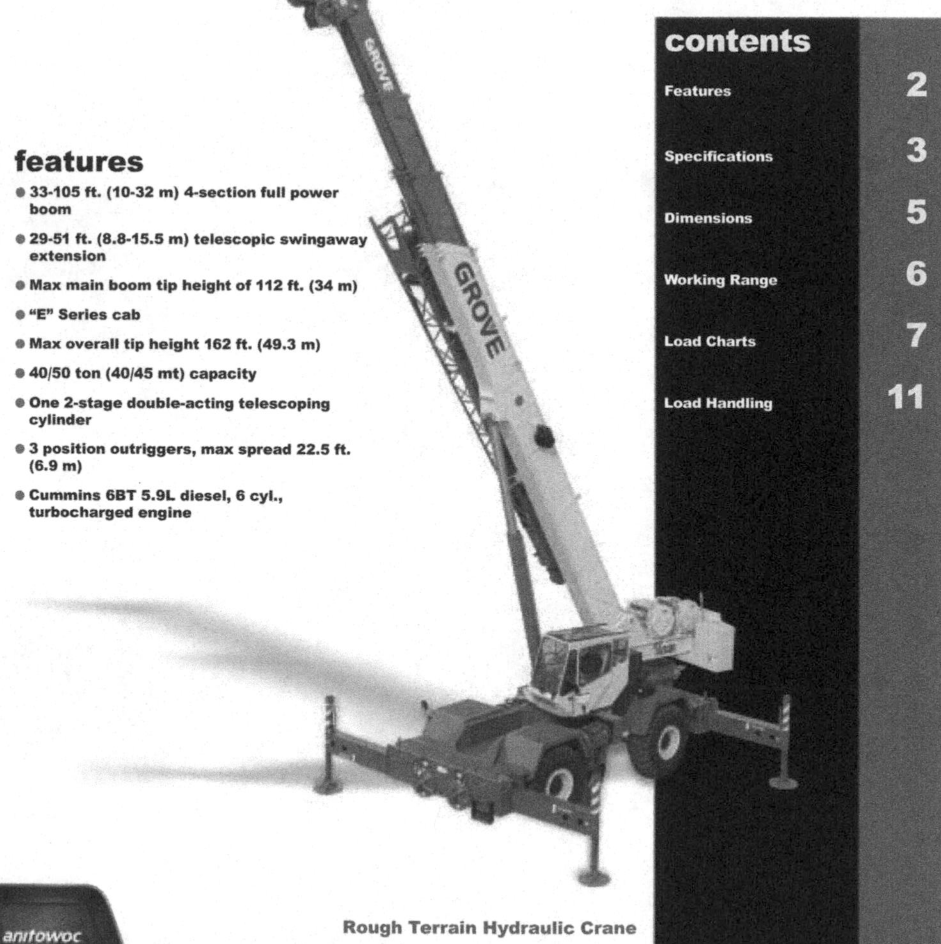
GROVE
RT600E
product guide
contents
Features 2
Specifications 3
Dimensions 5
Working Range 6
Load Charts 7
Load Handling 11
features
33-105 ft. (10-32 m) 4-section full power boom
29-51 ft. (8.8-15.5 m) telescopic swingaway extension
Max main boom tip height of 112 ft. (34 m)
"E" Series cab
Max overall tip height 162 ft. (49.3 m)
40/50 ton (40/45 mt) capacity
One 2-stage double-acting telescoping cylinder
3 position outriggers, max spread 22.5 ft. (6.9 m)
Cummins 6BT 5.9L diesel, 6 cyl., turbocharged engine
GROVE
GROVE
Manitowoc
Crane Group
Rough Terrain Hydraulic Crane

# features

2

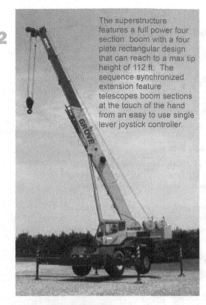

The superstructure features a full power four section boom with a four plate rectangular design that can reach to a max tip height of 112 ft. The sequence synchronized extension feature telescopes boom sections at the touch of the hand from an easy to use single lever joystick controller.

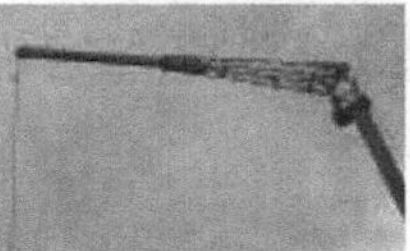

A telescopic swingaway lattice extension easily stows on the side of the base boom for easy transport. With a range of 29-51 ft. the max tip height reaches 162 ft. with a capacity of 6,000 lbs. An optional fixed lattice is also available, reaching a max height of 141 ft.

Optional full length aluminum decking is also available.

Features common to the Grove "E" Series cab include:

- hot water heater/defroster
- single axis joystick controllers
- sliding skylight and adjustable sunscreen
- engine instrumentation
- full accoustical lining

The PAT iFlex 5 graphic display LMI includes a work area definition system which allows the operator to define a preferred working area.

Large open stowage compartment for tools and rigging accessories.

The RT600E uses a 11,250 lbs. pinned-on counterweight. Cable power is provided through Grove model HO30G-16G grooved drum hoists with 16,800 lbs. permissible line pull. Max line speed is 593 fpm. Both the main and optional auxiliary hoists have cable capacity up to 450 ft.

# specifications

3

## Superstructure

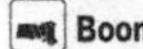

### Boom

33 ft. - 105 ft. (10.1 m - 32 m) four-section, full-power synchronized boom.
Maximum tip height: 112 ft. (34.1 m).

### *Optional Fixed Swingaway Extension

29 ft. (8.8 m) offsettable lattice swingaway extension. Offsettable at 0°, 25° and 45°. Stows alongside base boom section.
Maximum tip height: 141.5 ft. (43.1 m).

### *Optional Telescopic Swingaway Extension

29 ft. - 51 ft. (8.8 m - 15.5 m) telescoping lattice swingaway extension. Offsettable at 0°, 25° and 45°. Stows alongside base boom section. Maximum tip height: 162 ft. (49.3 m).

### Boom Nose

Three nylatron sheaves mounted on heavy-duty tapered roller bearings with removable pin-type rope guards. *(Four sheaves with optional 35 x 7 wire rope.) Quick-reeve type boom nose. *Optional removable auxiliary boom nose with removable pin type rope guard.

### Boom Elevation

One double-acting hydraulic cylinder with integral holding valve provides elevation from -2° to 78°.

### Load Moment & Anti-Two Block System

Standard "Graphic Display" load moment and anti-two block system with audio-visual warning and control lever lockout. These systems provide electronic display of boom angle, length, radius, tip height, relative load moment, maximum permissible load, load indication and warning of impending two-block condition. The system defaults to 360° on rubber chart. The standard Work Area Definition System allows the operator to pre-select and define working areas. If the crane approaches the pre-set limits, audio-visual warnings aid the operator in avoiding job-site obstructions.

### Cab

Full vision, all steel fabricated with acoustical lining and tinted safety glass throughout. Deluxe seat incorporates armrest-mounted hydraulic single-axis controllers. Dash panel incorporates gauges for all engine functions. Other standard features include: hot water heater/defroster, cab circulating air fan, sliding side and rear windows, sliding skylight with electric wiper and sunscreen, electric windshield wash/wipe, fire extinguisher, seat belt and circuit breakers.

### Swing

Planetary swing with foot-applied multi-disc brake. Spring applied, hydraulically-released swing brake and plunger-type, one position, mechanical house lock operated from cab. *Optional 360° mechanical swing lock. Maximum speed: 2.5 RPM.

### Counterweight

11,250 lbs. (5 103 kg) pinned to superstructure.

### Hydraulic System

Three main gear pumps with combined capacity of 103 GPM (391 L/min), 135 GPM (511 L/min) with optional air conditioning. Maximum operating pressure: 3500 psi (26.2 MPa)

Return line type filter with full flow by-pass protection and service indicator. Replaceable cartridge with micron filtration rating of 5/12/16. 134 gallon (509 L) reservoir. Hydraulic oil cooler. System pressure test ports.

### Hoist Specifications
### Main and Auxiliary Hoist: Grove Model HO30G-16G

Planetary reduction with automatic spring applied multi-disc brake. Grooved drum. Electronic hoist drum rotation indicator and hoist drum cable followers.

| | |
|---|---|
| Maximum Single Line Pull: | 18,180 lbs (8 246 kg) |
| Maximum Single Line Speed: | 588 FPM (179 m/min) |

Maximum Permissible Line Pull:
16,800 lbs. (7 620 kg) w/standard 6 x 37 class rope
16,800 lbs. (7 620 kg) w/optional 35 x 7 class rope

| | |
|---|---|
| Rope Diameter: | 3/4 in. (19 mm) |
| Rope Length: | 450 ft. (137 m) |
| Rope Type: | 6 x 37 Class EIPS IWRC |
| Optional: | 35 x 7 class rotation resistant |
| Maximum Rope Stowage: | 694 ft. (211 m) |

GROVE

# specifications

## Carrier

4

### Chassis

Box section frame fabricated from high-strength, low alloy steel. Integral outrigger housings and front/rear towing and tie down lugs.

### Outrigger System

Four hydraulic telescoping single-stage double box beam outriggers with inverted jacks and integral holding valves. Three position setting. All steel fabricated, quick-release type round outrigger floats, 24 in. (610 mm) diameter. Maximum outrigger pad load: 69,100 lbs. (31 344 kg).

### Outrigger Controls

Controls and crane level indicator located in cab.

### Engine

Cummins 6BT 5.9L diesel, six cylinders, turbocharged, 173 bhp (129 kW) (Gross) @ 2,500 rpm. Maximum torque: 530 ft. lbs. (719 Nm) @ 1,500 RPM.

### Fuel Tank Capacity

58 gallons (220 L)

### Transmission

Full powershift with 6 forward and 3 reverse speeds. Front axle disconnect for 4 x 2 travel.

### Electrical System

Two 12-volt maintenance free batteries. 12-volt starting and lighting, circuit breakers. *Optional battery disconnect switch.

### Drive

4 x 4

### Steering

Fully independent power steering:
Front: Full hydraulic, steering wheel controlled.
Rear: Full hydraulic, switch controlled.
Provides infinite variations of 4 main steering modes: front only, rear only, crab and coordinated.
"Rear steer centered" indicating light.
4 wheel turning radius - 21 ft. (6.4 m)

### Axles

Front: Drive/steer with differential and planetary reduction hubs rigid-mounted to frame.
Rear: Drive/steer with differential and planetary reduction hubs pivot-mounted to frame. Automatic full hydraulic lockouts on rear axle permit oscillation only with boom centered over the front.

### Brakes

Full hydraulic split circuit disc-type brakes operating on all wheels. Spring-applied, hydraulically released transmission-mounted parking brake.

### Tires

*23.5 x 25 - 20PR bias earthmover type.
*23.5R25 radial earthmover type.

### Lights

Full lighting package including turn indicators, head, tail, brake and hazard warning lights.

### Maximum Speed

24 MPH (39 km/h).

### Gradeability (Theoretical)

78% (Based on 75,000 lbs. [34 020 kg] GVW) 23.5 x 25 tires, pumps engaged, 105 ft. (32 m) boom, and tele-swingaway.

### Miscellaneous Standard Equipment

Full width steel fenders, dual rear view mirrors, hookblock tiedown, electronic back-up alarm, light package, front stowage well, tachometer, rear wheel position indicator, 36,000 BTU hot water heater, hoist mirrors, engine distress A/V warning system. Auxiliary hoist control valve arrangement (less hoist). Ether injection cold start aid (less canister) and immersion type engine block heater, 120V 1500 watt.

### *Optional Equipment

*VALUE PACKAGE: includes 29-51 ft. (8.8-15.5 m) offsettable telescoping swingaway, 360° NYC style swing lock, and auxiliary hoist package.
*AUXILIARY HOIST PACKAGE (includes Model HO30G-16G auxiliary hoist with electronic hoist drum rotation indicator, hoist drum cable follower, 450 ft. (137m) of 3/4 in.(19mm) 35 X 7 class wire rope, auxiliary single sheave boom nose.)
*AUXILIARY LIGHTING PACKAGE (includes cab mounted, 360° rotation spotlight, cab mounted amber flashing light, and dual base boom mounted floodlights.)
*CONVENIENCE PACKAGE (includes in cab LMI light bar)
*Air Conditioning
*Full-length aluminum decking
*Pintle hook - rear
*360 degree positive swing lock
*Battery disconnect switch
*Cab-controlled cross axle differential lock (front and rear)
*Manual hydraulic pump disconnect
*PAT datalogger
*Rubber mat for stowage trough
*Mounting hardware for gooseneck/trailer attachment

*Denotes optional equipment

RT600E

GROVE

# dimensions

5

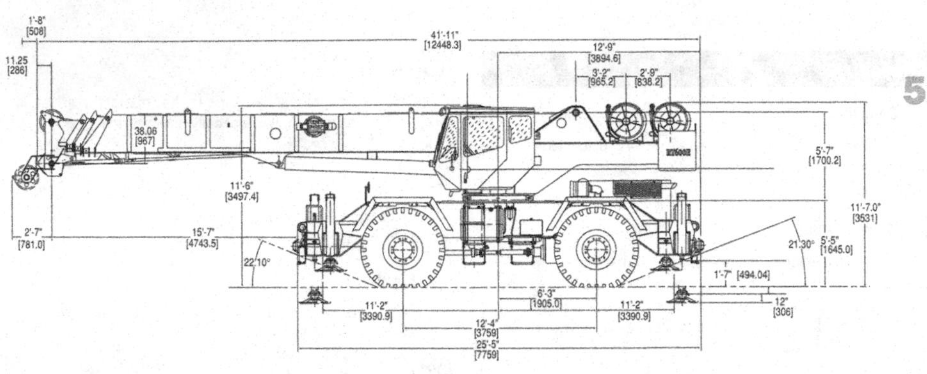


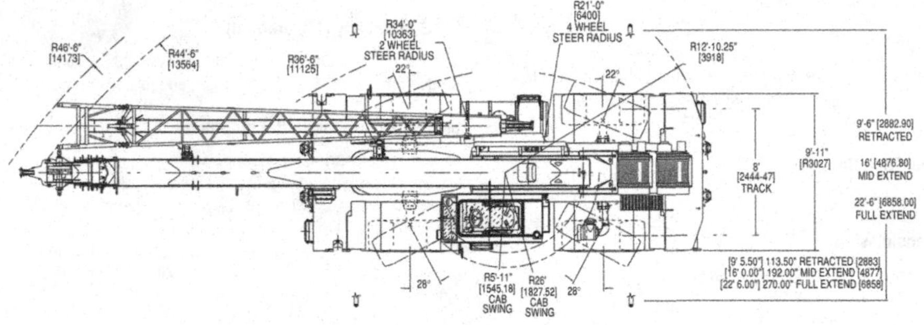


Note: [ ] Reference dimensions in mm

## Weights

| | GVW | | Front | | Rear | |
|---|---|---|---|---|---|---|
| | lb. | kg | lb. | kg | lb. | kg |
| **RT600E Basic Machine** | 71,691 | 32,519 | 32,934 | 14,939 | 38,757 | 17,580 |
| **ADD**: 29 - 51 ft. tele swingaway | 2,109 | 957 | 3,456 | 1,568 | -1,347 | -611 |
| **ADD**: 29 ft. swingaway | 1,493 | 677 | 2,506 | 1,137 | -1,013 | -459 |
| **ADD**: Auxiliary hoist cable | 563 | 255 | -213 | -97 | 775 | 352 |
| **ADD**: Auxiliary boom nose | 131 | 59 | 358 | 162 | -227 | -103 |
| **ADD**: 40 ton (35 mt) 3 sheave hookblock (stowed in trough) | 800 | 363 | 822 | 373 | -22 | -10 |
| **ADD**: 50 ton (45 mt) 3 sheave hookblock (stowed in trough) | 1,000 | 454 | 1,027 | 466 | -27 | -12 |
| **ADD**: 8.3 ton (7.5 mt) headache ball (hanging from aux. nose) | 370 | 168 | 643 | 292 | -273 | -124 |
| **Remove**: Counterweight | -11,250 | -5,103 | 4,570 | 2,073 | -15,820 | -7,176 |

RT600E

GROVE.

# working range

## Working range – 105 ft. Main Boom

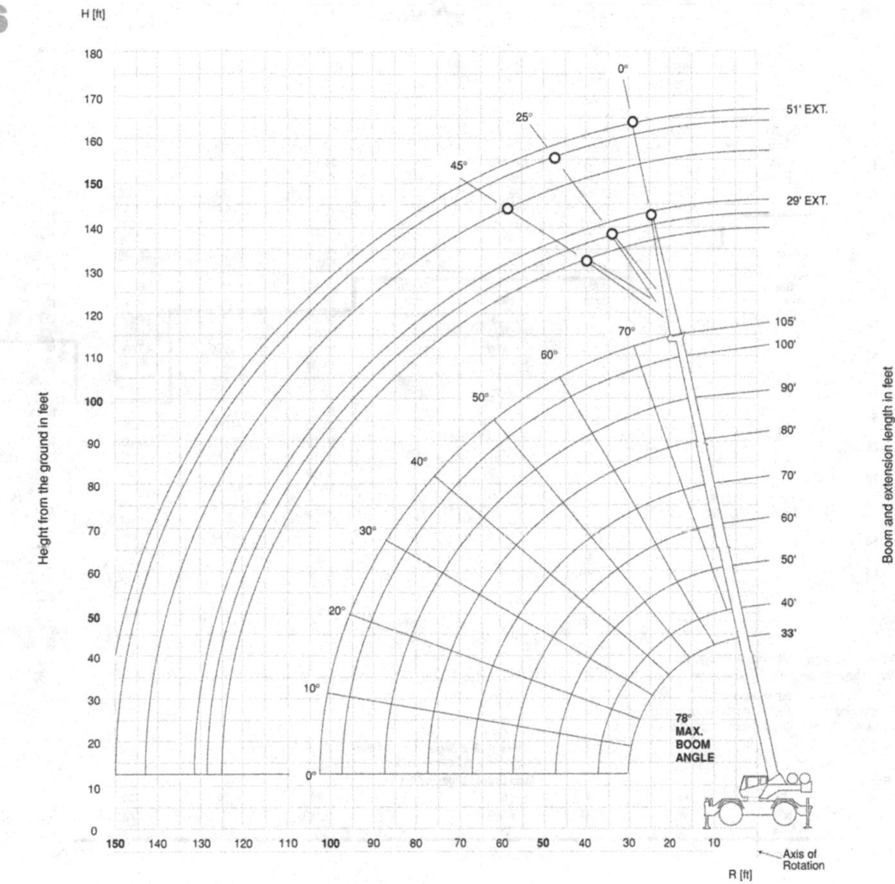


Operating Radius in Feet From Axis of Rotation

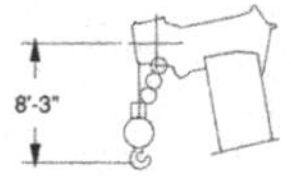


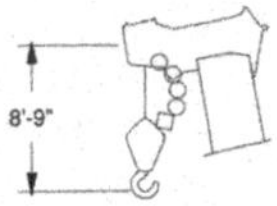


Dimensions are for Largest Grove furnished Hook Block and Headache Ball, with Anti-Two Block Activated.

RT600E

THIS CHART IS ONLY A GUIDE AND SHOULD NOT BE USED TO OPERATE THE CRANE. The individual crane's load chart, operating instructions and other instructional plates must be read and understood prior to operating the crane.

GROVE.

# RT650E load chart

7

33-105 ft. | 11,250 lbs | 100% 22' 6" spread | 360

Pounds

| Feet | 33 | 40 | 50 | 60 | 70 | 80 | 90 | 100 | 105 |
|---|---|---|---|---|---|---|---|---|---|
| 10 | 100,000 (69.5) | 80,550 (73.5) | 67,250 (77) | | | | | | |
| 12 | 87,100 (65.5) | 79,150 (70.5) | 64,200 (75) | *56,100 (78) | | | | | |
| 15 | 69,050 (59.5) | 69,550 (65.5) | 59,950 (71) | 51,800 (75) | 45,200 (77.5) | | | | |
| 20 | 50,500 (47.5) | 50,950 (57) | 51,400 (64.5) | 44,500 (69.5) | 38,550 (73) | 34,450 (75.5) | *31,400 (78) | | |
| 25 | 38,300 (32) | 38,850 (47) | 39,350 (58) | 39,650 (64.5) | 37,100 (68.5) | 29,850 (72) | 27,250 (74.5) | 21,000 (76.5) | 18,350 (77.5) |
| 30 | | 30,700 (34.5) | 31,200 (50.5) | 31,500 (58.5) | 31,700 (64) | 26,350 (68) | 24,100 (71) | 21,000 (73.5) | 18,350 (74.5) |
| 35 | | | 25,450 (41.5) | 25,750 (52.5) | 26,000 (59) | 23,650 (64) | 21,500 (67.5) | 19,150 (70) | 18,350 (71.5) |
| 40 | See Note 16 | | 20,850 (30.5) | 21,200 (46) | 21,600 (54) | 21,350 (59.5) | 19,400 (64) | 16,650 (67) | 17,300 (68.5) |
| 45 | | | | 17,100 (38) | 17,350 (48.5) | 17,300 (55) | 17,300 (60) | 14,650 (64) | 15,750 (65.5) |
| 50 | | | | 13,950 (28) | 14,150 (42.5) | 14,200 (50.5) | 14,200 (56) | 13,000 (60.5) | 14,300 (62.5) |
| 55 | | | | | 11,700 (35) | 11,750 (45.5) | 11,850 (52) | 11,900 (57) | 12,000 (59) |
| 60 | | | | | 9,730 (26) | 9,870 (39.5) | 9,980 (47.5) | 10,100 (53.5) | 10,150 (55.5) |
| 65 | | | | | | 8,300 (33) | 8,440 (42.5) | 8,600 (49.5) | 8,680 (52) |
| 70 | | | | | | 6,960 (24.5) | 7,170 (37.5) | 7,340 (45.5) | 7,430 (48.5) |
| 75 | | | | | | | 6,080 (31) | 6,290 (40.5) | 6,390 (44.5) |
| 80 | | | | | | | 5,130 (23) | 5,380 (35.5) | 5,490 (40) |
| 85 | | | | | | | | 4,580 (29.5) | 4,720 (35) |
| 90 | | | | | | | | 3,880 (22) | 4,020 (29) |
| 95 | | | | | | | | | 3,400 (21.5) |
| Minimum boom angle (∞) for indicated length (no load) | | | | | | | | | 0 |
| Maximum boom length (ft.) at 0∞ boom angle (no load) | | | | | | | | | 105 |

NOTE: ( ) Boom angles are in degrees.
#LMI operating code. Refer to LMI manual for operating instructions.
*This capacity is based on maximum boom angle.

**Lifting Capacities at Zero Degree Boom Angle On Outriggers Fully Extended - 360∞**

| Boom Angle | Main Boom Length in Feet 33 | 40 | 50 | 60 | 70 | 80 | 90 | 100 |
|---|---|---|---|---|---|---|---|---|
| 0∞ | 16,250 (28.2) | 12,500 (35) | 8,780 (45) | 6,290 (55) | 4,510 (65) | 3,160 (75) | 2,110 (85) | 1,260 (95) |

NOTE: ( ) Reference radii in feet.

A6-829-100936

THIS CHART IS ONLY A GUIDE AND SHOULD NOT BE USED TO OPERATE THE CRANE. The individual crane's load chart, operating instructions and other instructional plates must be read and understood prior to operating the crane.

GROVE.

# RT640E load chart

33-105 ft. | 11,250 lbs | 100% 22' 6" spread | 360

8

| Feet | 33 | 40 | 50 | 60 | 70 | 80 | 90 | 100 | 105 |
|---|---|---|---|---|---|---|---|---|---|
| | | | | | Pounds | | | | |
| 10 | 80,000 (69.5) | 73,500 (73.5) | 67,200 (77) | | | | | | |
| 12 | 77,750 (65.5) | 69,500 (70.5) | 62,300 (75) | *56,100 (78) | | | | | |
| 15 | 69,050 (59.5) | 65,550 (65.5) | 57,300 (71) | 51,800 (75) | 45,200 (77.5) | | | | |
| 20 | 50,500 (47.5) | 50,950 (57) | 51,400 (64.5) | 44,500 (69.5) | 38,550 (73) | 34,450 (75.5) | *31,400 (78) | | |
| 25 | 38,300 (32) | 38,850 (47) | 39,350 (58) | 39,650 (64.5) | 37,100 (68.5) | 29,850 (72) | 27,250 (74.5) | 21,000 (76.5) | 18,350 (77.5) |
| 30 | | 30,700 (34.5) | 31,200 (50.5) | 31,500 (58.5) | 31,700 (64) | 26,350 (68) | 24,100 (71) | 21,000 (73.5) | 18,350 (74.5) |
| 35 | | | 25,450 (41.5) | 25,750 (52.5) | 26,000 (59) | 23,650 (64) | 21,500 (67.5) | 19,150 (70) | 18,350 (71.5) |
| 40 | See Note 16 | | 20,850 (30.5) | 21,200 (46) | 21,600 (54) | 21,350 (59.5) | 19,400 (64) | 16,650 (67) | 17,300 (68.5) |
| 45 | | | | 17,100 (38) | 17,350 (48.5) | 17,300 (55) | 17,300 (60) | 14,650 (64) | 15,750 (65.5) |
| 50 | | | | 13,950 (28) | 14,150 (42.5) | 14,200 (50.5) | 14,200 (56) | 13,000 (60.5) | 14,300 (62.5) |
| 55 | | | | | 11,700 (35) | 11,750 (45.5) | 11,850 (52) | 11,900 (57) | 12,000 (59) |
| 60 | | | | | 9,730 (26) | 9,870 (39.5) | 9,980 (47.5) | 10,100 (53.5) | 10,150 (55.5) |
| 65 | | | | | | 8,300 (33) | 8,440 (42.5) | 8,600 (49.5) | 8,680 (52) |
| 70 | | | | | | 6,960 (24.5) | 7,170 (37.5) | 7,340 (45.5) | 7,430 (48.5) |
| 75 | | | | | | | 6,080 (31) | 6,290 (40.5) | 6,390 (44.5) |
| 80 | | | | | | | 5,130 (23) | 5,380 (35.5) | 5,490 (40) |
| 85 | | | | | | | | 4,580 (29.5) | 4,720 (35) |
| 90 | | | | | | | | 3,880 (22) | 4,020 (29) |
| 95 | | | | | | | | | 3,400 (21.5) |
| Minimum boom angle (∞) for indicated length (no load) | | | | | | | | | 0 |
| Maximum boom length (ft.) at 0∞ boom angle (no load) | | | | | | | | | 105 |

NOTE: ( ) Boom angles are in degrees.
#LMI operating code. Refer to LMI manual for operating instructions.
*This capacity is based on maximum boom angle.

**Lifting Capacities at Zero Degree Boom Angle On Outriggers Fully Extended - 360∞**

| Boom Angle | 33 | 40 | 50 | 60 | 70 | 80 | 90 | 100 | 105 |
|---|---|---|---|---|---|---|---|---|---|
| | Main Boom Length in Feet | | | | | | | | |
| 0∞ | 16,250 (28.2) | 12,500 (35) | 8,780 (45) | 6,290 (55) | 4,510 (65) | 3,160 (75) | 2,110 (85) | 1,260 (95) | |

NOTE: ( ) Reference radii in feet.

A6-829-100832A

THIS CHART IS ONLY A GUIDE AND SHOULD NOT BE USED TO OPERATE THE CRANE. The individual crane's load chart, operating instructions and other instructional plates must be read and understood prior to operating the crane.

GROVE.

# RT600E load charts

33-105 ft. | 29 - 51ft. | 11,250 lbs | 100% 22' 6" spread | 360

Pounds

| Feet | **29 ft. LENGTH | | | 51 ft. LENGTH | | |
|---|---|---|---|---|---|---|
| | #0021 0∞ OFFSET | #0022 25∞ OFFSET | #0023 45∞ OFFSET | #0041 0∞ OFFSET | #0042 25∞ OFFSET | #0043 45∞ OFFSET |
| 30 | *9,000 (78) | | | | | |
| 35 | 9,000 (77) | | | *6,000 (78) | | |
| 40 | 9,000 (74.5) | 8,000 (77.5) | | 6,000 (77) | | |
| 45 | 9,000 (72.5) | 7,560 (76) | *5,660 (78) | 6,000 (76) | | |
| 50 | 8,760 (70) | 7,170 (74) | 5,600 (76) | 6,000 (74) | | |
| 55 | 8,030 (67.5) | 6,820 (71.5) | 5,500 (73.5) | 6,000 (72) | *4,120 (78) | |
| 60 | 7,380 (65) | 6,500 (69) | 5,300 (71) | 6,000 (70) | 3,900 (77) | |
| 65 | 6,770 (62.5) | 6,210 (66.5) | 5,180 (68.5) | 6,000 (68) | 3,710 (75) | *2,740 (78) |
| 70 | 6,210 (60) | 5,950 (64) | 4,890 (66) | 5,620 (66) | 3,530 (72.5) | 2,660 (76.5) |
| 75 | 5,710 (57.5) | 5,710 (61.5) | 4,620 (63) | 5,210 (64) | 3,370 (70.5) | 2,580 (74) |
| 80 | 5,250 (55) | 5,500 (58.5) | 4,370 (60.5) | 4,860 (61.5) | 3,220 (68.5) | 2,520 (72) |
| 85 | 4,790 (52) | 5,300 (56) | 4,100 (57.5) | 4,540 (59.5) | 3,080 (66) | 2,460 (69.5) |
| 90 | 4,090 (49) | 4,650 (53) | 3,820 (54) | 4,260 (57) | 2,960 (63.5) | 2,410 (67) |
| 95 | 3,480 (46) | 3,960 (49.5) | | 4,000 (55) | 2,850 (61.5) | 2,360 (64.5) |
| 100 | 2,930 (42.5) | 3,350 (46) | | 3,770 (52.5) | 2,750 (59) | 2,330 (62) |
| 105 | 2,440 (39) | 2,810 (42.5) | | 3,360 (50) | 2,660 (56) | 2,300 (59) |
| 110 | 2,000 (35) | 2,320 (38.5) | | 2,910 (47.5) | 2,570 (53.5) | 2,280 (56) |
| 115 | 1,610 (30.5) | | | 2,500 (44.5) | 2,500 (50.5) | |
| 120 | 1,250 (25.5) | | | 2,120 (41.5) | 2,430 (47.5) | |
| 125 | | | | 1,780 (38.5) | 2,250 (44.5) | |
| 130 | | | | 1,470 (35) | 1,820 (40.5) | |
| 135 | | | | 1,180 (31) | 1,420 (36.5) | |
| Min. boom angle for indicated length (no load) | 24∞ | 32∞ | 45∞ | 25∞ | 35∞ | 45∞ |
| Max. boom length at 0∞ boom angle (no load) | | 90 ft. | | | 90 ft. | |

NOTE: ( ) Boom angles are in degrees. A6-829-100845A

#LMI operating code. Refer to LMI manual for instructions.

*This capacity based on maximum boom angle.

**29 ft. capacities are also applicable to fixed offsettable ext. However, the LMI codes will change to #0051, #0052 and #0053 for 0∞, 25∞ and 45∞ offset, respectively

33-105 ft. | 11,250 lbs | Stationary | 360

Pounds

| Feet | #9005 Main Boom Length in Feet | | | | |
|---|---|---|---|---|---|
| | 33 | 40 | 50 | 60 | 70 |
| 10 | 38,550 (69.5) | 38,550 (73.5) | | | |
| 12 | 32,550 (65.5) | 32,550 (70.5) | 32,550 (75) | | |
| 15 | 23,700 (59.5) | 23,700 (65.5) | 23,700 (71) | 23,700 (75) | |
| 20 | 14,450 (47.5) | 14,450 (57) | 14,450 (64.5) | 14,450 (69.5) | 14,450 (73) |
| 25 | 9,640 (32) | 9,640 (47) | 9,640 (58) | 9,640 (64.5) | 9,640 (68.5) |
| 30 | | 6,840 (34.5) | 6,840 (50.5) | 6,840 (58.5) | 6,840 (64) |
| 35 | | | 4,850 (41.5) | 4,850 (52.5) | 4,850 (59) |
| 40 | | | 3,450 (30.5) | 3,450 (46) | 3,450 (54) |
| 45 | | | | 2,410 (38) | 2,410 (48.5) |
| 50 | | | | 1,610 (28) | 1,610 (42.5) |
| Min. boom angle (∞) for indicated length (no load) | | | | | 30 |
| Max. boom length (ft.) at 0∞ boom angle (no load) | | | | | 60 |

NOTE: ( ) Boom angles are in degrees.

#LMI operating code. Refer to LMI manual for operating instructions.

| Lifting Capacities at Zero Degree Boom Angle On Rubber - 360∞ | | | |
|---|---|---|---|
| Boom Angle | Main Boom Length in Feet 33 | 40 | 50 |
| 0∞ | 7,580 (28.2) | 4,850 (35) | 2,410 (45) |

NOTE: ( ) Reference radii in feet. A6-829-100836B

9

**NOTES:**

1. All capacities above the bold line are based on structural strength of boom extension.
2. 29 ft. and 51 ft. boom extension lengths may be used for single line lifting service.
3. Radii listed are for a fully extended boom with the boom extension erected. For main boom lengths less than fully extended, the rated loads are determined by boom angle. Use only the column which corresponds to the boom extension length and offset for which the machine is configured. For boom angles not shown, use the rating of the next lower boom angle.
   **WARNING:** Operation of this machine with heavier loads than the capacities listed is strictly prohibited. Machine tipping with boom extension occurs rapidly and without advance warning.
4. Boom angle is the angle above or below horizontal of the longitudinal axis of the boom base section after lifting rated load.
5. Capacities listed are with outriggers fully extended and vertical jacks set.

THIS CHART IS ONLY A GUIDE AND SHOULD NOT BE USED TO OPERATE THE CRANE. The individual crane's load chart, operating instructions and other instructional plates must be read and understood prior to operating the crane.

GROVE.

# load charts

10

33-105 ft. | 11,250 lbs | Stationary | Defined arc over front

Pounds

#9005

| Feet | Main Boom Length in Feet 33 | 40 | 50 | 60 | 70 |
|---|---|---|---|---|---|
| 10 | 46,600 (69.5) | 40,800 (73.5) | 34,600 (77) | | |
| 12 | 40,800 (65.5) | 40,800 (70.5) | 34,600 (75) | | |
| 15 | 34,000 (59.5) | 34,000 (65.5) | 34,000 (71) | 26,650 (75) | 21,500 (77.5) |
| 20 | 26,050 (47.5) | 26,050 (57) | 26,050 (64.5) | 26,050 (69.5) | 21,500 (73) |
| 25 | 18,200 (32) | 18,200 (47) | 18,200 (58) | 18,200 (64.5) | 18,200 (68.5) |
| 30 | | 13,100 (34.5) | 13,100 (50.5) | 13,100 (58.5) | 13,100 (64) |
| 35 | | | 10,050 (41.5) | 10,050 (52.5) | 10,050 (59) |
| 40 | | | 7,900 (30.5) | 7,900 (46) | 7,900 (54) |
| 45 | | | | 6,290 (38) | 6,290 (48.5) |
| 50 | | | | 5,050 (28) | 5,050 (42.5) |
| 55 | | | | | 4,060 (35) |
| 60 | | | | | 3,260 (26) |
| Min. boom angle (∞) for indicated length (no load) | | | | | 0 |
| Max. boom length (ft.) at 0∞ boom angle (no load) | | | | | 70 |

NOTE: ( ) Boom angles are in degrees.
#LMI operating code. Refer to LMI manual for operating instructions.

Lifting Capacities at Zero Degree Boom Angle On Rubber - Defined Arc Over Front

| Boom Angle | Main Boom Length in Feet 33 | 40 | 50 | 60 | 70 |
|---|---|---|---|---|---|
| 0∞ | 14,550 (28.2) | 10,050 (35) | 6,290 (45) | 4,060 (55) | 2,590 (65) |

NOTE: ( ) Reference radii in feet. A6-829-100835B

33-105 ft. | 11,250 lbs | Pick & carry up to 2.5 mph | Boom centered over front

Pounds

#9006

| Feet | Main Boom Length in Feet 33 | 40 | 50 | 60 | 70 |
|---|---|---|---|---|---|
| 10 | 30,150 (69.5) | 30,150 (73.5) | 17,850 (77) | | |
| 12 | 30,150 (65.5) | 30,150 (70.5) | 17,850 (75) | | |
| 15 | 29,650 (59.5) | 29,650 (65.5) | 17,850 (71) | 17,850 (75) | 14,750 (77.5) |
| 20 | 22,650 (47.5) | 22,650 (57) | 17,850 (64.5) | 17,850 (69.5) | 14,750 (73) |
| 25 | 17,850 (32) | 17,850 (47) | 17,850 (58) | 17,850 (64.5) | 14,750 (68.5) |
| 30 | | 13,100 (34.5) | 13,100 (50.5) | 13,100 (58.5) | 13,100 (64) |
| 35 | | | 10,050 (41.5) | 10,050 (52.5) | 10,050 (59) |
| 40 | | | 7,340 (30.5) | 7,340 (46) | 7,340 (54) |
| 45 | | | | 6,020 (38) | 6,020 (48.5) |
| 50 | | | | 4,940 (28) | 4,940 (42.5) |
| 55 | | | | | 4,030 (35) |
| 60 | | | | | 3,260 (26) |
| Min. boom angle (∞) for indicated length (no load) | | | | | 0 |
| Max. boom length (ft.) at 0∞ boom angle (no load) | | | | | 70 |

NOTE: ( ) Boom angles are in degrees.
#LMI operating code. Refer to LMI manual for operating instructions.

Lifting Capacities at Zero Degree Boom Angle On Rubber - Pick & Carry

| Boom Angle | Main Boom Length in Feet 33 | 40 | 50 | 60 | 70 |
|---|---|---|---|---|---|
| 0∞ | 14,550 (28.2) | 10,050 (35) | 6,020 (45) | 4,030 (55) | 2,590 (65) |

NOTE: ( ) Reference radii in feet. A6-829-100837B

RT600E

THIS CHART IS ONLY A GUIDE AND SHOULD NOT BE USED TO OPERATE THE CRANE. The individual crane's load chart, operating instructions and other instructional plates must be read and understood prior to operating the crane.

GROVE.

# load handling

11

## Weight Reductions for Load Handling Devices

| 29 Ft. Offsettable Boom Extension | Pounds |
|---|---|
| *Erected – | 4,412 |

| 29 Ft. 51 ft. Tele. Boom Extension | Pounds |
|---|---|
| *Erected (Retracted) – | 6,611 |
| *Erected (Extended) – | 9,332 |

*Reduction of main boom capacities

| Auxiliary Boom Nose | Pounds |
|---|---|
| | 137 |

| Hookblocks and Headache Balls | Pounds |
|---|---|
| 50 Ton, 4 Sheave | 1075 |
| 50 Ton, 3 Sheave | 1000 |
| 40 Ton, 3 Sheave | 800 |
| 8.3 Ton Headache Ball (non-swivel) | 350 |
| 8.3 Ton Headache Ball (swivel)* | 370 |

+Refer to rating plate for actual weight.

When lifting over swingaway and/or jib combinations, deduct total weight of all load handling devices reeved over main boom nose directly from swingaway or jib capacity.

NOTE: All load handling devices and boom attachments are considered part of the load and suitable allowances MUST BE MADE for their combined weights. Weights are for Grove furnished equipment.

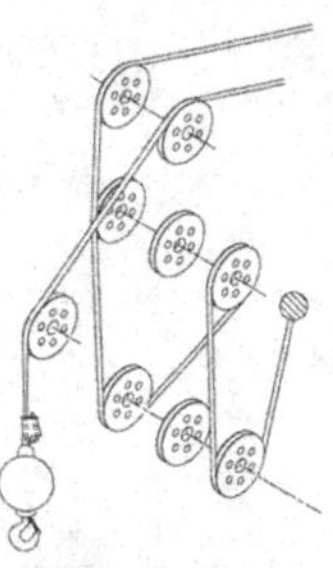

## Line Pulls and Reeving Information

| Hoists | Cable Specs | Permissible Line Pulls | Nominal Cable Length |
|---|---|---|---|
| Main | 3/4" (19 mm) 6x37 Class EIPS, IWRC Special Flexible Min. Breaking Str. 58,800 lb. | 16,800 lb. | 450 ft. |
| Main & Aux. | 3/4" (19 mm) Flex - X 35 Rotation Resistance (non-rotating) Min. Breaking Strength 85,500 lb. | 16,800 lb. | 450 ft. |

## Hoist Performance

| Wire Rope Layer | Hoist Line Pulls Two Speed Hoist Low Available lb.* | High Available lb.* | Drum Rope Capacity (ft.) Layer | Total |
|---|---|---|---|---|
| 1 | 18,134 | 9,067 | 78 | 78 |
| 2 | 16,668 | 8,334 | 85 | 164 |
| 3 | 15,420 | 7,710 | 92 | 256 |
| 4 | 14,347 | 7,174 | 99 | 356 |
| 5 | 13,413 | 6,707 | 106 | 462 |
| 6 | 12,594 | 6,297 | 113 | 575 |

*Max. lifting capacity: 6x37 or 35x7 class = 16,800 lb.

## Working Area Diagram

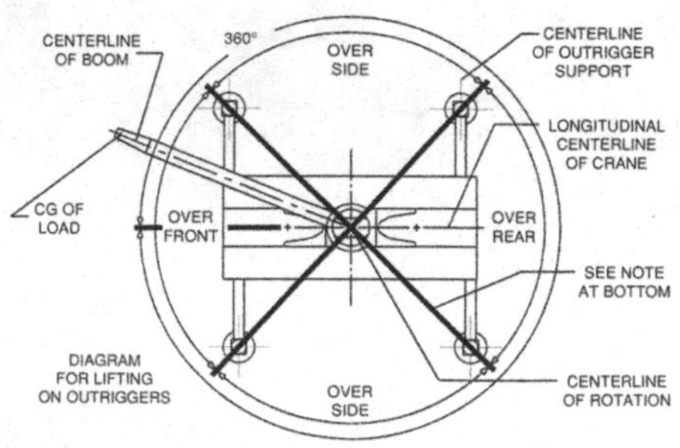


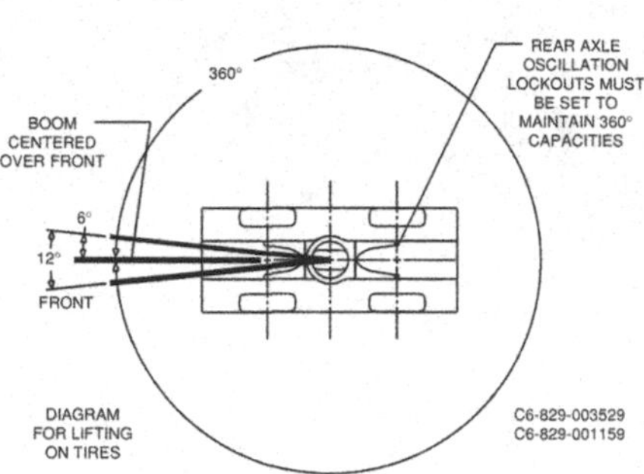


**Bold lines determine the limiting position of any load for operation within working areas indicated.**

THIS CHART IS ONLY A GUIDE AND SHOULD NOT BE USED TO OPERATE THE CRANE. The individual crane's load chart, operating instructions and other instructional plates must be read and understood prior to operating the crane.

RT600E

GROVE.

**GROVE**

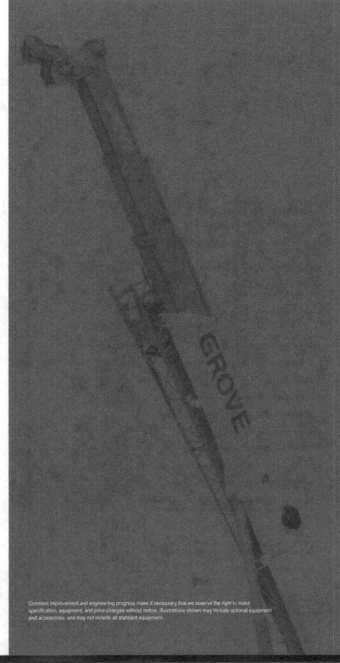

**Manitowoc Crane Group - Americas**
Manitowoc, Wisconsin Facility
Tel: [Int + 001] 920 684 6621
Fax: [Int + 001] 920 683 6277
Shady Grove, Pennsylvania Facility
Tel: [Int + 001] 717 597 8121
Fax: [Int + 001] 717 597 4062

**Manitowoc Crane Group - EMEA**
Europe Middle East & Africa
Tel: [Int + 33] (0) 4 72 18 20 20
Fax: [Int + 33] (0) 4 72 18 20 00

**Manitowoc Crane Group - UK**
Europe Middle East & Africa (Parts & Service)
Tel: [Int + 44] (0) 191 565-6281
Fax: [Int + 44] (0) 191 564-0442

**Manitowoc Crane Group - Germany**
(Sales, Parts & Service)
Tel: [Int + 49] (0) 2173 8909-0
Fax: [Int + 49] (0) 2173 8909-30

**Manitowoc Crane Group - France**
France & Africa (Sales, Parts & Service)
Tel: [Int + 33] (0) 1 303-13150
Fax: [Int + 33] (0) 1 303-86085

**Manitowoc Crane Group - Netherlands**
(Sales, Parts & Service)
Tel: [Int + 31] (0) 76 578 39 99
Fax: [Int + 31] (0) 76 578 39 78

**Manitowoc Crane Group - Italy**
Italy & Southern Europe (Sales, Parts & Service)
Tel: [Int + 39] (0) 331 49 33 11
Fax: [Int + 39] (0) 331 49 33 30

**Manitowoc Crane Group - Portugal**
Portugal & Spain (Sales, Parts & Service)
Tel: [Int + 351] (0) 22 968 08 89
Fax: [Int + 351] (0) 22 968 08 97

**Manitowoc Crane Group - Singapore**
Asia/Pacific excl China (Sales, Parts & Service)
Tel: [Int + 65] 6861-7133
Fax: [Int + 65] 6862-4040 / 4142

**Manitowoc Crane Group - Shanghai**
China (Sales, Parts & Service)
Tel: [Int + 86] (0) 21-64955555
Fax: [Int + 86] (0) 21-64852038

**Manitowoc Crane Group - Beijing**
China (Sales, Parts & Service)
Tel: [Int + 86] (0) 10 646-71690
Fax: [Int + 86] (0) 10 646-71691

**Manitowoc Crane Group - Middle East**
(Sales)
Tel: [Int + 971] (0) 4 348-4478
Fax: [Int + 971] (0) 4 348-4478
(Parts & Service)
Tel: [Int + 973] (0) 9 660-899
Fax: [Int + 973] (0) 2 707-740

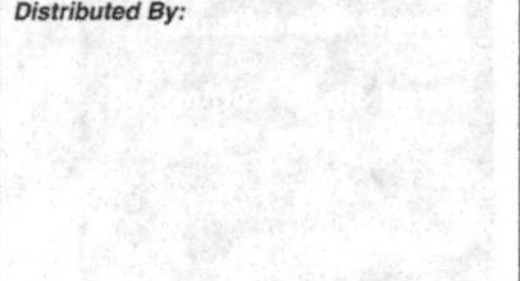

www.manitowoccranegroup.com



0704-2M Printed in USA Form No RT600E PG Part No. 03-473 Manitowoc Crane Group 2004

**product guide**

## features

- 181 mton (200 ton) capacity
- 307 mton-m (2,224 kips) Maximum Load Moment
- 82,3 m (270') Heavy-Lift Boom
- 91,4 m (300') Fixed Jib on Heavy-Lift Boom
- 106,7 m (350') Luffing Jib on Heavy-Lift Boom
- 125,0 m (410') Fixed Jib on Luffing Jib on Heavy-Lift Boom
- 253 kW (340 HP) engine standard
- EPIC® controls
- 135 m/min (442 fpm) line speed standard
- 205 m/min (674 fpm) line speed optional
- 131 kN (29,500 lb) line pull standard
- 196 kN (44,000 lb) line pull optional
- 13 380 kg (29,500 lb) Clamshell capacity
- 9 072 kg (20,000 lb) Dragline capacity
- Hydraulic-cylinder Boom Hoist for maximum performance with minimum maintenance
- Fast, efficient self-assembly
- Complete crane, maximum boom, fixed jib and counterweights ship on only 8 trucks
- Manitowoc CraneCARE®

## contents

# index

2

| | Specifications | Outline Dimensions/Weights | Wire Rope Lengths | Wire Rope Specifications | Boom Raising | Boom Combinations | Range Diagram/Capacity Charts |
|---|---|---|---|---|---|---|---|
| **No. 777 Basic Crane** | 3-5 | 6-7 | | | | | |
| Engine(s) | 3 | | | | | | |
| Hydraulic System | 3 | | | | | | |
| Drums & Laggings | 3, 17-18 | | | | | | |
| Swing System | 4 | | | | | | |
| Mast | 4 | | | | | | |
| Counterweights | 4 | 7-8 | | | | | |
| Operator's Cab | 4 | 7 | | | | | |
| Carbody | 5 | 7 | | | | | |
| Crawlers | 5 | 7 | | | | | |
| **Optional Equipment** | | | | | | | |
| Boom point | 4 | 10 | | | | | |
| Hook block/weight ball | 5 | 10 | | | | | |
| Miscellaneous | 5 | 10 | | | | | |
| Working Weights | 19 | | | | | | |
| Transport/Assembly | 11-12 | | | | | | |
| CraneCARE℠ | 35-36 | | | | | | |
| **No. 78 Boom** | | | | | | | |
| **Heavy-Lift No. 78** | 4 | 8 | 13 | 16 | 19 | 20 | 22-23 |
| Heavy-Lift No. 44 with **No. 134 Fixed Jib** | 4 | 8-9 | 13 | 16 | 19 | 20 | 24-26 |
| Heavy-Lift No. 44 with **No. 139 Luffing Jib** | 5 | 9-10 | 14-15 | 16 | 19 | 21 | 27-30 |
| Heavy-Lift No. 44 with **No. 138 Fixed Jib** on No. 139 Luffing Jib | 4-5 | 10 | 15 | 16 | 19 | 21 | 31-34 |
| **Clamshell** | 5 | | | | | | |

# specifications

## Upperworks

### Engine

Cummins Model QSC8.3 – C340 diesel, 6 cylinder, 253 kW (340 BHP) @ 2100 governed RPM.

- Optional: Caterpillar Model C-9 diesel, 6 cylinder, 253 kW (340 BHP) @ 2100 governed RPM.

Includes engine block heater (120 V), ether starting aid, manually operated disconnect clutch for cold weather starting, high silencing muffler, hydraulic oil cooler, radiator and fan.

Multiple hydraulic pump drive transmission provides independent power for all machine functions.

Two 12 volt maintenance-free, Group 8D batteries, 1400 CCA at -18°C (0° F), 24 volt starting and 120 amp alternator.

One 469 l (124 gal) capacity diesel fuel tank, mounted on rear of upperworks, with level indicator in operator's cab.

### Controls

Modulating electronic-over-hydraulic controls provide infinite speed response directly proportional to control lever movement. Controls include Manitowoc's exclusive EPIC® ***Electronically Processed Independent Control*** system providing microprocessor driven control logic, pump control, on-board diagnostics, and service information.

Block-up limit control is standard for hoist and auxiliary lines.

Integrated Load Moment Indicator system (LMI) is standard for main boom. "Function cut-out" or "warning only" operation is selected via a keyed switch on the LMI console. Includes travel and swing alarms.

- Optional: Upper boom point assembly with LMI.
- Optional: Anemometer (wind indicator).
- Optional: Foot controls for travel.

### Hydraulic System

Six high-pressure piston pumps are driven through a multi-hydraulic pump transmission. These six pumps provide independent "closed loop" hydraulic power for front drum, rear drum, boom hoist system, swing system, and both left and right crawler operation.

| System | kg/cm2 (psi) | lpm (gpm) |
|---|---|---|
| Front Drum | 422 (6,000) | 300 (79) |
| Rear Drum | 422 (6,000) | 300 (79) |
| Boom Hoist | 422 (6,000) | 300 (79) |
| Swing System | 386 (5,500) | 225 (59) |
| Left Crawler | 422 (6,000) | 225 (59) |
| Right Crawler | 422 (6,000) | 225 (59) |
| Auxiliary Pump* | 422 (6,000) | 390 (102) |

*Optional pump powers auxiliary drum on liftcrane.

Hydraulic reservoir capacity is 469 l (124 gal) and is equipped with breather, dipstick, clean out access, and internal diffuser.

Each function is equipped with relief valves to protect the hydraulic circuit from overload or shock.

Replaceable, spin on ten micron (absolute) full flow line filter is furnished in the hydraulic circuit. All oil is filtered prior to return to the hydraulic reservoir.

Hydraulic system also includes pump transmission disconnect clutch & hydraulic oil cooler.

### Drums

Two equal width winches 770 mm (30-1/8") wide and 495 mm (19-1/2") diameter are driven by independent variable displacement axial piston hydraulic motors through planetary reduction mounted on separate front and rear shafts with anti-friction bearings.

Powered hoisting/lowering operation is standard with automatic (spring applied, hydraulically released) multi-disc brakes, and drum rotation indicators.

- Optional: Free-fall operation for front and/or rear drum(s). External contracting band brake mounted on drum manually applied by foot pedal with locking latch in operator's cab. Operator may select free-fall or powered lowering mode using a selector switch.
- Optional: High line speed drum 205m/min (674 fpm) can be ordered in place of standard front or rear drum.
- Optional: High line pull drum 195 kN (44,000 lb) can be ordered in place of standard front or rear drum.
- Optional: Auxiliary (third) hydraulic powered drum rated at 89,0 kN (20,000 lb) line pull mounted in boom butt. Includes third drum control system.
- Optional: Auxiliary drum preparation includes electric wiring, controls, hydraulic pump and plumbing.
- Optional: Bolt-on liftcrane/clamshell laggings.
- Optional: Wire rope for various applications.

3

model 777

Manitowoc

# specifications

### Mast

4 Moving mast is 7,9 m (26') long and connects the boom hoist cylinders to the boom-support pendants.

Spring cushioned boom stop and automatic boom stop standard.

### Boom Hoist

Independent boom hoist is provided by two double-acting hydraulic cylinders connected to the mast. Boom hoist provides full range of boom angles from horizontal to 88 degrees, with or without load.

Boom hoist speed: raise 82,3 m (270') full main boom from 0°- 82° in 1 minute, 40 seconds.

### Swing System

High strength fabricated steel rotating bed is mounted on 2,15 m (84-1/2") diameter turntable single-row ball bearing.

Independent swing powered by a fixed displacement hydraulic motor coupled to a planetary gearbox with internal brake. 360° positive swing lock.

Swing system maximum speed: 2.7 rpm.

### Counterweight

| QTY. | ITEM | UNIT WEIGHT kg | UNIT WEIGHT lb | TOTAL WEIGHT kg | TOTAL WEIGHT lb |
|---|---|---|---|---|---|
| | **Series 1** | | | | |
| 1 | Tray | 15 876 | 35,000 | 15 876 | 35,000 |
| 4 | Upper Side Box | 7 938 | 17,500 | 31 752 | 70,000 |
| | | **SERIES 1 TOTAL** | | **47 628** | **105,000** |
| | **Series 2** | | | | |
| 2 | Upper Side Box | 7 938 | 17,500 | 15 876 | 35,000 |
| 2 | Upper Side Plate | 454 | 1,000 | 908 | 2,000 |
| 2 | Carbody Box | 9 979 | 22,000 | 19 958 | 44,000 |
| | | **SERIES 2 SUB-TOTAL** | | **36 742** | **81,000** |
| Optional: Add to **SERIES 1** for **SERIES 2 TOTAL** | | | | **84 370** | **186,000** |

Includes connecting pins, brackets, and stops.

### Operator's Cab

Fully enclosed and insulated steel module located at the left front corner of rotating bed. Module is equipped with sliding door, large safety glass windows on all sides and roof. Signal horn, cab space heater, front and roof windshield wipers, dome light, sun visor and shade, fire extinguisher and air circulating fan are standard.

- Optional: Air conditioner.
- Optional: Nylon protective window covers.

## Attachments

### No. 78 Heavy-Lift Main Boom

The liftcrane is equipped with a 18,3 m (60') No. 78 basic heavy-lift angle chord boom consisting of a 6,9 m (22' 6") butt and 11,4 m (37' 6") top with six 76,2 cm (30") diameter roller bearing sheaves on one shaft. Includes rope guides, and boom angle indicator. The No. 78 boom utilizes pendant rigging Manitowoc's patented, exclusive FACT™ connection system. The FACT connection system consists of two vertical pins, two horizontal connection pins and alignment pads for each boom connection location.

Luffing jib preparation included as standard.

- Optional: 3,05 m (10'), 6,1 m (20'), and 12,2 m (40') No. 78 boom inserts with pendant rigging and FACT™ connection system.
- Optional: No. 78 detachable upper boom point with one 76,2 cm (30") diameter tapered roller bearing steel sheave grooved for 1-1/8" rope with rope guard for liftcrane.

### No. 134 Fixed Jib

- Optional: No. 134 basic fixed jib 9,1 m (30') length consisting of 4,6 m (15') jib butt and 4,6 m (15') jib top with 3,7 m (12') jib strut, pendants and backstay. Includes LMI hardware.
- Optional: No. 134 fixed jib inserts 3,05 m (10') and 6,1 m (20') with pendants.

Utilize fixed jib inserts in combination with the No. 134 basic fixed jib for total lengths up to 24,4 m (80').

Note: Jib lengths greater than 18,3 m (60') require the use of at least one 6,1 m (20') No. 134 fixed jib insert.

### No. 138 Fixed Jib

- Optional: No. 138 basic fixed jib 9,1 m (30') length consisting of 4,6 m (15') jib butt and 4,6 m (15') jib top with 5,4 m (17' 9-7/16") jib strut, pendants and backstay. For use with No. 139 Luffing Jib.

# specifications

Optional: No. 138 fixed jib inserts 3,0 m (10') and 6,1 m (20') with pendants.

Utilize fixed jib inserts in combination with the No. 138 basic fixed jib for total lengths up to 18,3 m (60').

### No. 139 Luffing Jib

Optional: 21,3 m (70') basic No. 139 luffing jib assembly with LMI hardware consisting of 8,2 m (27') butt, 6,1 m (20') insert and 7,0 m (23') top with two 68,6 cm (27") straight roller bearing sheaves, pin connected jib sections, pendant rigging, fixed strut, jib strut, backstay straps, jib point wheel, 26 mm or (1") luffing jib hoist line, and 54 cm (21-1/4") grooved lagging for rear luffing drum.

Optional: 3,05 m (10'), 6,1 m (20'), and 12,2 m (40') No. 139 inserts with steel boom suspension straps.

## Lowerworks

### Carbody

Connects rotating bed to crawler assemblies. High strength fabricated steel assembly with FACT™ connection system for fast installation and removal of crawler assemblies.

### Crawlers

Crawler assemblies are 7,55 m (24' 9-1/4") long with 97 cm (38") wide cast steel crawler pads and sealed "low maintenance" intermediate rollers. Each crawler is powered independently by a variable displacement hydraulic motor and includes two hydraulically powered pin actuators for fast installation and removal from carbody. Carbody mounted drive motors are connected to crawler final reduction via telescoping shafts. This permits crawlers to be removed without opening their hydraulic circuits. Crawlers provide ample tractive effort that allows counter rotation with full rated load. Maximum ground speed of 1,7 kph (1.05 mph).

Optional: Self-Erect system includes: carbody jacking cylinders with pads, controls, 41 mton (45 ton) assembly block, boom-butt installation support, and crawler handling chains.

Optional: 122 cm (48") wide-cast steel crawler pads.

## Optional Equipment

Optional: Blocks and Hooks, each with 762 mm (30") roller-bearing sheaves for 26 mm or (1") wire rope, a roller-bearing swivel hook, a hook latch, and a swivel lock. 5

13,6 mton (15 ton) swivel hook and weight ball.

41,0 mton (45 ton) hook block with one 76,2 cm (30") sheave (assembly block).

54,4 mton (60 ton) hook block with two 76,2 cm (30") sheaves.

90,7 mton (100 ton) hook block with three 76,2 cm (30") sheaves .

160 mton (175 ton) hook block with six 76,2 cm (30") sheaves.

- Optional: Wire rope for various applications.
- Optional: Equipment and testing for special code compliance.
- Optional: Hydraulic Test Kit: required to properly analyze the performance of the EPIC® control system.
- Optional: Service Interval Kits: for the regularly scheduled maintenance of general crane operations.
- Optional: Lighting Packages: consult dealer for available options.
- Optional: Special paint colors other than Manitowoc standard red and black.
- Optional: Custom vinyl decals of customer name and/or logo from artwork supplied by customer.
- Optional: Export Packaging: basic crane, boom and jib sections.

## Optional Application

Optional: For limited clamshell work: guide bars for lower boom point; Rud-O-Matic® No. 1866 spring-powered three-barrel tagline with 76,2 cm (36") diameter wheel; and pressure rollers for the front and rear hoisting drums. Front drum is closing line. Rear drum is holding line. Manitowoc's EPIC® controls can be changed from liftcrane to clamshell mode with the flip of a switch.

model 777

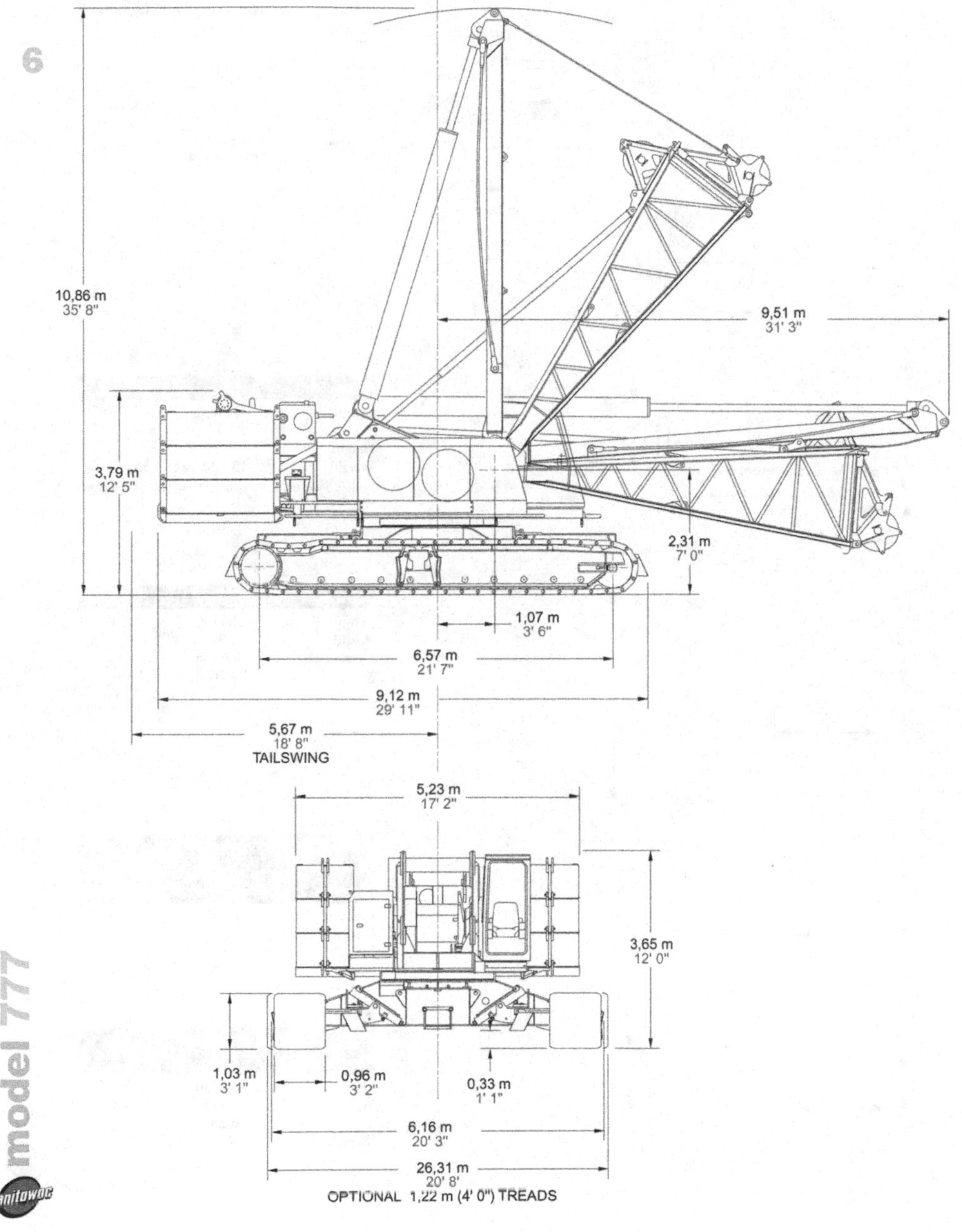
outline dimensions
6
10,86 m
35' 8"
9,51 m
31' 3"
3,79 m
12' 5"
2,31 m
7' 0"
1,07 m
3' 6"
6,57 m
21' 7"
9,12 m
29' 11"
5,67 m
18' 8"
TAILSWING
5,23 m
17' 2"
3,65 m
12' 0"
1,03 m
3' 1"
0,96 m
3' 2"
0,33 m
1' 1"
6,16 m
20' 3"
26,31 m
20' 8'
OPTIONAL 1,22 m (4' 0") TREADS
model 777
Manitowoc

# outline dimensions

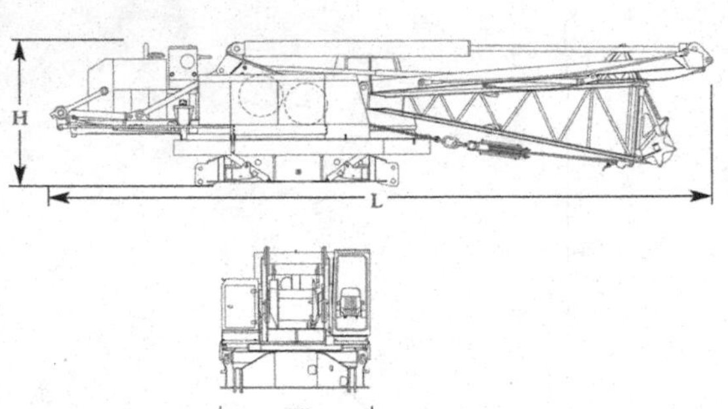


| Upperworks Module | | x 1 |
|---|---|---|
| Length | 15,17 m | 49' 9" |
| Width | 3,42 m | 11' 1" |
| Height | 3,36 m | 11' 0" |
| Weight | 39 065 kg | 86,125 lb |

Note: Weight includes carbody, upperworks, two full power drums with maximum length hoist and whip lines, diesel powerplant, operator's cab, boom hoist cylinders, mast, boom butt with integral wire rope guide, optional self assembly jacks, full hydraulic fluid reservoir, and half tank of fuel.

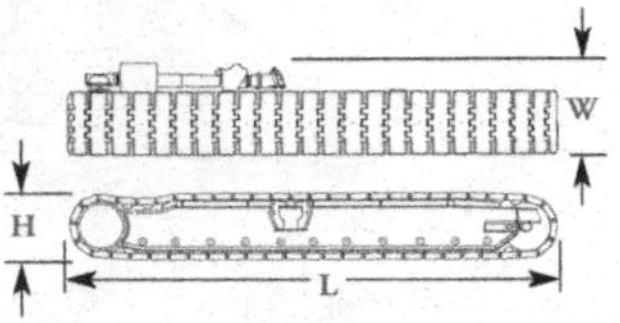


| Crawlers | | x 2 |
|---|---|---|
| Length | 7,56 m | 24' 10" |
| Width | 1,47 m | 4' 10" |
| Height | 1,04 m | 3' 5" |
| Weight | 15 320 kg | 33,775 lb |

Note: 970 mm (38") wide pad, 12 795 kg (28,210 lb) weight crawlers available.

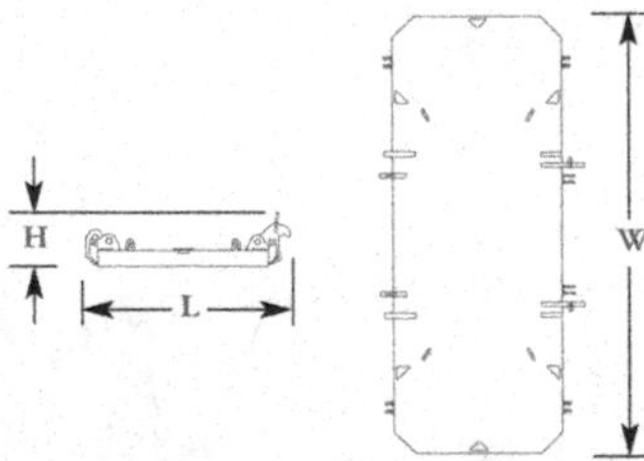


| Counterweight Tray | | x 1 |
|---|---|---|
| Length | 2,44 m | 8' 0" |
| Width | 6,99 m | 17' 1" |
| Height | 0,64 m | 2' 1" |
| Weight | 15 876 kg | 35,000 lb |

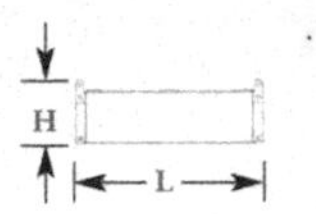


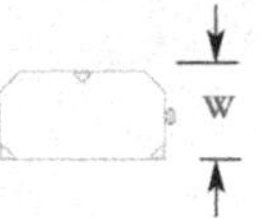


| Side Counterweight | | |
|---|---|---|
| Series 1 | | x 4 |
| Series 2 | | x 6 |
| Length | 2,24 m | 7' 4" |
| Width | 1,07 m | 3' 6" |
| Height | 0,69 m | 2' 3" |
| Weight | 7 938 kg | 17,500 lb |

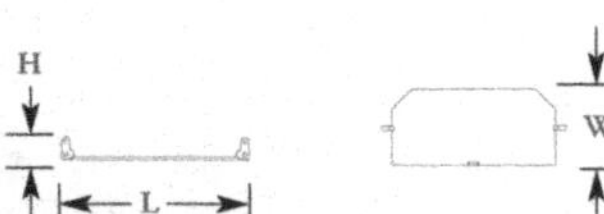


| Side Counterweight Plate | | |
|---|---|---|
| Series 2 | | x 2 |
| Length | 2,24 m | 7' 4" |
| Width | 0,90 m | 3' 0" |
| Height | 0,28 m | 0' 11" |
| Weight | 454 kg | 1,000 lb |

Option

# outline dimensions

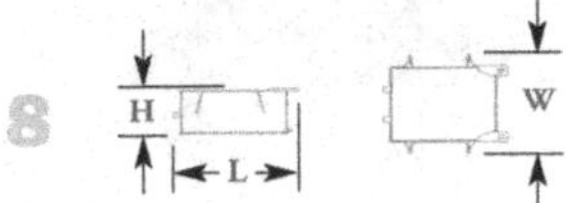


| Carbody Center Series 2 | | x 2 |
|---|---|---|
| Length | 1,90 m | 6' 3" |
| Width | 1,57 m | 5' 2" |
| Height | 0,72 m | 2' 4" |
| Weight | 9 979 kg | 22,000 lb |

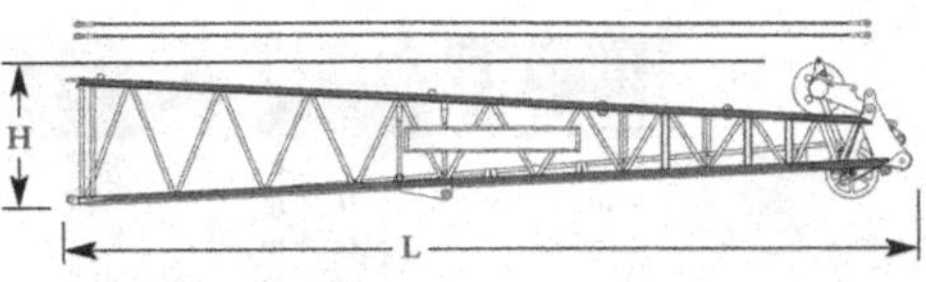


| No. 78 Boom Top 11,4 m (37' 6") & Wire Rope Guides, Pendants, Lower Point | | x 1 |
|---|---|---|
| Length | 12,38 m | 40' 8" |
| Width | 2,24 m | 7' 4" |
| Height | 2,04 m | 6' 11" |
| Weight | 3 567 kg | 7,865 lb |

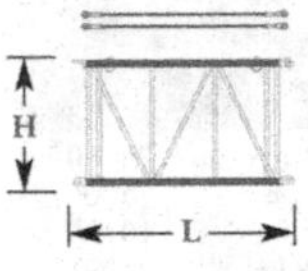


| No. 78 Boom Insert 3,0 m (10') & Pendants | | x 1, 2 |
|---|---|---|
| Length | 3,20 m | 10' 6" |
| Width | 2,24 m | 7' 4" |
| Height | 1,89 m | 6' 3" |
| Weight | 722 kg | 1,595 lb |

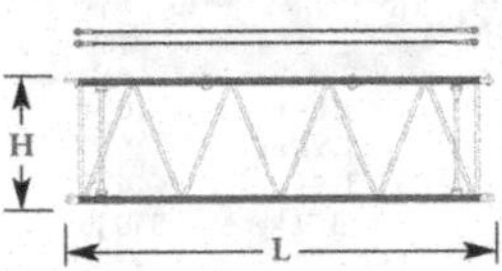


| No. 78 Boom Insert 6,1 m (20') & Pendants | | x 1, 2 |
|---|---|---|
| Length | 6,25 m | 20' 6" |
| Width | 2,24 m | 7' 4" |
| Height | 1,89 m | 6' 3" |
| Weight | 1 129 kg | 2,495 lb |

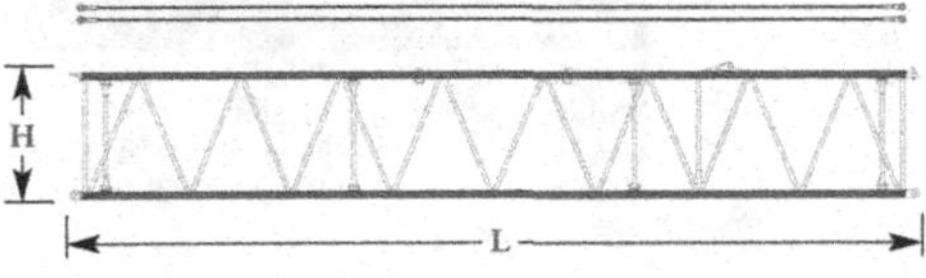


| No. 78 Boom Insert 12,3 m (40') & Pendants | | x 1, 2, 3, 4 |
|---|---|---|
| Length | 12,34 m | 40' 6" |
| Width | 2,24 m | 7' 4" |
| Height | 1,89 m | 6' 3" |
| Weight | 1 994 kg | 4,290 lb |

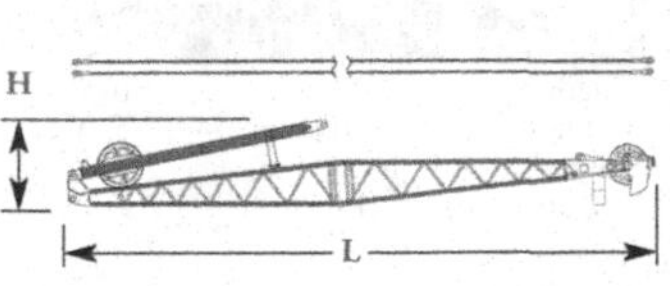


| No. 134 Fixed Jib 9,1 m (30') & Strut, Pendants | | x 1 |
|---|---|---|
| Length | 9,60 m | 31' 6" |
| Width | 0,86 m | 2' 10" |
| Height | 1,29 m | 4' 3" |
| Weight | 1 188 kg | 2,620 lb |

■ Option

model 777

Manitowoc

# outline dimensions

9

**No. 134 Jib Insert 3,0 m (10') & Pendants x 1, 2, 3**

| | | |
|---|---|---|
| Length | 3,12 m | 10' 3" |
| Width | 0,78 m | 2' 7" |
| Height | 0,78 m | 2' 7" |
| Weight | 213 kg | 480 lb |

**No. 134 Jib Insert 6,1 m (20') & Pendants x 1, 2**

| | | |
|---|---|---|
| Length | 6,16 m | 20' 3" |
| Width | 0,78 m | 2' 7" |
| Height | 0,78 m | 2' 7" |
| Weight | 339 kg | 750 lb |

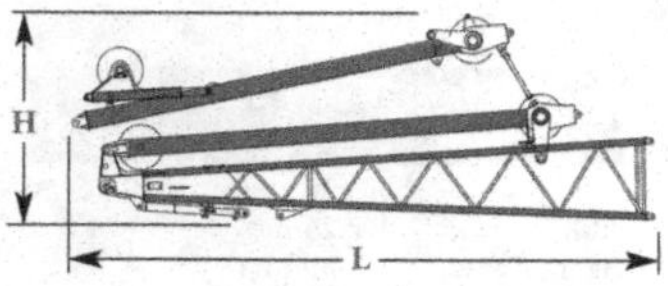


**No. 139 Luffing Jib 4,6 m (27') Butt & Main Strut, Jib Strut, Guides, Pendants x 1**

| | | |
|---|---|---|
| Length | 8,78 m | 28' 10" |
| Width | 1,51 m | 5' 0" |
| Height | 3,14 m | 10' 3" |
| Weight | 5 056 kg | 10,460 lb |

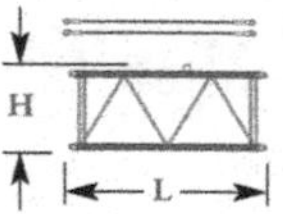


**No. 139 Jib Insert 3,0 m (10') & Pendants x 1**

| | | |
|---|---|---|
| Length | 3,15 m | 10' 4" |
| Width | 1,51 m | 5' 0" |
| Height | 1,34 m | 4' 5" |
| Weight | 379 kg | 840 lb |

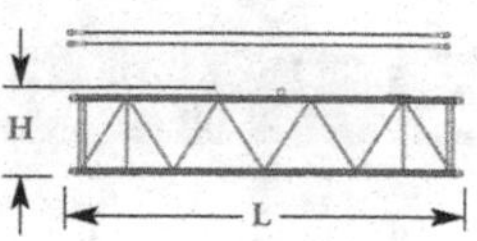


**No. 139 Jib Insert 6,1 m (20') & Pendants x 1, 2**

| | | |
|---|---|---|
| Length | 6,20 m | 20' 4" |
| Width | 1,51 m | 5' 0" |
| Height | 1,34 m | 4' 5" |
| Weight | 610 kg | 1,350 lb |

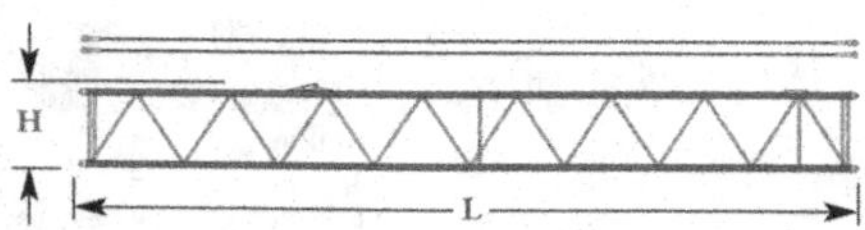


**No. 139 Jib Insert 12,3 m (40') & Pendants x 1, 2, 3**

| | | |
|---|---|---|
| Length | 12,40 m | 40' 4" |
| Width | 1,51 m | 5' 0" |
| Height | 1,34 m | 4' 5" |
| Weight | 1 048 kg | 2,315 lb |

Manitowoc

Option

# outline dimensions

10

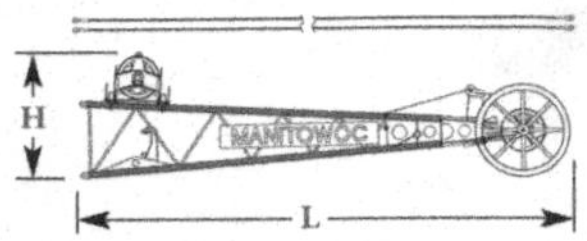


**No. 139 Jib Top 7,0 m (23') & Pendants x 1**

| | | |
|---|---|---|
| Length | 7,83 m | 25' 8" |
| Width | 1,51 m | 5' 0" |
| Height | 2,06 m | 6' 9" |
| Weight | 2 226 kg | 4,915 lb |

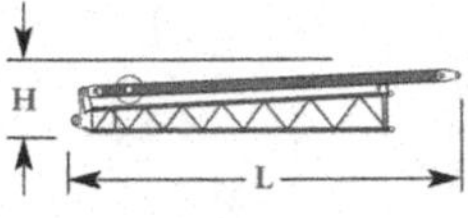


**No. 138 Fixed Jib 4,6 m (15') Butt & Strut x 1**

| | | |
|---|---|---|
| Length | 5,64 m | 18' 6" |
| Width | 0,75 m | 2' 6" |
| Height | 0,91 m | 3' 0" |
| Weight | 350 kg | 775 lb |

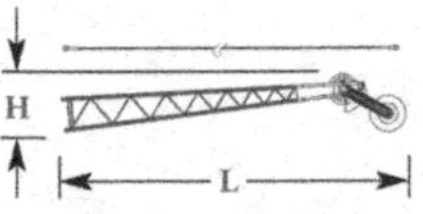


**No. 138 Fixed Jib 4,6 m (15') Top & Roller, Pendant x 1**

| | | |
|---|---|---|
| Length | 5,51 m | 18' 1" |
| Width | 0,76 m | 2' 6" |
| Height | 1,05 m | 3' 5" |
| Weight | 351 kg | 773 lb |

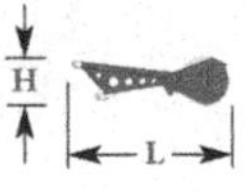


**No. 138 Jib Insert 3,0 m (10') & Pendants x 1, 2, 3**

| | | |
|---|---|---|
| Length | 3,12 m | 10' 3" |
| Width | 0,76 m | 2' 6" |
| Height | 0,58 m | 1' 11" |
| Weight | 98 kg | 215 lb |

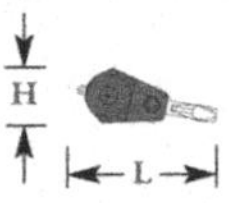


**No. 78 Upper Boom Point x 1**

| | | |
|---|---|---|
| Length | 2,64 m | 8' 8" |
| Width | 0,41 m | 1' 4" |
| Height | 0,81 m | 2' 8" |
| Weight | 420 kg | 925 lb |

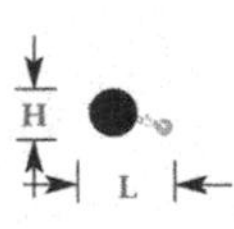


**Hook block for 26 mm or (1") wire rope**

| | | | | | |
|---|---|---|---|---|---|
| Capacity | 160 mt | 175 t | Length | 1,68 m | 5' 6" |
| Weight | 2 268 kg | 5,000 lb | Width | 1,29 m | 4' 0" |

**Weight Ball**

| | | | | | |
|---|---|---|---|---|---|
| Capacity/Swivel | 13,6 mt | 15 t | Diameter | 0,56 m | 1' 10" |
| Weight | 594 kg | 1 310 lb | Length | 1,07 m | 3' 6" |

■ Option

# transport data

11

## Trailer Load Out Summary

Model 777 Series 2
No. 78 Boom 82,3 m (270') and
No. 134 Fixed Jib 24,4 m (80')

| Item | Weight each Item kg (lb) | Quantity on Trailer Load # 1 | 2 | 3 | 4 | 5 | 6 | 7 | 8 |
|---|---|---|---|---|---|---|---|---|---|
| Upperworks, Carbody with No. 78 Boom Butt | 39 065 (86,125) | 1 | | | | | | | |
| Crawler Assembly | 15 320 (33,775) | | 1 | 1 | | | | | |
| Counterweight Tray | 15 876 (35,000) | | | | 1 | | | | |
| Upper Side Counterweight | 7 938 (17,500) | | | | | 2 | 2 | 1 | 1 |
| Upper Side Counterweight Plate | 454 (1,000) | | | | | | | 1 | 1 |
| Carbody Counterweight | 9 979 (22,000) | | | | | | | 1 | 1 |
| No. 78 Mast Top, Pendants | 3 567 (7,865) | | | | 1 | | | | |
| 3,0 m (10') No. 78 Mast Insert & Pendants | 722 (1,595) | | | | | | | | 1 |
| 6,1 m (20') No. 78 Mast Insert & Pendants | 1 129 (2,495) | | | | | | | 2 | |
| 12,2 m (40') No. 78 Mast Insert & Pendants | 1 944 (4,290) | | 1 | 1 | | 1 | 1 | | |
| 9,1 m (30') No. 134 Jib & Pendants | 1 586 (3,515) | | | | | | | | 1 |
| 3,0 m (10') No. 134 Jib Insert & Pendants | 216 (480) | | | | | | 1 | | |
| 6,1 m (20') No. 134 Jib Insert & Pendants | 339 (750) | | | | | 1 | 1 | | |
| 160-mton (175-ton) Hook Block | 2 268 (5,000) | | 1 | | | | | | |
| Miscellaneous* | | | | | | | | | |
| Approximate Total Shipping Weight Each Trailer Load kg (lb) | | 39 065 (86,125) | 19 534 (43,065) | 17 264 (38,065) | 19 443 (42,865) | 18 159 (40,040) | 18 375 (40,520) | 20 629 (45,490) | 20 679 (45,610) |

Trailer configurations - double drop (#2); step deck (#'s 4-6); flat bed (#'s 7,8).

*Miscellaneous weights vary and are not itemized for individual trailer load totals.

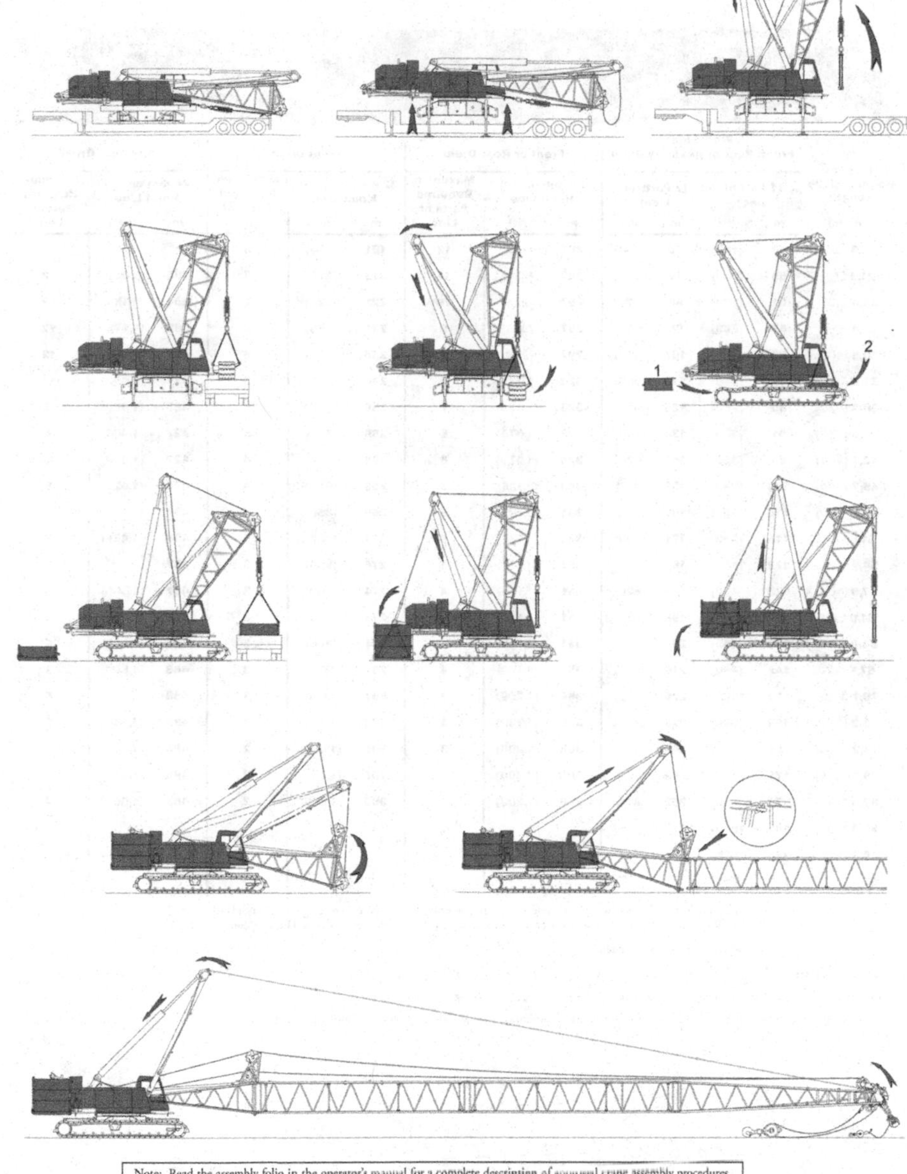
assembly
12
1
2
model 777
Manitowoc
Note: Read the assembly folio in the operator's manual for a complete description of approved crane assembly procedures.

# performance data

13

**Wire Rope Lengths**
**Boom No. 78**
**- or -**
**Fixed Jib No. 134 on Boom No. 78**

| Boom or Boom and Jib Length | Whip Line Front, Rear or Auxiliary Drum | | | | Front or Rear Drum | | | Front Drum | | | Auxiliary Drum | | |
|---|---|---|---|---|---|---|---|---|---|---|---|---|---|
| | (1 Part of Line) | | (2 Parts of Line) | | 26 mm or (1") Hoist Line | | Maximum Required Parts of Line | 32 mm or (1-1/4") Hoist Line | | Maximum Required Parts of Line | 26 mm or (1") Hoist Line | | Maximum Required Parts of Line |
| m (ft) | m | (ft) | m | (ft) | m | (ft) | | m | (ft) | | m | (ft) | |
| 18,3 (60) | 49 | (160) | 70 | (230) | 267 | (875) | 12 | 191 | (625) | 8 | – | – | – |
| 21,3 (70) | 55 | (180) | 79 | (260) | 282 | (925) | 11 | 213 | (700) | 8 | 305 | (1,000) | 12 |
| 24,4 (80) | 61 | (200) | 88 | (290) | 297 | (975) | 10 | 221 | (725) | 7 | 351 | (1,150) | 12 |
| 27,4 (90) | 67 | (220) | 98 | (320) | 297 | (975) | 9 | 221 | (725) | 6 | 389 | (1,275) | 12 |
| 30,5 (100) | 73 | (240) | 107 | (350) | 297 | (975) | 8 | 236 | (775) | 6 | 427 | (1,400) | 12 |
| 33,5 (110) | 79 | (260) | 116 | (380) | 328 | (1,075) | 8 | 236 | (775) | 5 | 427 | (1,400) | 11 |
| 36,6 (120) | 85 | (280) | 125 | (410) | 328 | (1,075) | 7 | 236 | (775) | 5 | 427 | (1,400) | 10 |
| 39,6 (130) | 91 | (300) | 134 | (440) | 328 | (1,075) | 6 | 259 | (850) | 5 | 427 | (1,400) | 9 |
| 42,7 (140) | 98 | (320) | 143 | (470) | 328 | (1,075) | 6 | 259 | (850) | 4 | 427 | (1,400) | 8 |
| 45,7 (150) | 104 | (340) | 152 | (500) | 343 | (1,125) | 6 | 259 | (850) | 4 | 434 | (1,425) | 8 |
| 48,8 (160) | 110 | (360) | 162 | (530) | 343 | (1,125) | 5 | 259 | (850) | 4 | 434 | (1,425) | 7 |
| 51,8 (170) | 116 | (380) | 171 | (560) | 343 | (1,125) | 5 | 274 | (900) | 4 | 434 | (1,425) | 7 |
| 54,9 (180) | 122 | (400) | 180 | (590) | 351 | (1,150) | 5 | 274 | (900) | 3 | 457 | (1,500) | 7 |
| 57,9 (190) | 128 | (420) | 189 | (620) | 351 | (1,150) | 4 | 274 | (900) | 3 | 457 | (1,500) | 6 |
| 61,0 (200) | 134 | (440) | 198 | (650) | 351 | (1,150) | 4 | 274 | (900) | 3 | 457 | (1,500) | 6 |
| 64,0 (210) | 140 | (460) | 207 | (680) | 351 | (1,150) | 4 | 274 | (900) | 3 | 465 | (1,525) | 6 |
| 67,1 (220) | 146 | (480) | 216 | (710) | 351 | (1,150) | 4 | 282 | (975) | 3 | 488 | (1,600) | 6 |
| 70,1 (230) | 152 | (500) | 226 | (740) | 366 | (1,200) | 3 | 297 | (1,000) | 3 | 488 | (1,600) | 5 |
| 73,2 (240) | 158 | (520) | 235 | (770) | 366 | (1,200) | 3 | 305 | (1,000) | 3 | 488 | (1,600) | 5 |
| 76,2 (250) | 165 | (540) | 244 | (800) | 366 | (1,200) | 3 | 305 | (1,000) | 2 | 488 | (1,600) | 5 |
| 79,2 (260) | 171 | (560) | 253 | (830) | 366 | (1,200) | 3 | 305 | (1,000) | 2 | 488 | (1,600) | 4 |
| 82,3 (270) | 177 | (580) | 262 | (860) | 366 | (1,200) | 3 | 305 | (1,000) | 2 | 488 | (1,600) | 4 |
| 85,3 (280) | 183 | (600) | 271 | (890) | | | | | | | | | |
| 88,4 (290) | 189 | (620) | | | | | | | | | | | |
| 91,4 (300) | 195 | (640) | | | | | | | | | | | |

**Note:** Line lengths are based on single part lead line. Hoist line and whip line lengths given in table will allow hook to touch ground. When block travel below ground is required, add additional rope equal to parts of line times added travel distance. Hoisting distance or line pull may be limited when block travel below ground is required.

Auxiliary Drum cannot be used with 18,3 m (60') boom length.

Maximum block travel with maximum parts of line may be restricted when hoist line length exceeds 213 m (700') using 32 mm (1-1/4") wire rope.

Maximum capacity with 2 part whip line using the 32 mm or (1-1/4") wire rope is 26 760 kg (59,000 lb).

Capacity chart restrictions will occur when auxiliary drum is used. Maximum capacity is 108 860 kg (240,000 lb) with 12 parts line.

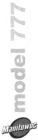

Manitowoc

# performance data

14

## Wire Rope Lengths
## Luffing Jib No. 139 on Boom No. 78

| Boom and Luffing Jib Length | Luffing Jib Hoist Line Front Drum | | | | | | | |
|---|---|---|---|---|---|---|---|---|
| | (4 Parts of Line) | | (3 Parts of Line) | | (2 Parts of Line) | | (1 Part of Line) | |
| m (ft) | m | (ft) | m | (ft) | m | (ft) | m | (ft) |
| 42,7 (140) | 229 | (750) | | | | | 95 | (310) |
| 45,7 (150) | 244 | (800) | | | | | 101 | (330) |
| 48,8 (160) | 259 | (850) | 213 | (700) | | | 107 | (350) |
| 51,8 (170) | 274 | (900) | 221 | (725) | | | 113 | (370) |
| 54,9 (180) | 290 | (950) | 236 | (775) | | | 119 | (390) |
| 57,9 (190) | 305 | (1,000) | 244 | (800) | | | 125 | (410) |
| 61,0 (200) | 320 | (1,050) | 259 | (850) | 198 | (650) | 131 | (430) |
| 64,0 (210) | 335 | (1,100) | 274 | (900) | 206 | (675) | 137 | (450) |
| 67,1 (220) | – | – | 282 | (925) | 213 | (700) | 143 | (470) |
| 70,1 (230) | – | – | 297 | (975) | 229 | (750) | 149 | (490) |
| 73,2 (240) | – | – | 305 | (1,000) | 236 | (775) | 155 | (510) |
| 76,2 (250) | – | – | 320 | (1,050) | 244 | (800) | 162 | (530) |
| 79,2 (260) | – | – | 335 | (1,100) | 251 | (825) | 168 | (550) |
| 82,3 (270) | – | – | 343 | (1,125) | 259 | (850) | 174 | (570) |
| 85,3 (280) | – | – | – | – | 267 | (875) | 180 | (590) |
| 88,4 (290) | – | – | – | – | 282 | (925) | 186 | (610) |
| 91,4 (300) | – | – | – | – | 290 | (950) | 192 | (630) |
| 94,5 (310) | – | – | – | – | 297 | (975) | 198 | (650) |
| 97,5 (320) | – | – | – | – | 305 | (1,000) | 204 | (670) |
| 100,6 (330) | – | – | – | – | 312 | (1,025) | 210 | (690) |
| 103,6 (340) | – | – | – | – | 320 | (1,050) | 216 | (710) |
| 106,7 (350) | – | – | – | – | – | – | 223 | (730) |

**Note:** Line lengths are based on single part lead line. Hoist line and whip line lengths given in table will allow hook to touch ground. When block travel below ground is required, add additional rope equal to parts of line times added travel distance. Hoisting distance or line pull may be limited when block travel below ground is required.

Capacity chart restrictions will occur when auxiliary drum is used. Maximum capacity is 36 290 kg (80,000 lb) with four parts of line.

# performance data

15

**Wire Rope Lengths**
**Luffing Jib No. 139 on Boom No. 78**

| Boom and Luffing Jib Length | Luffing Jib Hoist Line Auxiliary Drum | | | | | | | |
|---|---|---|---|---|---|---|---|---|
| | (4 Parts of Line) | | (3 Parts of Line) | | (2 Parts of Line) | | (1 Part of Line) | |
| m (ft) | m | (ft) | m | (ft) | m | (ft) | m | (ft) |
| 42,7 (140) | 229 | (750) | | | | | 95 | (310) |
| 45,7 (150) | 244 | (800) | | | | | 101 | (330) |
| 48,8 (160) | 259 | (850) | | | | | 107 | (350) |
| 51,8 (170) | 274 | (900) | | | | | 113 | (370) |
| 54,9 (180) | 290 | (950) | | | | | 119 | (390) |
| 57,9 (190) | 305 | (1,000) | | | | | 125 | (410) |
| 61,0 (200) | 320 | (1,050) | 259 | (850) | | | 131 | (430) |
| 64,0 (210) | 335 | (1,100) | 274 | (900) | | | 137 | (450) |
| 67,1 (220) | 351 | (1,150) | 282 | (925) | | | 143 | (470) |
| 70,1 (230) | 366 | (1,200) | 297 | (975) | 229 | (750) | 149 | (490) |
| 73,2 (240) | 381 | (1,250) | 305 | (1,000) | 236 | (775) | 155 | (510) |
| 76,2 (250) | 396 | (1,300) | 320 | (1,050) | 244 | (800) | 162 | (530) |
| 79,2 (260) | 411 | (1,350) | 335 | (1,100) | 251 | (825) | 168 | (550) |
| 82,3 (270) | 427 | (1,400) | 343 | (1,125) | 259 | (850) | 174 | (570) |
| 85,3 (280) | – | – | 358 | (1,175) | 267 | (875) | 180 | (590) |
| 88,4 (290) | – | – | 366 | (1,200) | 282 | (925) | 186 | (610) |
| 91,4 (300) | – | – | 373 | (1,225) | 290 | (950) | 192 | (630) |
| 94,5 (310) | – | – | – | – | 297 | (975) | 198 | (650) |
| 97,5 (320) | – | – | – | – | 305 | (1,000) | 204 | (670) |
| 100,6 (330) | – | – | – | – | 312 | (1,025) | 210 | (690) |
| 103,6 (340) | – | – | – | – | 320 | (1,050) | 216 | (710) |

**Note:** Line lengths are based on single part lead line. Hoist line and whip line lengths given in table will allow hook to touch ground. When block travel below ground is required, add additional rope equal to parts of line times added travel distance. Hoisting distance or line pull may be limited when block travel below ground is required.

Drums each provide 000 kN (00,000 lb) maximum single line pull.

**Wire Rope Lengths**
**Fixed Jib No. 138 on Luffing Jib No. 139 on Boom No. 80**

| Boom, Luffing Jib and Fixed Jib Length | Fixed Jib Whip Line Front Drum or Auxiliary Drum | |
|---|---|---|
| | (1 Part of Line) | |
| m (ft) | m | (ft) |
| 94,5 (310) | 198 | (650) |
| 97,5 (320) | 204 | (670) |
| 100,6 (330) | 210 | (690) |
| 103,6 (340) | 216 | (710) |
| 106,7 (350) | 223 | (730) |
| 109,7 (360) | 229 | (750) |
| 112,8 (370) | 235 | (770) |
| 115,8 (380) | 241 | (790) |
| 118,9 (390) | 247 | (810) |
| 121,9 (400) | 253 | (830) |
| 125,0 (410) | 259 | (850) |

**Note:** Line lengths are based on single part lead line. Hoist line and whip line lengths given in table will allow hook to touch ground. When block travel below ground is required, add additional rope equal to parts of line times added travel distance. Hoisting distance or line pull may be limited when block travel below ground is required.

Maximum load on 26 mm or (1") wire rope is 12 400 kg (27,500 lb) per line. When auxiliary drum is used maximum load is 9 070 kg (20,000 lb).

model 777

Manitowoc

# performance data

16

## Wire Rope Specifications 5:1 Safety Factor
Boom No. 78
- or -
Fixed Jib No. 134 on Boom No. 78
- or -
Luffing Jib No. 139 on Boom No. 78
- or -
Fixed Jib No. 138 on Luffing Jib No. 139 on Boom No. 78

5:1 Safety Factor
2 160 N/nm²

| Part Number / Function | No. 719379 Hoist Line Whip Line | No. 719378 Auxiliary Line | No. 719413* Hoist Line Whip Line (High Line Pull) | No. 719392 Hoist Line Whip Line | No. 719393 Auxiliary Line |
|---|---|---|---|---|---|
| Size Wire Rope | 26 mm – | 26 mm – | 32 mm – | – (1") | – (1") |
| Minimum Breaking Strength | 656,1 kN (147,500 lb) | 650,6 kN (146,200 lb) | 1 085,3 kN (244,000 lb) | 684,2 kN (153,800 lb) | 620,9 kN (139,600 lb) |
| Maximum Load Per Line | 13 380 kg (29,500 lb) | 9 070 kg (20,000 lb) | 19 960 kg (44,000 lb) | 13 380 kg (29,500 lb) | 9 070 kg (20,000 lb) |
| Approximate Weight | 3,17 kg/m (2.13 lb/ft) | 3,17 kg/m (2.13 lb/ft) | 4,381 kg/m (3.23 lb/ft) | 3,02 kg/m (2.03 lb/ft) | 3,02 kg/m (2.03 lb/ft) |

* Right Regular Lay

## Wire Rope Specifications 3.5:1 Safety Factor
Boom No. 78
- or -
Fixed Jib No. 134 on Boom No. 78
- or -
Luffing Jib No. 139 on Boom No. 78
- or -
Fixed Jib No. 138 on Luffing Jib No. 139 on Boom No. 78

3.5:1 Safety Factor
Right Regular Lay, 6x25 Filler Wire,
Extra Improved Plow Steel, IWRC

| Part Number / Function | No. 719387 Hoist Line Whip Line Auxiliary Line | No. 719060 Hoist Line Whip Line | No. 719278 Hoist Line Whip Line (High Line Pull) | No. 719073 Auxiliary Line |
|---|---|---|---|---|
| Size Wire Rope | 26 mm – | – (1") | – (1-1/4") | – (1") |
| Minimum Breaking Strength | 483,1 kN (108,600 lb) | 460,0 kN (103,400 lb) | 72 480 kg (159,800 lb) | 40 730 kg (89,890 lb) |
| Maximum Load Per Line | 13 380 kg (29,500 lb) | 13 380 kg (29,500 lb) | 19 960 kg (44,000 lb) | 9 070 kg (20,000 lb) |
| Approximate Weight | 2,89 kg/m (1.94 lb/ft) | 2,75 kg/m (1.85 lb/ft) | 4,30 kg/m (2.89 lb/ft) | 2,75 kg/m (1.85 lb/ft) |

model 777

Manitowoc

# performance data

17

## Drums & Laggings

| | Application | Drum Location | Drum Part Number | Drum Type | Drum Diameter | Drum Width | Optional Grooved Lagging Part Number | Lagging Diameter | Wire Rope Size |
|---|---|---|---|---|---|---|---|---|---|
| | | Drums | | | | | | | |
| Basic Liftcrane | Hoist | Front | 178272 | Bare | 495 mm (19-1/2") | 765 mm (30-1/8") | 502392 | 540 mm (21-1/4") | 26 mm<br>– |
| | | | | | | | 502391 | 540 mm (21-1/4") | –<br>1" |
| | Hoist (Optional Dual Input Drive) | Front | 178273 | Bare | 495 mm (19-1/2") | 765 mm (30-1/8") | 502392 | 540 mm (21-1/4") | 26 mm<br>– |
| | | | | | | | 502391 | 540 mm (21-1/4") | –<br>1" |
| | | | | | | | 502395 | 641 mm (25-1/4") | 32 mm<br>OR 1-1/4" |
| | Whip | Rear | 178272 | Bare | 495 mm (19-1/2") | 765 mm (30-1/8") | 502392 | 540 mm (21-1/4") | 26 mm<br>– |
| | | | | | | | 502391 | 540 mm (21-1/4") | –<br>1" |
| | Whip (Optional Dual Input Drive) | Rear | 178273 | Bare | 495 mm (19-1/2") | 765 mm (30-1/8") | 502392 | 540 mm (21-1/4") | 26 mm<br>– |
| | | | | | | | | 540 mm (21-1/4") | –<br>1" |
| | Hoist (Auxiliary) | Boom Butt | 177379 | Bare | 495 mm (19-1/2") | 940 mm (37") | 502391 | 540 mm (21-1/4") | 26 mm<br>– |
| | | | | | | | | 540 mm (21-1/4") | –<br>1" |
| Luffing Jib Liftcrane | Hoist | Front | 178272 | Bare | 495 mm (19-1/2") | 765 mm (30-1/8") | 502392 | 540 mm (21-1/4") | 26 mm<br>– |
| | | | | | | | 502391 | 540 mm (21-1/4") | –<br>1" |
| | Hoist (Optional Dual Input Drive) | Front | 178273 | Bare | 495 mm (19-1/2") | 765 mm (30-1/8") | 502392 | 540 mm (21-1/4") | 26 mm<br>– |
| | | | | | | | 502391 | 540 mm (21-1/4") | –<br>1" |
| | Hoist (Auxiliary) | Boom Butt | 177379 | Bare | 495 mm (19-1/2") | 940 mm (37") | 502372 | 540 mm (21-1/4") | 26 mm<br>– |
| | | | | | | | 502382 | 540 mm (21-1/4") | –<br>1" |
| Clamshell | Closing | Front | 178272 | | | 765 mm (30-1/8") | 502410 | 540 mm (21-1/4") | 26 mm<br>OR 1" |
| | Holding | Rear | 178272 | | | 765 mm (30-1/8") | 502410 | 540 mm (21-1/4") | 26 mm<br>OR 1" |
| Dragline | Drag | Front | 178272 | | | 765 mm (30-1/8") | 502410 | 540 mm (21-1/4") | 26 mm<br>OR 1" |
| | Hoist | Rear | 178272 | | | 765 mm (30-1/8") | 502410 | 540 mm (21-1/4") | 26 mm<br>OR 1" |

# performance data

18

## Drum Capacities - Wire Rope

| | Maximum Length | |
|---|---|---|
| | **No Lagging** | **With Lagging** |
| **Front Drum (Hoist or Whip)** | | |
| **26 mm Wire Rope** | **327 m**<br>**6 Layers** | **280 m**<br>**5 Layers *** |
| **(1") Wire Rope** | **(1,074 ft)**<br>**6 Layers** | **(918 ft)**<br>**5 Layers*** |
| **32 mm Wire Rope** | – | **145 m**<br>**3 Layers**** |
| **(1-1/4") Wire Rope** | – | **(475 ft)**<br>**3 Layers**** |
| **Left Rear Drum (Whip)** | | |
| **29 mm Wire Rope** | **327 m**<br>**6 Layers** | **280 m**<br>**5 Layers *** |
| **(1-1/8") Wire Rope** | **(1,074 ft)**<br>**6 Layers** | **(918 ft)**<br>**5 Layers*** |
| **Auxiliary Drum (Whip)** | | |
| **26 mm Wire Rope** | **403 m**<br>**6 Layers** | **431 m**<br>**8 Layers*** |
| **(1") Wire Rope** | **(1,323 ft)**<br>**6 Layers** | **(1,415 ft)**<br>**8 Layers*** |

**Note:** 5 m (17') is deducted from maximum spooling capacities for 3 dead wraps per drum or lagging.

*Lagging diameter 540 mm (21-1/4").

**Lagging diameter 641 mm (25-1/4").

## Main & Whip Drums - 131 kN (29,500 lb)

**Full Power Drum - Continuous Duty**
**Standard Pull/Single Line Speed**

| Layer | 1 | 2 | 3 | 4 | 5 |
|---|---|---|---|---|---|
| **Line Pull kg (lb)** | **m/min (ft/min)** | | | | |
| **0** (0) | **99** (325) | **108** (354) | **117** (383) | **126** (413) | **135** (442) |
| **2 268** (5,000) | **93** (306) | **101** (331) | **109** (357) | **116** (382) | **124** (406) |
| **4 536** (10,000) | **88** (287) | **94** (309) | **101** (330) | **107** (351) | **111** (363) |
| **6 803** (15,000) | **75** (245) | **76** (249) | **77** (252) | **78** (256) | **79** (260) |
| **9 072** (20,000) | **59** (194) | **60** (197) | **61** (201) | **63** (205) | **64** (208) |
| **11 340** (25,000) | **50** (163) | **51** (167) | **52** (170) | **53** (174) | **54** (177) |
| **13 380** (29,500) | **44** (144) | **45** (148) | **46** (151) | **47** (155) | **48** (158) |

**NOTE:** Line pull is infinitely variable. With 540 mm (21-1/4") lagging for 26 mm or (1") wire rope.

## Main & Whip Drums - 131 kN (29,500 lb)

**High Speed Drum - Continuous Duty**
**Single Line Pull/Single Line Speed**

| Layer | 1 | 2 | 3 | 4 | 5 |
|---|---|---|---|---|---|
| **Line Pull kg (lb)** | **m/min (ft/min)** | | | | |
| **0** (0) | **151** (496) | **165** (541) | **178** (585) | **192** (630) | **205** (674) |
| **2 268** (5,000) | **141** (462) | **152** (500) | **164** (537) | **175** (574) | **186** (611) |
| **4 536** (10,000) | **130** (427) | **140** (459) | **144** (471) | **145** (474) | **145** (477) |
| **6 803** (15,000) | **98** (321) | **99** (324) | **99** (326) | **100** (329) | **101** (332) |
| **9 072** (20,000) | **76** (249) | **77** (251) | **77** (254) | **78** (257) | **79** (260) |
| **11 340** (25,000) | **63** (205) | **63** (208) | **64** (211) | **65** (214) | **66** (217) |
| **13 380** (29,500) | **55** (179) | **56** (182) | **56** (185) | **57** (187) | **58** (190) |

**NOTE:** Line pull is infinitely variable. With 540 mm (21-1/4") lagging for 26 mm or (1") wire rope.

## Main & Whip Drums - 196 kN (44,000 lb)

**High Pull Drum - Continuous Duty**
**Single Line Pull/Single Line Speed**

| Layer | 1 | 2 | 3 |
|---|---|---|---|
| **Line Pull kg (lb)** | **m/min (ft/min)** | | |
| **0** (0) | **119** (389) | **130** (425) | **141** (462) |
| **2 268** (5,000) | **112** (368) | **122** (400) | **132** (432) |
| **4 536** (10,000) | **106** (347) | **114** (375) | **123** (403) |
| **9 072** (20,000) | **74** (242) | **74** (244) | **75** (246) |
| **13 608** (30,000) | **52** (169) | **52** (172) | **53** (174) |
| **18 143** (40,000) | **41** (133) | **42** (136) | **42** (138) |
| **19 958** (44,000) | **38** (123) | **38** (126) | **39** (128) |

**NOTE:** Line pull is infinitely variable. With 641 mm (25-1/4") lagging for 32 mm or (1-1/4") wire rope.

model 777

# performance data

19

## Maximum Length – Unassisted Raising

No. 134 Fixed Jib on No. 78 Main Boom Series 2

| Method | Main Boom | Fixed Jib |
|---|---|---|
| m (ft) | | |
| Over front of blocked crawlers | 82,3 (270) | – – |
| | 79,2 (260) | – – |
| | 76,2 (250) | – – |
| | 73,2 (240) | 9,1 (30) |
| | 70,1 (230) | 15,2 (50) |
| | 67,1 (220) | 24,4 (80) |
| Over rear of blocked crawlers | 79,2 (260) | – – |
| | 76,2 (250) | – – |
| | 73,2 (240) | – – |
| | 70,1 (230) | 15,2 (50) |
| | 67,1 (220) | 67,1 (70) |
| | 64,0 (210) | 24,4 (80) |
| Over side of blocked crawlers | 70,1 (230) | – – |
| | 67,1 (220) | – – |
| | 64,0 (210) | 15,2 (50) |
| | 61,0 (200) | 67,1 (70) |
| | 57,9 (190) | 24,4 (80) |

Note: Load block(s), hook(s) and weight ball(s) on ground at start.

## Working Weight

| Configuration | kg (lb) 777 Series 1 | 777 Series 2 |
|---|---|---|
| 18,3 m (60') No. 78 Main Boom | 118 655 (261,590) | 155 450 (342,710) |
| 82,3 m (270') No. 78 Main Boom combined with 24,4 m (80') No. 134 Fixed Jib | 133 034 (293,290) | 169 830 (374,410) |

Typical working weight consists of: hydraulic reservoirs full, fuel half-full, drums loaded with standard lengths of wire rope, upper boom point, 160 mt (175 t) hook block, and standard weight ball.

Note: Upper boom point not used with fixed jib.

## Maximum Length – Unassisted Raising

No. 139 Luffing Jib on No. 78 Main Boom Series 2

| Method | In-Line Procedure: Main Boom | In-Line Procedure: Luffing Jib | Jack-Knife Procedure: Main Boom | Jack-Knife Procedure: Luffing Jib |
|---|---|---|---|---|
| Over end of blocked crawlers m (ft) | 16,8 (55) | 15,2 - 36,6 (50 - 120) | 16,8 (55) | 39,6 - 45,7 (130 - 150) |
| | 19,8 (65) | 15,2 - 33,5 (50 - 110) | 19,8 (65) | 36,6 - 45,7 (120 - 150) |
| | 22,9 (75) | 15,2 - 30,5 (50 - 100) | 22,9 (75) | 33,5 - 45,7 (110 - 150) |
| | 25,9 (85) | 15,2 - 24,4 (50 - 80) | 25,9 (85) | 27,4 - 45,7 (90 - 150) |
| | 29,0 (95) | 15,2 - 21,3 (50 - 70) | 29,0 (95) | 24,4 - 45,7 (80 - 150) |
| | 32,0 (105) | 15,2 - 18,3 (50 - 60) | 32,0 (105) | 21,3 - 45,7 (70 - 150) |
| | – – | – – | 35,1 (115) | 15,2 - 45,7 (50 - 150) |
| | – – | – – | 38,1 (125) | 15,2 - 42,7 (50 - 140) |
| | – – | – – | 41,1 (135) | 15,2 - 30,5 (50 - 100) |
| | – – | – – | 44,2 (145) | 18,3 (60) |

NOTE: Load block(s), hook(s) and weight ball(s) on ground at start.

## Maximum Length – Unassisted Raising

No. 10 Fixed Jib on No. 222 Luffing Jib on No. 260 Main Boom
222, 222EX Series B
Jack-Knife Procedure

| Method | Main Boom | Luffing Jib | Fixed Jib |
|---|---|---|---|
| Over end of blocked crawlers m (ft) | 35,1 (115) | 30,5 - 39,6 (100 - 130) | 9,1 - 18,3 (30 - 60) |
| | 38,1 (125) | 30,5 - 39,6 (100 - 130) | 9,1 - 18,3 (30 - 60) |
| | 41,1 (135) | 30,5 (100) | 9,1 - 18,3 (30 - 60) |

NOTE: Load block(s), hook(s) and weight ball(s) on ground at start.

# boom combinations

20

**No. 78 Main Boom Combinations**

| Boom Length m (ft) | Boom Inserts 3,0 m (10 ft) | 6,1 m (20 ft) | 12,2 m (40 ft) |
|---|---|---|---|
| 18,3 (60) | – | – | – |
| 21,3 (70) | 1 | – | – |
| 24,4 (80) | – | 1 | – |
| 27,4 (90) | 1 | 1 | – |
| 30,5 (100) | – | – | 1 |
| 33,5 (110) | 1 | – | 1 |
| 36,6 (120) | – | 1 | 1 |
| 39,6 (130) | 1 | 1 | 1 |
| 42,7 (140) | – | – | 2 |
| 45,7 (150) | 1 | – | 2 |
| 48,8 (160) | – | 1 | 2 |
| 51,8 (170) | 1 | 1 | 2 |
| 54,9 (180) | – | – | 3 |
| 57,9 (190) | 1 | – | 3 |
| 61,0 (200) | – | 1 | 3 |
| 64,0 (210) | 1 | 1 | 3 |
| 67,1 (220) | – | – | 4 |
| 70,1 (230) | 1 | – | 4 |
| 73,2 (240) | – | 1 | 4 |
| 76,2 (250) | 1 | 1 | 4 |
| 79,2 (260) | – | – | 5 |
| 82,3 (270) | 1 | 2 | 4 |

**Note**: Intermediate suspension required for 57,9 m (190') and longer boom lengths.

**No. 134 Fixed Jib Combinations**

| Jib Length m (ft) | Fixed Jib Inserts 3,0 m (10 ft) | 6,1 m (20 ft) |
|---|---|---|
| 9,1 (30) | – | – |
| 12,2 (40) | 1 | – |
| 15,2 (50) | – | 1 |
| 18,3 (60) | 1 | 1 |
| 21,3 (70) | – | 2 |
| 24,4 (80) | 1 | 2 |

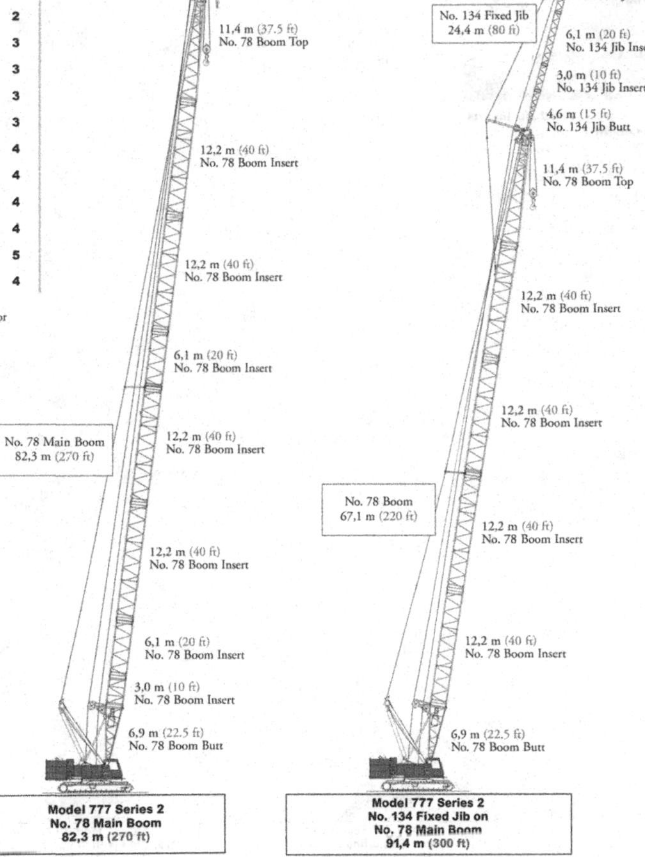


model 777

Manitowoc

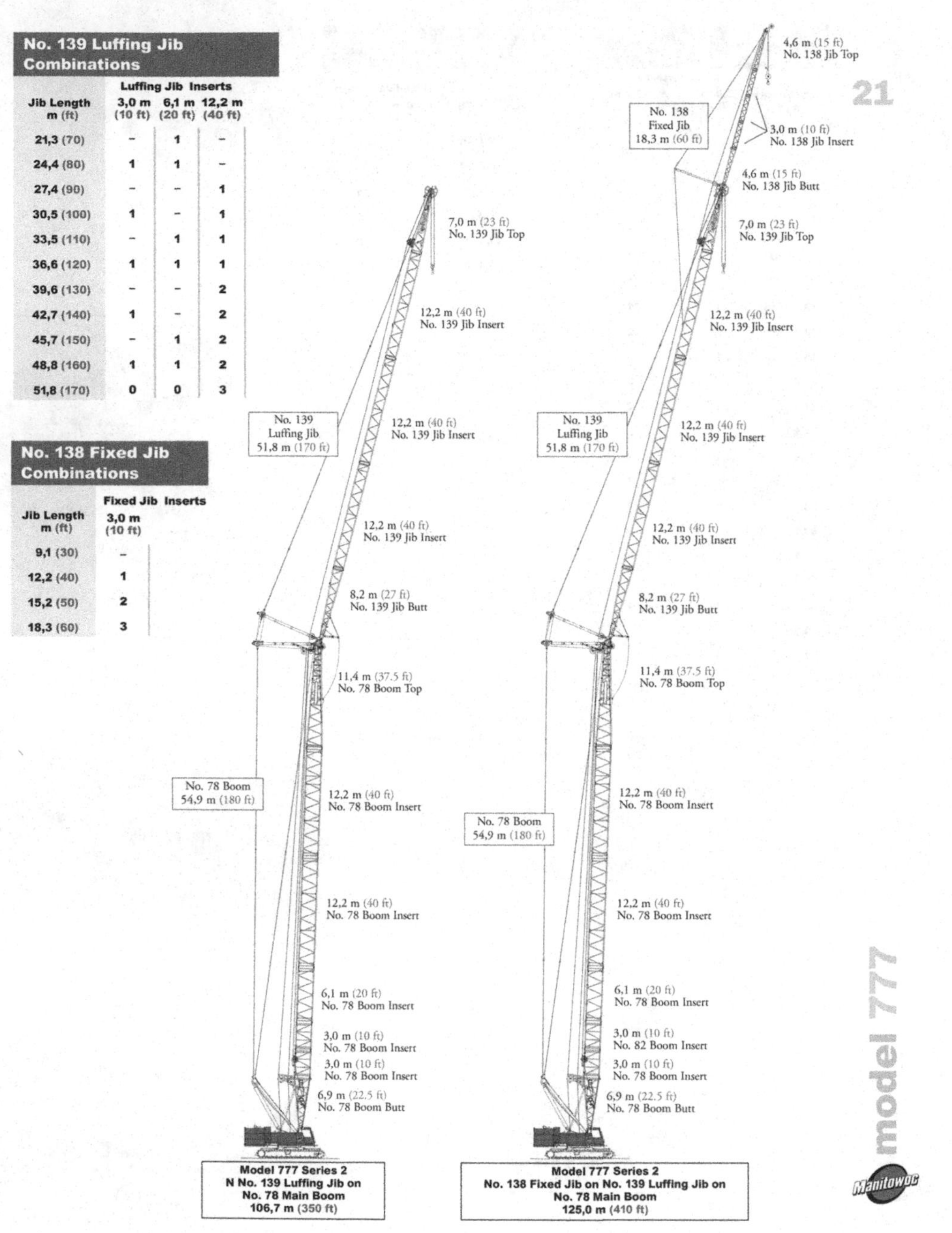

# boom combinations

**No. 139 Luffing Jib Combinations**

| Jib Length m (ft) | Luffing Jib Inserts 3,0 m (10 ft) | 6,1 m (20 ft) | 12,2 m (40 ft) |
|---|---|---|---|
| 21,3 (70) | – | 1 | – |
| 24,4 (80) | 1 | 1 | – |
| 27,4 (90) | – | – | 1 |
| 30,5 (100) | 1 | – | 1 |
| 33,5 (110) | – | 1 | 1 |
| 36,6 (120) | 1 | 1 | 1 |
| 39,6 (130) | – | – | 2 |
| 42,7 (140) | 1 | – | 2 |
| 45,7 (150) | – | 1 | 2 |
| 48,8 (160) | 1 | 1 | 2 |
| 51,8 (170) | 0 | 0 | 3 |

**No. 138 Fixed Jib Combinations**

| Jib Length m (ft) | Fixed Jib Inserts 3,0 m (10 ft) |
|---|---|
| 9,1 (30) | – |
| 12,2 (40) | 1 |
| 15,2 (50) | 2 |
| 18,3 (60) | 3 |

# heavy-lift boom range diagram

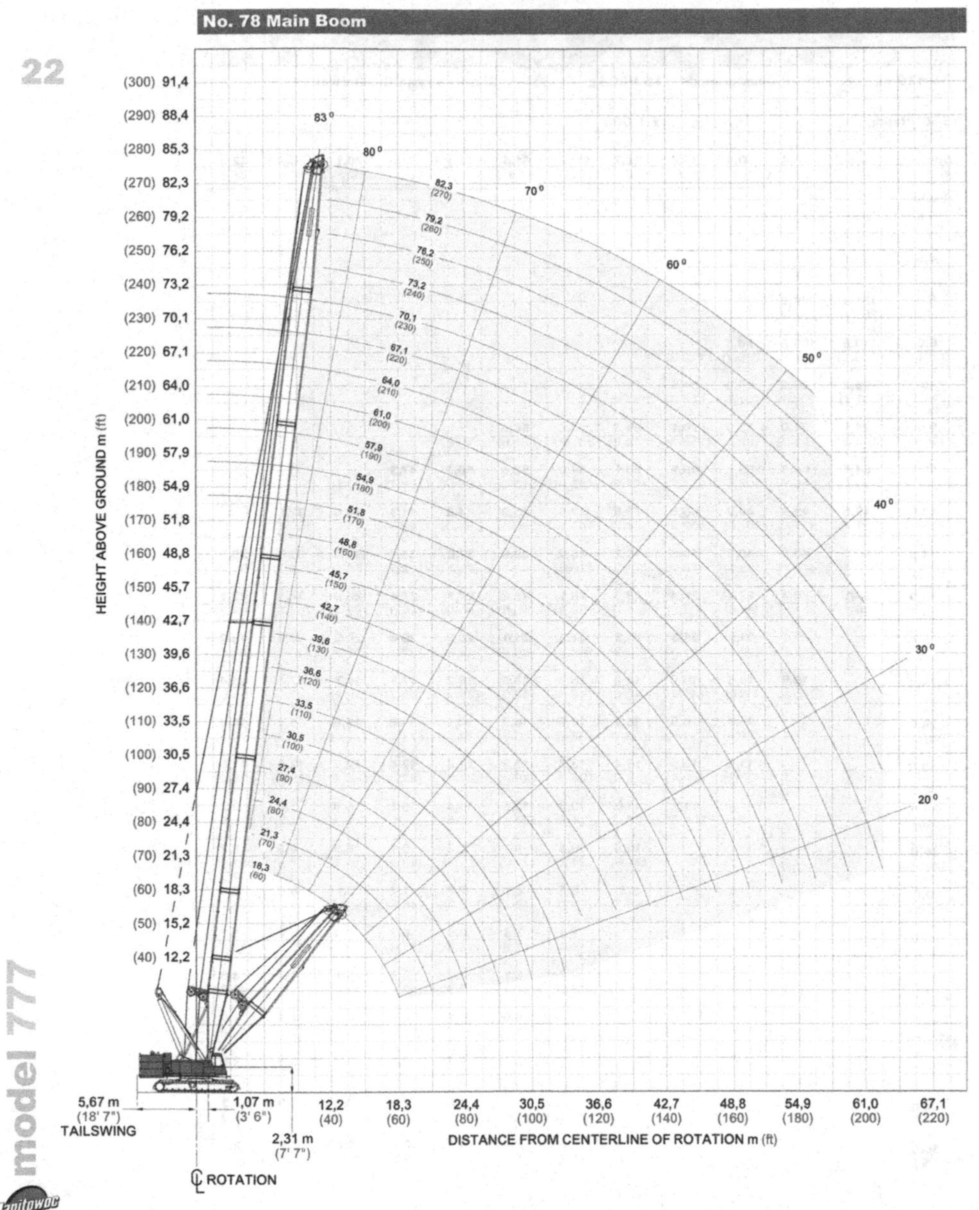

# heavy-lift load charts

23

**Liftcrane Boom Capacities - Series 2**
**Boom No. 78**

**64 410 kg** (142,000 lb) **Counterweight 19 960 kg** (44,000 lb) **Crawler Frame Counterweight**

**360° Rating** kg (lb) x 1 000

| Boom m (ft) | 18,3 (60) | 24,4 (80) | 30,5 (100) | 36,6 (120) | 42,7 (140) | 48,8 (160) | 51,8 (170) | 57,9 (190) | 64,0 (210) | 70,1 (230) | 76,2 (250) | 82,3 (270) |
|---|---|---|---|---|---|---|---|---|---|---|---|---|
| **Radius** | | | | | | | | | | | | |
| 4,0 (13) | 160,0 (352.8) | | | | | | | | | | | |
| 4,5 (15) | 146,1 (316.7) | - (275.5) | | | | | | | | | | |
| 5,0 (17) | 131,9 (281.3) | 119,9 (259.9) | | | | | | | | | | |
| 6,0 (20) | 110,8 (240.6) | 109,6 (239.9) | 96,3 (210.7) | - (192.2) | | | | | | | | |
| 7,0 (24) | 95,2 (201.2) | 95,2 (201.0) | 89,0 (191.7) | 81,5 (175.9) | - (157.9) | | | | | | | |
| 9,0 (30) | 74,1 (160.9) | 74,0 (160.7) | 73,9 (160.2) | 70,1 (152.2) | 64,3 (140.6) | 58,6 (128.1) | 56,3 (123.0) | - (116.6) | | | | |
| 10,0 (34) | 64,5 (135.0) | 65,0 (135.9) | 65,1 (136.0) | 63,2 (134.7) | 60,1 (127.5) | 54,9 (118.4) | 52,6 (113.4) | 50,7 (109.9) | 47,3 (102.7) | - (97.1) | | |
| 12,0 (40) | 49,6 (107.1) | 49,9 (107.8) | 49,9 (107.8) | 50,0 (107.9) | 49,8 (107.7) | 47,5 (103.2) | 46,2 (100.5) | 44,2 (96.0) | 41,3 (89.7) | 39,0 (84.7) | 37,1 (80.5) | |
| 14,0 (50) | 40,0 (78.4) | 40,3 (79.1) | 40,3 (79.1) | 40,3 (79.1) | 40,2 (78.8) | 40,2 (78.5) | 39,6 (78.3) | 37,8 (76.5) | 35,6 (72.5) | 33,5 (68.0) | 31,6 (64.0) | 29,6 (59.6) |
| 18,0 (60) | 28,1 (60.6) | 28,4 (61.5) | 28,4 (61.5) | 28,4 (61.5) | 28,3 (61.1) | 28,1 (60.8) | 28,0 (60.6) | 27,7 (59.8) | 27,4 (59.3) | 26,0 (56.4) | 24,5 (53.3) | 22,7 (49.3) |
| 22,0 (70) | | 21,4 (49.5) | 21,5 (49.6) | 21,5 (49.6) | 21,3 (49.2) | 21,2 (48.9) | 21,0 (48.6) | 20,7 (47.8) | 20,4 (47.3) | 20,2 (46.7) | 19,6 (44.9) | 18,2 (41.8) |
| 24,0 (80) | | 18,9 (40.8) | 19,0 (41.0) | 19,0 (41.0) | 18,8 (40.6) | 18,6 (40.3) | 18,5 (40.0) | 18,2 (39.2) | 17,9 (38.7) | 17,7 (38.1) | 17,4 (37.4) | 16,4 (35.5) |
| 26,0 (90) | | | 16,9 (34.5) | 16,9 (34.5) | 16,7 (34.1) | 16,6 (33.8) | 16,5 (33.5) | 16,1 (32.7) | 15,8 (32.2) | 15,6 (31.5) | 15,2 (30.9) | 14,8 (30.1) |
| 30,0 (100) | | | 13,5 (29.2) | 13,6 (29.4) | 13,4 (29.0) | 13,3 (28.7) | 13,2 (28.4) | 12,8 (27.6) | 12,6 (27.1) | 12,3 (26.4) | 12,0 (25.8) | 11,6 (25.0) |
| 34,0 (110) | | | | 11,2 (25.3) | 11,0 (25.0) | 10,9 (24.6) | 10,8 (24.4) | 10,4 (23.5) | 10,1 (23.0) | 9,8 (22.3) | 9,5 (21.7) | 9,2 (20.9) |
| 36,0 (120) | | | | | 10,0 (21.6) | 9,8 (21.2) | 9,7 (21.0) | 9,3 (20.1) | 9,1 (19.6) | 8,8 (18.9) | 8,5 (18.3) | 8,2 (17.5) |
| 40,0 (130) | | | | | 8,3 (18.7) | 8,2 (18.4) | 8,1 (18.2) | 7,7 (17.3) | 7,4 (16.8) | 7,1 (16.1) | 6,8 (15.5) | 6,5 (14.7) |
| 42,0 (140) | | | | | | 7,4 (16.0) | 7,3 (15.8) | 6,9 (14.9) | 6,7 (14.4) | 6,4 (13.7) | 6,1 (13.1) | 5,7 (12.3) |
| 46,0 (150) | | | | | | 6,2 (13.9) | 6,1 (13.7) | 5,7 (12.8) | 5,5 (12.4) | 5,1 (11.6) | 4,9 (11.0) | 4,5 (10.2) |
| 48,0 (160) | | | | | | | 5,6 (11.9) | 5,1 (11.0) | 4,9 (10.6) | 4,6 (9.9) | 4,3 (9.2) | 4,0 (8.4) |
| 52,0 (170) | | | | | | | | 4,2 (9.4) | 4,0 (9.0) | 3,7 (8.3) | 3,4 (7.6) | 2,9 (6.7) |
| 54,0 (180) | | | | | | | | 3,8 (8.0) | 3,6 (7.6) | 3,3 (6.9) | 2,9 (6.2) | 2,5 (5.2) |
| 56,0 (185) | | | | | | | | 3,3 (7.3) | 3,2 (7.0) | 2,8 (6.2) | 2,6 (5.6) | 2,0 (4.4) |

Meets ANSI B30.5 Requirements - Capacities do not exceed 75% of static tipping load.
NOTICE: This capacity chart is for reference only and must not be used for lifting purposes.

Manitowoc

# fixed jib range diagram

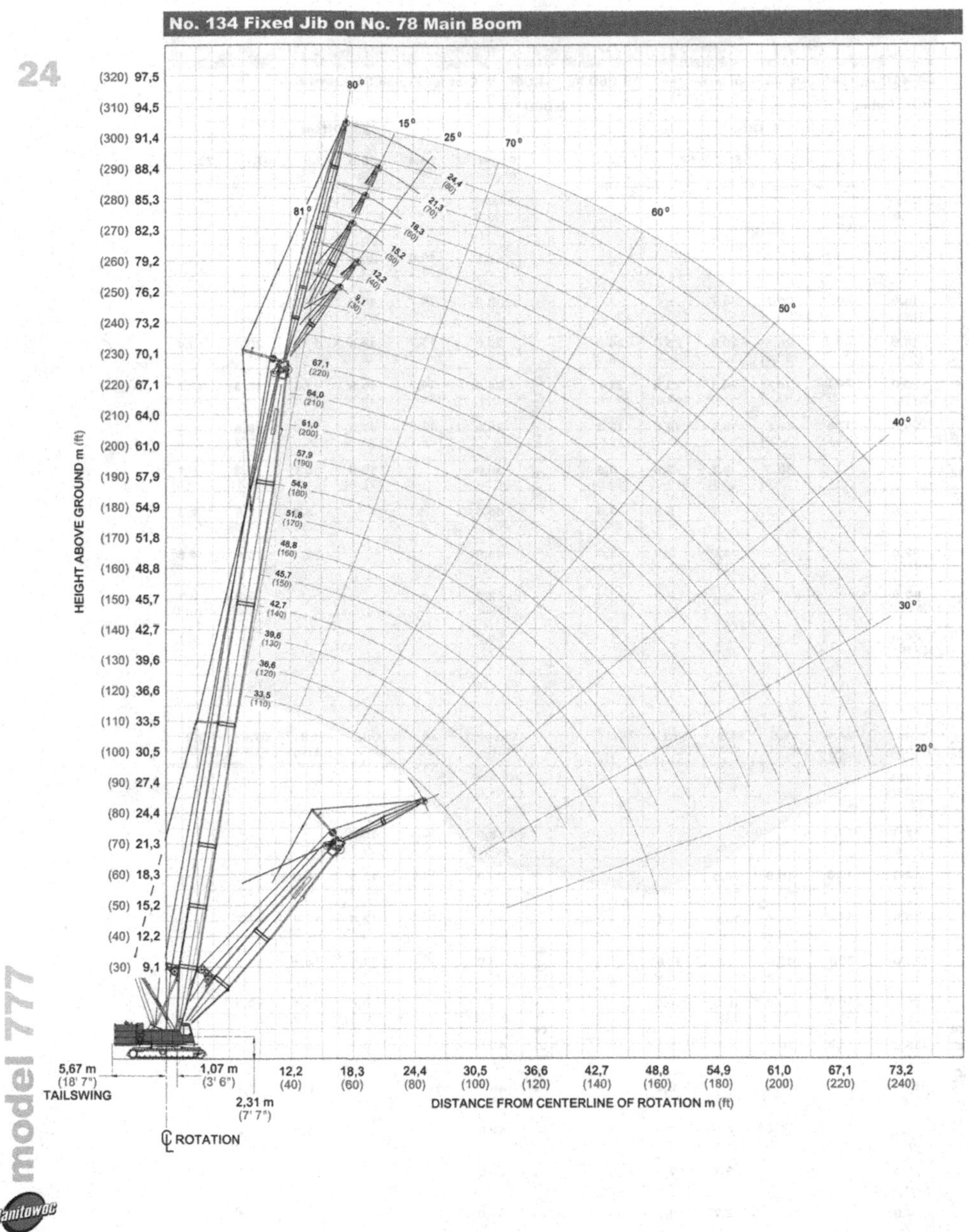

# fixed jib load charts

**Liftcrane Jib Capacities - Series 2**
**Jib No. 134 with 3 810 mm (12'6") Strut on Boom No. 78**

**64 410 kg (142,000 lb) Counterweight 19 960 kg (44,000 lb) Crawler Frame Counterweight**

**360° Rating kg (lb) x 1 000**

25

### 5° Offset — Jib 9,1 m (30 ft)

| Boom m (ft) | 24,4 (80) | 39,6 (130) | 51,8 (170) | 67,1 (220) | 73,2 (240) |
|---|---|---|---|---|---|
| **Radius** | | | | | |
| 7,6<br>(25) | 26,7<br>(59.0) | | | | |
| 10,0<br>(35) | 26,7<br>(59.0) | –<br>(59.0) | | | |
| 14,0<br>(50) | 26,6<br>(58.2) | 26,7<br>(59.0) | 26,7<br>(59.0) | –<br>(58.3) | –<br>(56.8) |
| 18,0<br>(60) | 25,3<br>(55.6) | 26,5<br>(58.5) | 26,6<br>(58.8) | 25,8<br>(56.8) | 24,5<br>(53.5) |
| 20,0<br>(70) | 24,5<br>(51.2) | 24,8<br>(50.0) | 24,3<br>(48.8) | 23,4<br>(47.0) | 21,9<br>(45.1) |
| 26,0<br>(90) | 17,6<br>(36.1) | 17,1<br>(34.8) | 16,5<br>(33.6) | 15,6<br>(31.7) | 15,3<br>(31.0) |
| 36,0<br>(120) | | 10,3<br>(22.2) | 9,7<br>(21.0) | 8,8<br>(19.0) | 8,5<br>(18.3) |
| 44,0<br>(150) | | 7,2<br>(14.9) | 6,7<br>(13.7) | 5,8<br>(11.7) | 5,5<br>(11.0) |
| 52,0<br>(180) | | | 4,6<br>(8.9) | 3,7<br>(6.9) | 3,4<br>(6.2) |
| 60,0<br>(200) | | | | 2,2<br>(4.5) | |
| 64,0<br>(210) | | | | | |

### 25° Offset — Jib 9,1 m (30 ft)

| Boom m (ft) | 24,4 (80) | 39,6 (130) | 51,8 (170) | 67,1 (220) | 73,2 (240) |
|---|---|---|---|---|---|
| **Radius** | | | | | |
| 10,0<br>(35) | –<br>(47.7) | | | | |
| 14,0<br>(50) | 19,0<br>(40.2) | 20,9<br>(44.7) | –<br>(47.0) | | |
| 18,0<br>(60) | 16,7<br>(36.6) | 18,8<br>(41.4) | 20,1<br>(44.1) | –<br>(46.5) | –<br>(45.1) |
| 20,0<br>(70) | 15,7<br>(33.7) | 18,0<br>(38.6) | 19,2<br>(41.5) | 20,4<br>(44.1) | 19,4<br>(41.2) |
| 24,0<br>(80) | 14,3<br>(31.4) | 16,5<br>(36.3) | 17,9<br>(39.3) | 18,6<br>(40.2) | 17,3<br>(37.9) |
| 26,0<br>(90) | 13,8<br>(29.7) | 15,9<br>(34.3) | 16,9<br>(34.9) | 16,5<br>(33.4) | 16,2<br>(32.8) |
| 36,0<br>(120) | | 10,6<br>(22.8) | 10,1<br>(21.8) | 9,3<br>(20.1) | 9,1<br>(19.5) |
| 40,0<br>(140) | | | 8,4<br>(16.3) | 7,6<br>(14.6) | 7,3<br>(14.0) |
| 48,0<br>(160) | | | 5,7<br>(12.3) | 5,0<br>(10.6) | 4,6<br>(9.9) |
| 52,0<br>(180) | | | | 3,9<br>(7.4) | 3,7<br>(6.8) |
| 60,0<br>(200) | | | | | 2,1<br>(4.3) |

### 5° Offset — Jib 15,2 m (50 ft)

| Boom m (ft) | 24,4 (80) | 39,6 (130) | 51,8 (170) | 67,1 (220) | 73,2 (240) |
|---|---|---|---|---|---|
| **Radius** | | | | | |
| 8,0<br>(30) | –<br>(43.3) | | | | |
| 12,0<br>(40) | 18,5<br>(40.7) | –<br>(43.0) | | | |
| 14,0<br>(50) | 17,8<br>(38.4) | 18,9<br>(41.1) | 19,5<br>(42.5) | | |
| 18,0<br>(60) | 16,6<br>(36.5) | 17,9<br>(39.4) | 18,6<br>(41.0) | 18,1<br>(40.0) | |
| 20,0<br>(70) | 16,1<br>(34.8) | 17,4<br>(37.8) | 18,1<br>(39.6) | 17,8<br>(39.0) | |
| 26,0<br>(90) | 14,7<br>(31.9) | 16,2<br>(35.2) | 16,9<br>(34.4) | 16,0<br>(32.5) | |
| 36,0<br>(120) | 11,3<br>(24.5) | 10,6<br>(23.0) | 10,0<br>(21.6) | 9,1<br>(19.7) | |
| 44,0<br>(150) | | 7,6<br>(15.7) | 7,0<br>(14.3) | 6,0<br>(12.3) | |
| 52,0<br>(180) | | 5,4<br>– | 4,9<br>(9.5) | 3,9<br>(7.5) | |
| 60,0<br>(200) | | | 3,3<br>(7.1) | 2,4<br>(5.1) | |
| 64,0<br>(220) | | | 2,7<br>– | 1,8<br>– | |

### 25° Offset — Jib 15,2 m (50 ft)

| Boom m (ft) | 24,4 (80) | 39,6 (130) | 51,8 (170) | 67,1 (220) | 73,2 (240) |
|---|---|---|---|---|---|
| **Radius** | | | | | |
| 14,0<br>(50) | –<br>(30.9) | | | | |
| 18,0<br>(60) | 12,6<br>(27.7) | 14,0<br>(30.7) | | | |
| 20,0<br>(70) | 11,8<br>(25.1) | 13,3<br>(28.3) | 14,0<br>(30.2) | –<br>(31.9) | |
| 24,0<br>(80) | 10,5<br>(23.0) | 12,0<br>(26.4) | 12,9<br>(28.3) | 13,7<br>(30.2) | |
| 26,0<br>(90) | 10,0<br>(21.3) | 11,5<br>(24.7) | 12,4<br>(26.7) | 13,3<br>(28.7) | |
| 36,0<br>(120) | | 9,5<br>(20.9) | 10,4<br>(22.9) | 10,0<br>(21.5) | |
| 40,0<br>(140) | | 9,0<br>(18.5) | 8,9<br>(17.5) | 8,2<br>(15.9) | |
| 48,0<br>(160) | | | 6,2<br>(13.3) | 5,4<br>(11.6) | |
| 52,0<br>(180) | | | 5,2<br>– | 4,4<br>(8.4) | |
| 60,0<br>(200) | | | | 2,8<br>(5.8) | |
| 64,0<br>(210) | | | | 2,1<br>(4.7) | |

Meets ANSI B30.5 Requirements - Capacities do not exceed 75% of static tipping load.
NOTICE: This capacity chart is for reference only and must not be used for lifting purposes.

model 777

Manitowoc

# fixed jib load charts

26

**Liftcrane Jib Capacities - Series 2**
**Jib No. 134 with 3 810 mm (12'6") Strut on Boom No. 78**

**64 410 kg (142,000 lb) Counterweight    19 960 kg (44,000 lb) Crawler Frame Counterweight**

**360° Rating    kg (lb) x 1 000**

**5° Offset** — Jib 18,3 m (60 ft)

| Boom m (ft) | 24,4 (80) | 39,6 (130) | 51,8 (170) | 67,1 (220) | 73,2 (240) |
|---|---|---|---|---|---|
| **Radius** | | | | | |
| 10,0 (35) | – (35.3) | | | | |
| 14,0 (50) | 15,0 (32.5) | 15,9 (34.5) | 16,4 (35.6) | | |
| 18,0 (60) | 14,0 (30.9) | 15,0 (33.1) | 15,6 (34.4) | 15,5 (34.1) | |
| 24,0 (80) | 12,7 (27.9) | 13,9 (30.7) | 14,6 (32.2) | 14,7 (32.5) | |
| 28,0 (100) | 11,9 (25.5) | 13,2 (28.4) | 14,0 (29.4) | 14,1 (27.5) | |
| 36,0 (120) | 9,9 (21.5) | 10,7 (23.2) | 10,1 (21.8) | 9,2 (19.9) | |
| 40,0 (140) | 8,8 – | 9,1 (18.0) | 8,4 (16.6) | 7,5 (14.6) | |
| 48,0 (160) | | 6,6 (14.1) | 5,9 (12.7) | 5,0 (10.6) | |
| 56,0 (190) | | 4,7 – | 4,1 (8.4) | 3,2 (6.4) | |
| 64,0 (210) | | | 2,8 (6.2) | 1,9 (4.2) | |
| 68,0 (230) | | | | | |

**25° Offset** — Jib 18,3 m (60 ft)

| Boom m (ft) | 24,4 (80) | 39,6 (130) | 51,8 (170) | 67,1 (220) | 73,2 (240) |
|---|---|---|---|---|---|
| **Radius** | | | | | |
| 16,0 (55) | – (25.4) | | | | |
| 18,0 (65) | 11,3 (24.0) | – (25.2) | | | |
| 22,0 (75) | 10,0 (21.6) | 11,2 (24.4) | 11,4 (25.1) | – (25.5) | |
| 26,0 (90) | 8,9 (19.0) | 10,2 (21.8) | 11,0 (23.6) | 11,3 (24.7) | |
| 32,0 (110) | 7,7 (16.4) | 8,9 (19.2) | 9,7 (21.0) | 10,5 (22.8) | |
| 36,0 (130) | 7,0 – | 8,3 (17.3) | 9,1 (19.0) | 9,8 (18.9) | |
| 44,0 (150) | | 7,3 (15.8) | 7,5 (15.7) | 7,0 (14.2) | |
| 48,0 (170) | | 6,8 – | 6,4 (12.1) | 5,7 (10.4) | |
| 56,0 (190) | | | 4,5 (9.1) | 3,8 (7.5) | |
| 64,0 (210) | | | | 2,3 (5.1) | |
| – (220) | | | | – (4.1) | |

**5° Offset** — Jib 24,4 m (80 ft)

| Boom m (ft) | 24,4 (80) | 39,6 (130) | 51,8 (170) | 67,1 (220) | 73,2 (240) |
|---|---|---|---|---|---|
| **Radius** | | | | | |
| 10,0 (35) | – (20.6) | | | | |
| 14,0 (50) | 9,1 (20.1) | 9,3 (20.5) | – (20.6) | | |
| 18,0 (65) | 8,9 (19.6) | 9,1 (20.1) | 9,2 (20.3) | 9,3 (20.5) | |
| 24,0 (80) | 8,4 (18.4) | 8,9 (19.7) | 9,0 (20.0) | 9,1 (20.2) | |
| 28,0 (100) | 7,8 (16.6) | 8,7 (18.6) | 8,9 (19.6) | 9,0 (19.8) | |
| 36,0 (130) | 6,8 (14.3) | 7,8 (16.5) | 8,3 (17.8) | 8,4 (17.2) | |
| 48,0 (160) | 5,8 (12.7) | 6,6 (14.4) | 6,0 (12.9) | 5,0 (10.8) | |
| 56,0 (190) | | 4,9 (10.2) | 4,2 (8.7) | 3,3 (6.6) | |
| 64,0 (210) | | | 2,9 (6.6) | 2,0 (4.5) | |
| 68,0 (230) | | | 2,4 (4.8) | | |
| 72,0 (240) | | | 1,9 (4.0) | | |

**25° Offset** — Jib 24,4 m (80 ft)

| Boom m (ft) | 24,4 (80) | 39,6 (130) | 51,8 (170) | 67,1 (220) | 73,2 (240) |
|---|---|---|---|---|---|
| **Radius** | | | | | |
| – (65) | – (17.2) | | | | |
| 20,0 (70) | 7,7 (16.8) | – (17.4) | | | |
| 24,0 (80) | 7,3 (16.0) | 7,6 (16.8) | – (17.2) | | |
| 26,0 (90) | 7,1 (15.2) | 7,4 (16.2) | 7,7 (16.7) | 7,8 (17.1) | |
| 32,0 (110) | 6,0 (12.8) | 6,9 (15.0) | 7,2 (15.8) | 7,4 (16.3) | |
| 36,0 (130) | 5,4 (11.1) | 6,4 (13.2) | 6,9 (14.5) | 7,2 (15.6) | |
| 44,0 (150) | 4,6 – | 5,5 (11.8) | 6,1 (13.1) | 6,7 (14.3) | |
| 48,0 (170) | | 5,1 (10.7) | 5,7 (12.0) | 5,8 (11.0) | |
| 56,0 (190) | | | 4,6 (9.6) | 4,0 (8.0) | |
| 64,0 (210) | | | 3,3 (7.3) | 2,4 (5.4) | |
| 68,0 (230) | | | | | |

model 777

Meets ANSI B30.5 Requirements - Capacities do not exceed 75% of static tipping load.
NOTICE: This capacity chart is for reference only and must not be used for lifting purposes.

# luffing jib range diagram

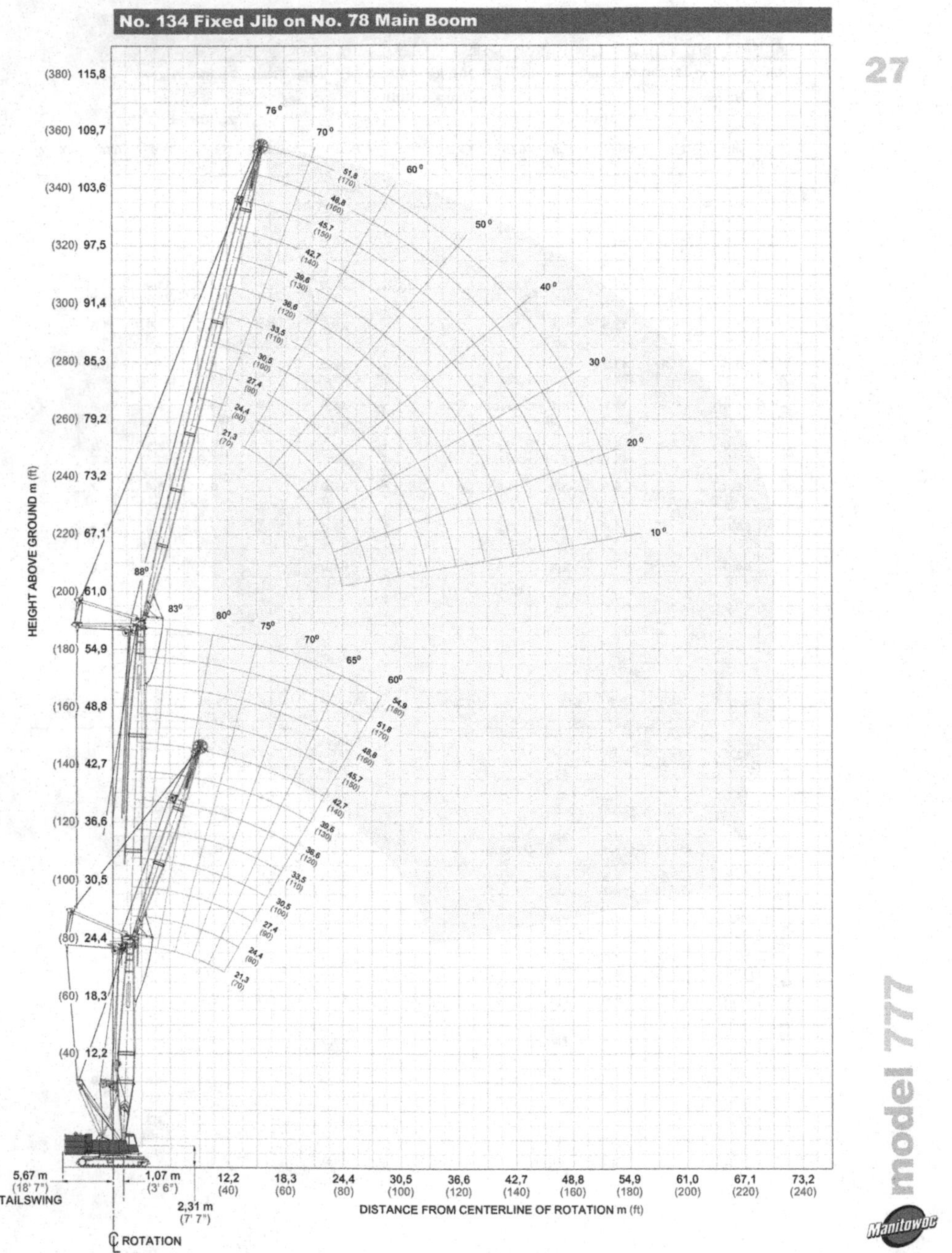

# luffing jib load charts

28

**Liftcrane Luffing Jib Capacities - Series 2**
**Luffing Jib No. 139 on Boom No. 78**

**64 410 kg (142,000 lb) Counterweight 19 960 kg (44,000 lb) Carbody Counterweight**

**360° Rating** **kg (lb) x 1 000**

**88° Boom Angle**

**Luffing Jib Length 21,3 m (70 ft)**

| Boom m (ft) | 21,3 (70) | 30,5 (100) | 39,6 (130) | 45,7 (150) | 54,9 (180) |
|---|---|---|---|---|---|
| **Radius** | | | | | |
| 9,1 (30) | 47,4 (104.6) | | | | |
| 12,0 (40) | 43,2 (94.7) | 42,9 (94.0) | 41,2 (90.3) | 37,2 (81.4) | – (71.4) |
| 14,0 (50) | 37,4 (73.5) | 38,7 (76.2) | 37,7 (78.1) | 34,2 (71.9) | 30,3 (63.9) |
| 20,0 (70) | 23,1 (46.6) | 23,8 (48.0) | 24,4 (49.2) | 25,0 (50.2) | 24,2 (49.4) |
| 26,0 (90) | | | | | |
| 32,0 (110) | | | | | |
| 38,0 (130) | | | | | |
| 44,0 (150) | | | | | |
| 50,0 (170) | | | | | |
| 54,0 (180) | | | | | |

**Luffing Jib Length 30,5 m (100 ft)**

| Boom m (ft) | 21,3 (70) | 30,5 (100) | 39,6 (130) | 45,7 (150) | 54,9 (180) |
|---|---|---|---|---|---|
| **Radius** | | | | | |
| 9,1 (30) | | | | | |
| 12,0 (40) | 33,3 (73.0) | 32,7 (71.9) | 31,6 (69.6) | 30,6 (67.4) | – (51.7) |
| 14,0 (50) | 31,1 (66.2) | 31,1 (66.1) | 30,4 (65.4) | 29,6 (63..9) | 22,9 (49.8) |
| 20,0 (70) | 23,3 (47.3) | 24,0 (48.6) | 24,6 (49.8) | 24,6 (50.2) | 20,6 (43.6) |
| 26,0 (90) | 16,3 (33.4) | 16,7 (34.2) | 17,1 (34.9) | 17,3 (35.4) | 16,5 (34.5) |
| 32,0 (110) | 11,3 – | 11,7 – | 12,2 – | 12,4 – | 12,8 (23.3) |
| 38,0 (130) | | | | | |
| 44,0 (150) | | | | | |
| 50,0 (170) | | | | | |
| 54,0 (180) | | | | | |

**Luffing Jib Length 39,6 m (130 ft)**

| Boom m (ft) | 21,3 (70) | 30,5 (100) | 39,6 (130) | 45,7 (150) | 54,9 (180) |
|---|---|---|---|---|---|
| **Radius** | | | | | |
| 9,1 (30) | | | | | |
| 12,0 (40) | – (51.7) | – (49.9) | | | |
| 14,0 (50) | 22,7 (49.0) | 22,1 (48.0) | 21,1 (46.0) | 20,4 (44.5) | 18,8 (40.7) |
| 20,0 (70) | 19,3 (41.0) | 19,4 (41.2) | 19,2 (41.1) | 19,1 (40.9) | 16,8 (35.9) |
| 26,0 (90) | 16,2 (33.6) | 16,4 (34.4) | 16,5 (35.1) | 16,5 (35.2) | 14,3 (30.4) |
| 32,0 (110) | 12,1 (25.0) | 12,4 (25.6) | 12,7 (26.1) | 12,8 (26.4) | 11,9 (25.0) |
| 38,0 (130) | 9,2 (19.1) | 9,5 (19.6) | 9,7 (19.9) | 9,7 (20.1) | 9,8 (20.5) |
| 44,0 (150) | | | | | |
| 50,0 (170) | | | | | |
| 54,0 (180) | | | | | |

**Luffing Jib Length 51,8 m (170 ft)**

| Boom m (ft) | 21,3 (70) | 30,5 (100) | 39,6 (130) | 45,7 (150) | 54,9 (180) |
|---|---|---|---|---|---|
| **Radius** | | | | | |
| 9,1 (30) | | | | | |
| 12,0 (40) | | | | | |
| 14,0 (50) | – (35.6) | – (34.3) | | | |
| 20,0 (70) | 13,8 (29.2) | 13,8 (29.3) | 13,7 (29.1) | 12,2 (26.0) | 11,8 (25.6) |
| 26,0 (90) | 11,2 (23.6) | 11,3 (23.9) | 11,4 (24.1) | 10,2 (21.6) | 10,2 (21.7) |
| 32,0 (110) | 9,0 (18.9) | 9,2 (19.3) | 9,3 (19.5) | 8,4 (17.6) | 8,5 (17.9) |
| 38,0 (130) | 7,2 (15.1) | 7,4 (15.4) | 7,5 (15.7) | 6,8 (14.3) | 7,0 (14.6) |
| 44,0 (150) | 5,7 (12.0) | 5,9 (12.3) | 6,0 (12.5) | 5,5 (11.5) | 5,6 (11.8) |
| 50,0 (170) | 4,4 (8.8) | 4,5 (9.1) | 4,6 (9.3) | 4,2 (8.5) | 4,3 (8.8) |
| 54,0 (180) | | | | | 3,5 (7.5) |

model 777

Manitowoc

**Meets ANSI B30.5 Requirements - Capacities do not exceed 75% of static tipping load.**
**NOTICE: This capacity chart is for reference only and must not be used for lifting purposes.**

# luffing jib load charts

29

**Liftcrane Luffing Jib Capacities - Series 2**
**Luffing Jib No. 139 on Boom No. 78**

**64 410 kg (142,000 lb) Counterweight 19 960 kg (44,000 lb) Carbody Counterweight**

**360° Rating** **kg (lb) x 1 000**

**75° Boom Angle**

**Luffing Jib Length 21,3 m (70 ft)**

| Boom m (ft) / Radius | 21,3 (70) | 30,5 (100) | 39,6 (130) | 45,7 (150) | 54,9 (180) |
|---|---|---|---|---|---|
| 16,0 (55) | – (71.7) | | | | |
| 18,0 (65) | 29,8 (58.2) | – (55.9) | | | |
| 26,0 (90) | 18,8 (38.8) | 18,0 (37.3) | 17,1 (35.3) | 16,4 (33.9) | 15,3 (31.6) |
| 32,0 (110) | | | 13,1 – | 12,6 (26.2) | 11,7 (24.4) |
| 38,0 (130) | | | | | |
| 44,0 (150) | | | | | |
| 50,0 (170) | | | | | |
| 56,0 (190) | | | | | |
| 62,0 (210) | | | | | |
| 66,0 (220) | | | | | |

**Luffing Jib Length 30,5 m (100 ft)**

| Boom m (ft) / Radius | 21,3 (70) | 30,5 (100) | 39,6 (130) | 45,7 (150) | 54,9 (180) |
|---|---|---|---|---|---|
| 16,0 (55) | | | | | |
| 18,0 (65) | | | | | |
| 26,0 (90) | 18,4 (38.0) | 17,6 (36.3) | 16,6 (34.3) | – (32.8) | |
| 32,0 (110) | 14,1 (29.4) | 13,5 (28.1) | 12,7 (26.4) | 12,1 (25.3) | 11,2 (23.3) |
| 38,0 (130) | | 10,7 (22.4) | 10,1 (21.1) | 9,6 (20.2) | 8,9 (18.6) |
| 44,0 (150) | | | | | 7,2 (15.1) |
| 50,0 (170) | | | | | |
| 56,0 (190) | | | | | |
| 62,0 (210) | | | | | |
| 66,0 (220) | | | | | |

**Luffing Jib Length 39,6 m (130 ft)**

| Boom m (ft) / Radius | 21,3 (70) | 30,5 (100) | 39,6 (130) | 45,7 (150) | 54,9 (180) |
|---|---|---|---|---|---|
| 16,0 (55) | | | | | |
| 18,0 (65) | | | | | |
| 26,0 (90) | 18,1 (37.4) | – (35.6) | | | |
| 32,0 (110) | 13,8 (28.8) | 13,2 (27.4) | 12,4 (25.7) | 11,8 (24.5) | |
| 38,0 (130) | 11,0 (23.1) | 10,5 (21.9) | 9,8 (20.5) | 9,3 (19.5) | 8,6 (17.8) |
| 44,0 (150) | 8,9 (18.3) | 8,5 (17.9) | 8,0 (16.7) | 7,6 (15.9) | 6,9 (14.5) |
| 50,0 (170) | | | 6,6 – | 6,2 (13.1) | 5,7 (11.9) |
| 56,0 (190) | | | | | |
| 62,0 (210) | | | | | |
| 66,0 (220) | | | | | |

**Luffing Jib Length 51,8 m (170 ft)**

| Boom m (ft) / Radius | 21,3 (70) | 30,5 (100) | 39,6 (130) | 45,7 (150) | 54,9 (180) |
|---|---|---|---|---|---|
| 16,0 (55) | | | | | |
| 18,0 (65) | | | | | |
| 26,0 (90) | | | | | |
| 32,0 (110) | 10,7 (22.6) | – (24.3) | | | |
| 38,0 (130) | 8,7 (18.2) | 9,5 (19.9) | 9,3 (19.4) | – (18.4) | – (16.7) |
| 44,0 (150) | 6,9 (14.5) | 7,7 (16.0) | 7,5 (15.7) | 7,1 (14.8) | 6,4 (13.3) |
| 50,0 (170) | 5,5 (11.4) | 6,1 (12.8) | 6,1 (12.9) | 5,8 (12.1) | 5,2 (10.8) |
| 56,0 (190) | 4,1 (8.3) | 4,7 (9.5) | 5,1 (10.7) | 4,8 (10.0) | 4,2 (8.7) |
| 62,0 (210) | | | 4,0 – | 3,9 (8.0) | 3,3 (6.9) |
| 66,0 (220) | | | | | 2,9 (6.2) |

model 777

Manitowoc

Meets ANSI B30.5 Requirements - Capacities do not exceed 75% of static tipping load.
NOTICE: This capacity chart is for reference only and must not be used for lifting purposes.

# luffing jib load charts

30

**Liftcrane Luffing Jib Capacities - Series 2**
**Luffing Jib No. 139 on Boom No. 78**

**64 410 kg (142,000 lb) Counterweight 19 960 kg (44,000 lb) Carbody Counterweight**

**360° Rating** **kg (lb) x 1 000**

## 60° Boom Angle

**Luffing Jib Length 21,3 m (70 ft)**

| Boom m (ft) / Radius | 21,3 (70) | 30,5 (100) | 39,6 (130) | 45,7 (150) | 54,9 (180) |
|---|---|---|---|---|---|
| 26,0 (90) | – (34.7) | | | | |
| 30,0 (100) | 14,0 (30.3) | | | | |
| 32,0 (110) | 12,9 (26.7) | 11,6 (24.1) | | | |
| 36,0 (120) | | 9,9 (21.5) | – (18.7) | | |
| 38,0 (130) | | 9,2 – | 8,0 (16.7) | – (14.8) | |
| 44,0 (150) | | | | 5,7 (11.9) | – (9.3) |
| 50,0 (170) | | | | | 3,5 – |
| 56,0 (190) | | | | | |
| 64,0 (210) | | | | | |
| 70,0 (230) | | | | | |

**Luffing Jib Length 30,5 m (100 ft)**

| Boom m (ft) / Radius | 21,3 (70) | 30,5 (100) | 39,6 (130) | 45,7 (150) | 54,9 (180) |
|---|---|---|---|---|---|
| 26,0 (90) | | | | | |
| 30,0 (100) | | | | | |
| 32,0 (110) | – (25.9) | | | | |
| 36,0 (120) | 10,6 (23.1) | | | | |
| 38,0 (130) | 9,9 (20.7) | – (18.3) | | | |
| 44,0 (150) | | 7,1 (14.8) | 6,0 (12.5) | – (10.8) | |
| 50,0 (170) | | | 4,9 (10.2) | 4,2 (8.7) | – (6.3) |
| 56,0 (190) | | | | | 2,4 (4.9) |
| 64,0 (210) | | | | | |
| 70,0 (230) | | | | | |

## 70° Boom Angle

**Luffing Jib Length 39,6 m (130 ft)**

| Boom m (ft) / Radius | 21,3 (70) | 30,5 (100) | 39,6 (130) | 45,7 (150) | 54,9 (180) |
|---|---|---|---|---|---|
| 26,0 (90) | | | | | |
| 30,0 (100) | – (31.2) | | | | |
| 32,0 (110) | 13,2 (27.6) | – (25.6) | | | |
| 36,0 (120) | 11,3 (24.6) | 10,5 (22.8) | – (20.7) | | |
| 38,0 (130) | 10,5 (22.0) | 9,7 (20.4) | 8,8 (18.5) | – (17.1) | |
| 44,0 (150) | 8,6 (18.0) | 7,9 (16.6) | 7,2 (15.0) | 6,6 (13.8) | 5,7 (11.9) |
| 50,0 (170) | | 6,5 – | 5,9 (12.4) | 5,4 (11.3) | 4,6 (9.7) |
| 56,0 (190) | | | | 4,5 – | 3,8 (7.9) |
| 64,0 (210) | | | | | |
| 70,0 (230) | | | | | |

**Luffing Jib Length 51,8 m (170 ft)**

| Boom m (ft) / Radius | 21,3 (70) | 30,5 (100) | 39,6 (130) | 45,7 (150) | 54,9 (180) |
|---|---|---|---|---|---|
| 26,0 (90) | | | | | |
| 30,0 (100) | | | | | |
| 32,0 (110) | | | | | |
| 36,0 (120) | – (21.4) | | | | |
| 38,0 (130) | 9,2 (19.3) | – (19.3) | | | |
| 44,0 (150) | 7,4 (15.5) | 7,4 (15.6) | 6,6 (13.9) | – (12.7) | |
| 50,0 (170) | 5,9 (12.3) | 6,1 (12.8) | 5,4 (11.3) | 4,9 (10.2) | 4,1 (8.5) |
| 56,0 (190) | 4,5 (9.1) | 5,0 (10.6) | 4,4 (9.3) | 4,0 (8.3) | 3,3 (6.8) |
| 64,0 (210) | | | 3,4 (7.6) | 3,0 (6.8) | 2,4 (5.4) |
| 70,0 (230) | | | | | 1,9 (4.2) |

model 777

Manitowoc

**Meets ANSI B30.5 Requirements - Capacities do not exceed 75% of static tipping load.**
**NOTICE: This capacity chart is for reference only and must not be used for lifting purposes.**

# fixed jib on luffing jib range diagram

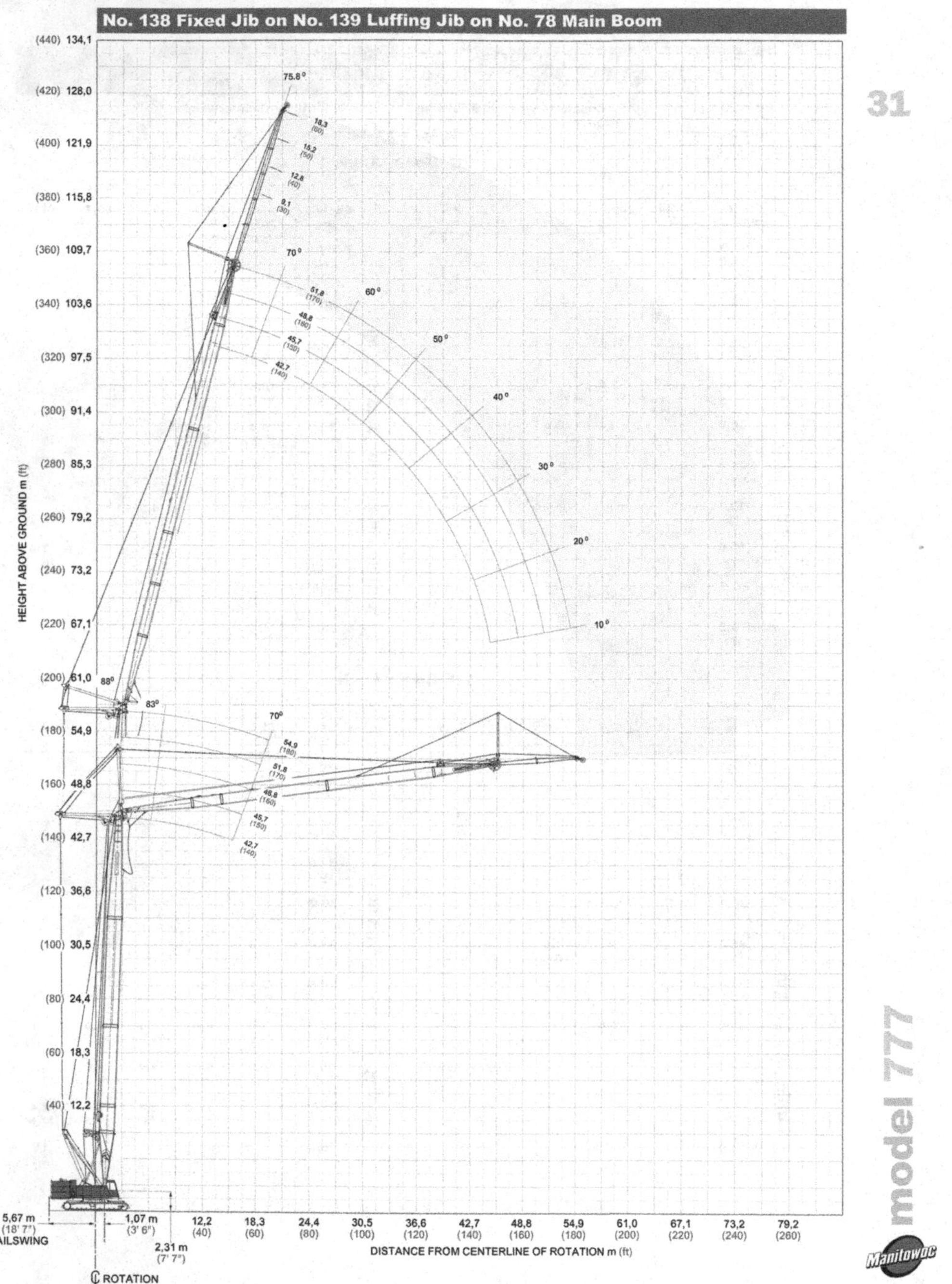

# fixed jib on luffing jib load charts

32

**Liftcrane Fixed Jib on Luffing Capacities - Series 2**
**Fixed Jib No. 138 at 3 Degree Angle on Luffing Jib No. 139 on Boom No. 78**

**64 410 kg** (142,000 lb) **Upperworks Counterweight 19 960 kg** (44,000 lb) **Carbody Counterweight**

**360° Rating** kg (lb) x 1 000

**88° Boom Angle**

**Fixed Jib Length 9,1 m (30 ft)**

| Luffing Jib m (ft) | 42,7 (140) | | | 45,7 (150) | | | 48,8 (160) | | | 51,8 (170) | | |
|---|---|---|---|---|---|---|---|---|---|---|---|---|
| Boom m (ft) | 42,7 (140) | 48,8 (160) | 54,9 (180) | 42,7 (140) | 48,8 (160) | 54,9 (180) | 42,7 (140) | 48,8 (160) | 54,9 (180) | 42,7 (140) | 48,8 (160) | 54,9 (180) |
| Radius | | | | | | | | | | | | |
| 16,8 (55) | 12,2 (27.0) | 12,0 (26.6) | | | | | | | | | | |
| 20,0 (70) | 12,0 (26.2) | 11,8 (25.9) | 11,6 (25.4) | 11,7 (25.7) | 11,2 (24.7) | 10,6 (23.4) | 10,5 (23.0) | 10,1 (22.1) | 9,6 (21.0) | 9,4 (20.6) | 9,0 (19.8) | 8,6 (18.8) |
| 24,0 (80) | 11,6 (25.7) | 11,5 (25.4) | 11,0 (24.3) | 11,4 (25.1) | 10,9 (24.1) | 10,2 (22.5) | 10,2 (22.5) | 9,8 (21.6) | 9,3 (20.6) | 9,1 (20.1) | 8,7 (19.3) | 8,3 (18.4) |
| 30,0 (100) | 10,0 (21.9) | 10,0 (22.0) | 9,8 (21.5) | 9,8 (21.5) | 9,8 (21.5) | 9,2 (20.2) | 9,6 (21.0) | 9,2 (20.2) | 8,5 (18.7) | 8,6 (19.0) | 8,3 (18.3) | 7,9 (17.3) |
| 36,0 (120) | 8,7 (19.1) | 8,8 (19.2) | 8,5 (18.6) | 8,6 (18.8) | 8,6 (18.8) | 8,1 (17.7) | 8,4 (18.5) | 8,2 (18.0) | 7,6 (16.7) | 8,1 (17.8) | 7,7 (16.8) | 7,1 (15.6) |
| 42,0 (140) | 7,7 (16.9) | 7,8 (17.0) | 7,2 (15.7) | 7,6 (16.7) | 7,6 (16.5) | 7,0 (15.2) | 7,5 (16.5) | 7,2 (15.8) | 6,7 (14.6) | 7,3 (16.0) | 6,8 (15.0) | 6,3 (13.8) |
| 48,0 (160) | 6,6 (14.2) | 6,5 (14.2) | 6,1 (13.2) | 6,6 (14.1) | 6,4 (14.0) | 6,0 (13.0) | 6,4 (13.9) | 6,3 (13.7) | 5,8 (12.6) | 6,3 (13.6) | 6,0 (13.1) | 5,5 (12.1) |
| 54,0 (180) | 4,9 (9.7) | 5,0 (10.0) | 5,0 (10.3) | 5,2 (11.1) | 5,2 (11.2) | 5,0 (11.0) | 5,1 (10.9) | 5,1 (11.0) | 5,0 (10.8) | 5,0 (10.7) | 5,0 (10.8) | 4,9 (10.6) |
| 60,0 (200) | | | | | | | 3,8 (7.2) | 3,8 (7.5) | 3,9 (7.7) | 3,8 (8.2) | 3,9 (8.3) | 3,9 (8.4) |
| 64,0 (210) | | | | | | | | | | 2,7 (6.1) | 2,8 (6.3) | 2,9 (6.5) |

**Fixed Jib Length 18,3 m (60 ft)**

| Luffing Jib m (ft) | 42,7 (140) | | | 45,7 (150) | | | 48,8 (160) | | | 51,8 (170) | | |
|---|---|---|---|---|---|---|---|---|---|---|---|---|
| Boom m (ft) | 42,7 (140) | 48,8 (160) | 54,9 (180) | 42,7 (140) | 48,8 (160) | 54,9 (180) | 42,7 (140) | 48,8 (160) | 54,9 (180) | 42,7 (140) | 48,8 (160) | 54,9 (180) |
| Radius | | | | | | | | | | | | |
| 20,0 (65) | 7,1 (15.7) | 6,9 (15.4) | 6,8 (15.1) | 6,8 (15.1) | 6,6 (14.8) | | | | | | | |
| 24,0 (80) | 6,7 (14.8) | 6,6 (14.6) | 6,5 (14.3) | 6,5 (14.3) | 6,3 (14.0) | 6,2 (13.7) | 6,2 (13.7) | 6,1 (13.4) | 5,9 (13.0) | 5,8 (12.9) | 5,7 (12.6) | 5,6 (12.3) |
| 30,0 (100) | 6,2 (13.6) | 6,1 (13.5) | 6,0 (13.3) | 6,0 (13.2) | 5,9 (13.0) | 5,8 (12.8) | 5,7 (12.7) | 5,6 (12.5) | 5,5 (12.2) | 5,5 (12.1) | 5,4 (11.9) | 5,2 (11.6) |
| 36,0 (120) | 5,7 (12.6) | 5,6 (12.5) | 5,6 (12.4) | 5,6 (12.3) | 5,5 (12.2) | 5,4 (12.0) | 5,4 (11.9) | 5,3 (11.7) | 5,2 (11.5) | 5,1 (11.4) | 5,0 (11.2) | 4,9 (10.9) |
| 42,0 (140) | 5,1 (11.2) | 5,2 (11.3) | 5,2 (11.4) | 5,0 (11.0) | 5,1 (11.1) | 5,1 (11.2) | 5,0 (10.9) | 5,0 (10.9) | 4,9 (10.9) | 4,8 (10.6) | 4,8 (10.6) | 4,7 (10.4) |
| 48,0 (160) | 4,5 (9.9) | 4,6 (10.0) | 4,6 (10.1) | 4,5 (9.8) | 4,5 (9.8) | 4,5 (9.9) | 4,4 (9.6) | 4,4 (9.7) | 4,4 (9.7) | 4,3 (9.4) | 4,3 (9.5) | 4,3 (9.5) |
| 54,0 (180) | 4,1 (8.9) | 4,1 (9.0) | 4,1 (9.0) | 4,0 (8.8) | 4,0 (8.8) | 4,0 (8.9) | 3,9 (8.7) | 3,9 (8.7) | 4,0 (8.8) | 3,9 (8.5) | 3,9 (8.6) | 3,9 (8.6) |
| 58,0 (200) | 3,8 (8.1) | 3,8 (8.2) | 3,8 (8.1) | 3,8 (8.0) | 3,8 (8.1) | 3,8 (8.1) | 3,7 (7.9) | 3,7 (7.9) | 3,8 (8.0) | 3,6 (7.8) | 3,6 (7.8) | 3,7 (7.8) |
| 64,0 (220) | 3,3 – | 3,4 – | 3,4 – | 3,4 (6.4) | 3,4 (6.7) | 3,4 (6.9) | 3,4 (7.3) | 3,4 (7.3) | 3,4 (7.0) | 3,3 (7.1) | 3,4 (7.2) | 3,3 (6.9) |
| 72,0 (240) | | | | | | | | | | 2,4 (4.4) | 2,5 (4.7) | 2,5 (4.8) |

model 777

Manitowoc

Meets ANSI B30.5 Requirements - Capacities do not exceed 75% of static tipping load.
NOTICE: This capacity chart is for reference only and must not be used for lifting purposes.

# fixed jib on luffing jib load charts

## Liftcrane Fixed Jib on Luffing Capacities - Series 2
## Fixed Jib No. 138 at 3 Degree Angle on Luffing Jib No. 139 on Boom No. 78

33

**64 410 kg** (142,000 lb) **Upperworks Counterweight 19 960 kg** (44,000 lb) **Carbody Counterweight**

**360° Rating** **kg** (lb) **x 1 000**

**83° Boom Angle**

**Fixed Jib Length 9,1 m (30 ft)**

| Luffing Jib m (ft) | 42,7 (140) | | | 45,7 (150) | | | 48,8 (160) | | | 51,8 (170) | | |
|---|---|---|---|---|---|---|---|---|---|---|---|---|
| **Boom m (ft)** | 42,7 (140) | 48,8 (160) | 54,9 (180) | 42,7 (140) | 48,8 (160) | 54,9 (180) | 42,7 (140) | 48,8 (160) | 54,9 (180) | 42,7 (140) | 48,8 (160) | 54,9 (180) |
| **Radius** | | | | | | | | | | | | |
| 24,0 (80) | – (26.4) | | | | | | | | | | | |
| 26,0 (90) | 11,8 (26.0) | 11,7 (25.7) | 11,1 (24.6) | 11,1 (24.5) | – (23.4) | – (22.0) | – (21.7) | – (20.7) | | – (19.3) | | |
| 30,0 (100) | 11,2 (24.5) | 11,4 (25.0) | 11,0 (24.4) | 10,9 (23.9) | 10,4 (23.1) | 9,9 (21.9) | 9,7 (21.4) | 9,3 (20.5) | 8,8 (19.5) | 8,6 (19.0) | 8,2 (18.2) | 7,9 (17.4) |
| 36,0 (120) | 9,6 (21.0) | 9,8 (21.5) | 10,0 (21.9) | 9,4 (20.6) | 9,6 (21.0) | 9,7 (21.4) | 9,2 (20.2) | 9,0 (19.9) | 8,6 (19.0) | 8,3 (18.2) | 8,0 (17.6) | 7,6 (16.9) |
| 42,0 (140) | 8,4 (18.4) | 8,6 (18.8) | 8,2 (17.8) | 8,3 (18.1) | 8,4 (18.5) | 8,1 (17.5) | 8,1 (17.8) | 8,3 (18.1) | 7,9 (17.1) | 7,8 (17.2) | 7,7 (17.0) | 7,4 (16.4) |
| 48,0 (160) | 7,5 (16.4) | 7,1 (15.4) | 6,6 (14.3) | 7,4 (16.2) | 7,0 (15.2) | 6,5 (14.1) | 7,3 (15.8) | 6,8 (14.7) | 6,3 (13.7) | 7,0 (15.4) | 6,7 (14.4) | 6,2 (13.3) |
| 54,0 (180) | 6,1 (13.0) | 5,9 (12.6) | 5,4 (11.7) | 6,1 (13.0) | 5,8 (12.4) | 5,3 (11.5) | 5,9 (12.7) | 5,5 (12.0) | 5,1 (11.0) | 5,8 (12.5) | 5,4 (11.6) | 5,0 (10.7) |
| 60,0 (200) | | | | 4,7 (10.1) | 4,7 (10.2) | 4,4 (9.4) | 4,7 (10.0) | 4,5 (9.8) | 4,1 (8.9) | 4,6 (9.8) | 4,4 (9.4) | 4,0 (8.6) |
| 62,0 (210) | | | | | 4,3 – | 4,1 – | 4,3 (8.7) | 4,2 (8.8) | 3,9 (8.1) | 4,2 (8.6) | 4,1 (8.5) | 3,7 (7.7) |
| 64,0 (220) | | | | | | | 3,9 – | 3,9 – | 3,6 – | 3,9 (7.4) | 3,8 (7.6) | 3,4 (6.9) |

**Fixed Jib Length 18,3 m (60 ft)**

| Luffing Jib m (ft) | 42,7 (140) | | | 45,7 (150) | | | 48,8 (160) | | | 51,8 (170) | | |
|---|---|---|---|---|---|---|---|---|---|---|---|---|
| **Boom m (ft)** | 42,7 (140) | 48,8 (160) | 54,9 (180) | 42,7 (140) | 48,8 (160) | 54,9 (180) | 42,7 (140) | 48,8 (160) | 54,9 (180) | 42,7 (140) | 48,8 (160) | 54,9 (180) |
| **Radius** | | | | | | | | | | | | |
| 28,0 (95) | – (14.3) | – (14.1) | | – (13.8) | | | | | | | | |
| 32,0 (105) | 6,2 (13.8) | 6,2 (13.7) | 6,1 (13.5) | 6,0 (13.3) | 5,9 (13.2) | 5,8 (12.9) | 5,7 (12.7) | 5,6 (12.5) | 5,5 (12.2) | 5,4 (12.0) | 5,3 (11.8) | |
| 36,0 (120) | 5,9 (13.1) | 5,9 (13.0) | 5,8 (12.9) | 5,7 (12.7) | 5,7 (12.6) | 5,6 (12.4) | 5,5 (12.2) | 5,4 (12.0) | 5,3 (11.8) | 5,2 (11.6) | 5,1 (11.4) | 5,0 (11.1) |
| 42,0 (140) | 5,5 (12.2) | 5,5 (12.2) | 5,5 (12.1) | 5,4 (11.9) | 5,3 (11.8) | 5,3 (11.7) | 5,2 (11.5) | 5,1 (11.4) | 5,0 (11.2) | 5,0 (11.0) | 4,9 (10.8) | 4,8 (10.6) |
| 48,0 (160) | 4,9 (10.8) | 5,0 (11.0) | 5,1 (11.3) | 4,9 (10.6) | 5,0 (10.9) | 5,0 (11.1) | 4,8 (10.5) | 4,8 (10.7) | 4,8 (10.7) | 4,7 (10.2) | 4,7 (10.3) | 4,6 (10.2) |
| 54,0 (180) | 4,4 (9.6) | 4,5 (9.8) | 4,6 (10.0) | 4,3 (9.5) | 4,4 (9.7) | 4,5 (9.8) | 4,2 (9.3) | 4,3 (9.5) | 4,4 (9.7) | 4,2 (9.2) | 4,3 (9.3) | 4,3 (9.5) |
| 62,0 (200) | 3,8 (8.7) | 3,9 (8.8) | 4,0 (9.0) | 3,8 (8.6) | 3,8 (8.7) | 3,9 (8.9) | 3,7 (8.4) | 3,8 (8.6) | 3,8 (8.7) | 3,7 (8.3) | 3,7 (8.4) | 3,7 (8.5) |
| 68,0 (220) | – (8.0) | 3,6 (8.1) | 3,6 (8.2) | 3,5 (7.8) | 3,5 (7.9) | 3,5 (8.1) | 3,4 (7.7) | 3,5 (7.8) | 3,3 (7.6) | 3,3 (7.6) | 3,4 (7.7) | 3,2 (7.3) |
| 72,0 (240) | | | | | | | 3,2 (6.7) | 3,2 (6.9) | 2,9 (6.2) | 3,1 (6.5) | 3,0 (6.6) | 2,7 (5.8) |
| 76,0 (250) | | | | | | | | | | 2,5 (5.6) | 2,6 (5.9) | 2,3 (5.2) |

Meets ANSI B30.5 Requirements - Capacities do not exceed 75% of static tipping load.
NOTICE: This capacity chart is for reference only and must not be used for lifting purposes.

model 777

Manitowoc

# fixed jib on luffing jib load charts

34

**Liftcrane Fixed Jib on Luffing Capacities - Series 2**
**Fixed Jib No. 138 at 3 Degree Angle on Luffing Jib No. 139 on Boom No. 78**

**64 410 kg (142,000 lb) Upperworks Counterweight 19 960 kg (44,000 lb) Carbody Counterweight**

**360° Rating** kg (lb) x 1 000

## 70° Boom Angle

### Fixed Jib Length 9,1 m (30 ft)

| Luffing Jib m (ft) | 42,7 (140) | | | 45,7 (150) | | | 48,8 (160) | | | 51,8 (170) | | |
|---|---|---|---|---|---|---|---|---|---|---|---|---|
| Boom m (ft) | 42,7 (140) | 48,8 (160) | 54,9 (180) | 42,7 (140) | 48,8 (160) | 54,9 (180) | 42,7 (140) | 48,8 (160) | 54,9 (180) | 42,7 (140) | 48,8 (160) | 54,9 (180) |
| Radius | | | | | | | | | | | | |
| – (150) | – (13.6) | | | | | | | | | | | |
| 48,0 (160) | 5,6 (12.2) | 5,1 (11.0) | – (9.3) | 5,6 (12.1) | – (10.9) | | – (11.7) | | | | | |
| 50,0 (170) | 5,3 (11.1) | 4,8 (10.0) | 4,0 (8.5) | 5,2 (10.9) | 4,7 (9.8) | – (8.2) | 5,0 (10.6) | – (9.4) | | – (10.3) | – (9.1) | |
| 54,0 (180) | 4,6 (10.0) | 4,2 (9.0) | 3,6 (7.8) | 4,6 (9.9) | 4,1 (8.8) | 3,5 (7.5) | 4,4 (9.5) | 3,9 (8.5) | 3,2 (7.0) | 4,2 (9.2) | 3,8 (8.2) | – (6.7) |
| 56,0 (190) | 4,3 (9.1) | 3,9 (8.1) | 3,4 (7.0) | 4,3 (9.0) | 3,8 (8.0) | 3,2 (6.9) | 4,1 (8.6) | 3,6 (7.6) | 3,0 (6.3) | 4,0 (8.3) | 3,5 (7.3) | 2,9 (6.1) |
| 60,0 (200) | 3,8 (8.3) | 3,4 (7.3) | 2,9 (6.3) | 3,8 (8.2) | 3,3 (7.2) | 2,8 (6.2) | 3,6 (7.8) | 3,1 (6.8) | 2,6 (5.7) | 3,5 (7.5) | 3,0 (6.5) | 2,5 (5.4) |
| 64,0 (210) | 3,4 (7.5) | 2,9 (6.6) | 2,5 (5.7) | 3,3 (7.4) | 2,9 (6.5) | 2,4 (5.5) | 3,1 (7.0) | 2,7 (6.1) | 2,3 (5.1) | 3,0 (6.8) | 2,6 (5.9) | 2,1 (4.8) |
| 68,0 (230) | | 2,6 (5.4) | 2,2 (4.5) | 2,9 (6.1) | 2,5 (5.3) | 2,1 (4.4) | 2,7 (5.7) | 2,4 (4.9) | 2,0 (4.0) | 2,6 (5.5) | 2,2 (4.7) | 1,8 – |
| 72,0 (240) | | | | | 2,2 (4.7) | | 2,4 (5.1) | 2,0 (4.4) | | 2,3 (4.9) | 1,9 (4.1) | |
| 76,0 (250) | | | | | | | | | | 2,0 (4.4) | | |

### Fixed Jib Length 18,3 m (60 ft)

| Luffing Jib m (ft) | 42,7 (140) | | | 45,7 (150) | | | 48,8 (160) | | | 51,8 (170) | | |
|---|---|---|---|---|---|---|---|---|---|---|---|---|
| Boom m (ft) | 42,7 (140) | 48,8 (160) | 54,9 (180) | 42,7 (140) | 48,8 (160) | 54,9 (180) | 42,7 (140) | 48,8 (160) | 54,9 (180) | 42,7 (140) | 48,8 (160) | 54,9 (180) |
| Radius | | | | | | | | | | | | |
| 50,0 (165) | – (11.8) | | | | | | | | | | | |
| 52,0 (175) | 5,0 (10.7) | 4,5 (9.6) | | 5,0 (10.5) | – (9.4) | | – (10.2) | | | | | |
| 56,0 (190) | 4,4 (9.3) | 3,9 (8.3) | 3,3 (6.9) | 4,4 (9.2) | 3,9 (8.1) | – (6.6) | 4,2 (8.8) | – (7.7) | – (6.1) | 4,0 (8.5) | – (7.4) | |
| 60,0 (200) | 3,9 (8.5) | 3,5 (7.5) | 2,9 (6.3) | 3,8 (8.3) | 3,4 (7.4) | 2,8 (6.1) | 3,7 (8.0) | 3,2 (7.0) | 2,5 (5.6) | 3,6 (7.7) | 3,1 (6.7) | 2,4 (5.3) |
| 62,0 (210) | 3,7 (7.8) | 3,3 (6.8) | 2,7 (5.8) | 3,6 (7.6) | 3,2 (6.7) | 2,6 (5.5) | 3,5 (7.2) | 3,0 (6.3) | 2,4 (5.1) | 3,3 (7.0) | 2,9 (6.0) | 2,3 (4.8) |
| 64,0 (230) | 3,5 (6.5) | 3,0 (5.6) | 2,6 (4.7) | 3,4 (6.3) | 3,0 (5.5) | 2,4 (4.5) | 3,2 (6.0) | 2,8 (5.1) | 2,3 (4.1) | 3,1 (5.7) | 2,7 (4.8) | 2,1 – |
| 68,0 (240) | 3,1 (5.9) | 2,7 (5.1) | 2,2 (4.2) | 3,0 (5.8) | 2,6 (5.0) | 2,2 (4.0) | 2,9 (5.4) | 2,5 (4.6) | 1,9 – | 2,7 (5.1) | 2,3 (4.3) | 1,8 – |
| 72,0 (250) | 2,7 (5.4) | 2,4 (4.6) | 2,0 – | 2,6 (5.2) | 2,3 (4.5) | 1,9 – | 2,5 (4.9) | 2,1 (4.1) | | 2,4 (4.6) | 2,0 – | |
| 76,0 (260) | 2,4 – | 2,1 (4.1) | | 2,3 (4.7) | 2,0 (4.0) | | 2,2 (4.4) | 1,8 – | | 2,1 (4.2) | | |
| 80,0 (270) | | | | | | | 1,9 (4.0) | | | | | |

model 777

Manitowoc

Meets ANSI B30.5 Requirements - Capacities do not exceed 75% of static tipping load.
NOTICE: This capacity chart is for reference only and must not be used for lifting purposes.

# CraneCARE℠

**CraneCARE** is Manitowoc's comprehensive service and support program. It includes classroom and on-site training, prompt parts availability, expert field service, technical support and documentation — for every one of the more than 7,000 Manitowoc cranes currently in use throughout the world.

That's commitment you won't find anywhere else.

That's **CraneCARE**.

## Service Training

Manitowoc specialists work with you in our training center and in the field to make sure you know how to get maximum performance, reliability, and life from your cranes.

Manitowoc Cranes Technical Training Center provides valuable multi-level training, which is available for all models and attachments, in the following format:

- **Basic** – Provides technicians with the basic skills required in our Level I and II classes covering hydraulic and electrical theory and schematics, pump, motor, control, and LMI operation, and the use of meters and gauges.

- **Level 1** – This model-specific class covers theory and offers hands-on training and trouble shooting for all crane systems.

- **Level 2** – This model-specific class provides in depth coverage of all crane systems and components, and advanced troubleshooting of simulated faults. (Requires Level 1.)

- **Level 3/Masters** – Covering all EPIC models and the 4100W, this class stresses high level system knowledge and trouble shooting of simulated faults. (Requires Level 2.)

## Parts Availability

Genuine Manitowoc replacement parts are accessible through your distributor 24 hours a day, 7 days a week, 365 days a year.

**Service Interval Kits**
Provides all the parts required by Manitowoc's Preventative Maintenance Checklist.

**Hydraulic Filter Kit – Part No. 495138-0**
Consists of the following:

- Filter Element - No substitutions allowed (1x Part No. 427283-0)
- Filter Element (1x Part No. 427292-0)
- Filter Element (Spin on) - No substitutions allowed (1x Part No. 427282-0)

**Cummins Model QSC8.3 – C340 Diesel – Part No. 414176-0 Service Interval Kits** 35

**200 Hour Kit – Part No. 495198-0**
Consists of the following:

**Engine**

- Air Cleaner - Disposable (1x Part No. 347066-0)
- Oil Filter (1x Part No. 413653-0)
- Water Filter (1x Part No. 413654-0)
- Separator - Fuel/Water (1x Part No. 413652-0)

**1,000 Hour Kit – Part No. 495199-0**
Consists of the following:

**Engine**

- Air Cleaner - Disposable (1x Part No. 347066-0)
- Oil Filter (1x Part No. 413653-0)
- Water Filter (1x Part No. 413654-0)
- Separator - Fuel/Water (1x Part No. 413652-0)

**Hydraulic**

- Filter Element - No substitutions allowed (1x Part No. 427283-0)
- Filter Element (1x Part No. 427292-0)
- Filter Element (Spin on) - No substitutions allowed (1x Part No. 427282-0)

**2,000 Hour Kit – Part No. 495202-0**
Consists of the following:

**Engine**

- Air Cleaner - Disposable (1x Part No. 347066-0)
- Oil Filter (1x Part No. 413653-0)
- Water Filter (1x Part No. 413654-0)
- Separator - Fuel/Water (1x Part No. 413652-0)
- Belt - Alternator & Fan (1x Part No. 413650-0)
- Belt - Compressor (1x Part No. 413651-0)

**Hydraulic**

- Filter Element - No substitutions allowed (1x Part No. 427283-0)
- Filter Element (1x Part No. 427292-0)
- Filter Element (Spin on) - No substitutions allowed (1x Part No. 427282-0)
- Filter Element (Spin on) - Hydraulic (3x Part No. 427278-0)
- Filter Element (Spin on) - Hydraulic (1x Part No. 427291-0)
- Filter Element (Spin on) - Hydraulic (1x Part No. 427279-0)

# CraneCARE℠

**Hydraulic Test Kit – Part No. 499791-6**
Protect your investment by demanding Genuine Manitowoc Parts Service Kits. The Hydraulic Service Kit consist of the following:



- All hydraulic fittings to access all pressures and flows
- Hydraulic flow meters and pressure gauges to record hydraulic data.
- Electrical "Break out" harnesses to access voltages on all electrical circuits on all machines.
- Fluke® Digital volt ohm meter, as used in all Manitowoc service literature.

**Hydraulic Test Kit with case – Part No. 499792-9**
The above kit (Part No. 4299791-6) plus a custom heavy-duty carrying case.

**U.S. Standard Tools Kit – Part No. 22205-1**
All standard tools needed to properly maintain and service your crane. (Does not include torque wrench.)

## Field Service

Factory-trained service experts are always ready to help maintain your crane's peak performance.

For a worldwide listing of dealer locations, please consult our website at: **www.manitowoccranes.com**

## Technical Support

Manitowoc's dealer network and factory personnel are available 24 hours a day, 7 days a week, 365 days a year to answer your technical questions and more, with the help of computerized programs that simplify crane selection, lift planning, and ground-bearing calculations.

For a worldwide listing of dealer locations, please consult our website at: **www.manitowoccranes.com**

## Technical Documentation

Manitowoc has the industry's most extensive documentation, and the easiest to understand, available in major languages and formats that include print, disk, and videotape.

A complete set of Operator's, Parts, Capacity, Vendor, and Service Technician's Manuals are shipped with each crane.

Additional copies available through your Authorized Manitowoc Distributor.

- **Crane Operator's Manual – Part No. 899721**
- **Crane Parts Manual – Part No. 899720**
- **Crane Capacity Chart Manual – Part No. 899794**
- **Attachment Capacity Chart Manual – Part No. 899795**

- **Crane Vendor Manual – Part No. 899722**
- **Service Technician's Manual (EPIC) – Part No. 899732**

CD rom versions of the Operator's and Parts Manuals are shipped with each crane.

Also available are the following CDs:

- **777 Service – Yearly subscription CD – Part No. 899760-0**
- **Ground Bearing Pressure Estimator CD – Part No. 899765-0**
- **Crane Selection and Planning Software (CompuCRANE©) CD – Part No. 899766-0**

- **EPIC® Crane Library** consisting of capacity charts, range diagrams,wire rope specifications, travel specifications, crane weights, counter weight arrangement, luffing jib raising procedures, operating range diagrams, drum and lagging charts, and wind condition charts. **CD – Part No. 899801-0**

Available from your Authorized Manitowoc Cranes Distributor, these VHS videos are available in NTSC, PAL and SECAM formats.

- **Model 777 Assembly Video – Part No. 899826-0**
- **Model 777 Operation Video – Part No. 899827-0**
- **Model 777 Lubrication Video – Part No. 899828-0**
- **Your Capacity Chart Video – Part No. 899737**
- **Respect the Limits Video – Part No. 899734**
- **Crane Safety Video – Part No. 899736**
- **Boom Inspection/Repair Video – Part No. 899738**

## CraneCARE Package

Manitowoc has assembled all of the available literature, CDs, and videos listed above plus several Manitowoc premiums into one complete CraneCARE Package.

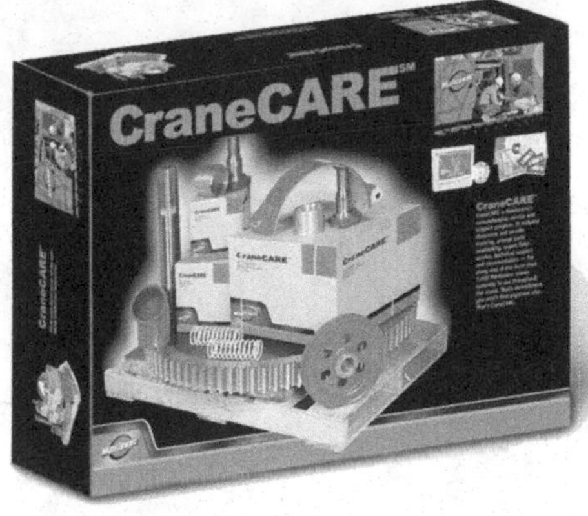

# Notes

37

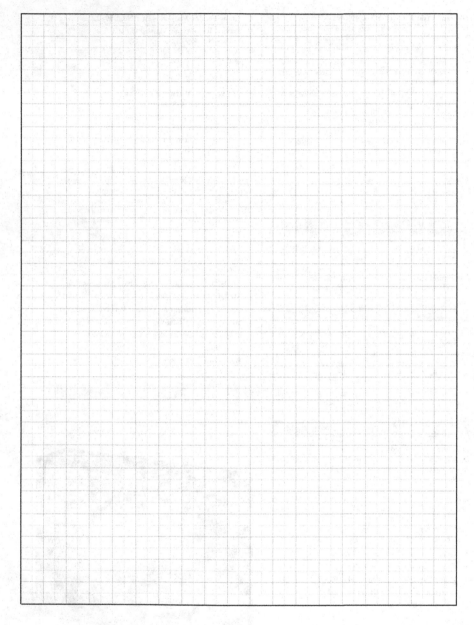

model 777

Manitowoc

## Notes

38

## Notes

**Backed by Manitowoc**

# CraneCARE℠

CraneCARE is Manitowoc's comprehensive service and support program. It includes classroom and on-site training, prompt parts availability, expert field service, technical support and documentation — for every one of the more than 7,000 Manitowoc cranes currently in use throughout the world. That's commitment you won't find anywhere else. That's CraneCARE.

Manitowoc Cranes, Inc.
P.O. Box 70
Manitowoc, WI, USA 54221-0070
Telephone 920-684-6621
Facsimile 920-683-6277
www.manitowoccranes.com

PRODUCT IMPROVEMENTS MAY CHANGE SPECIFICATIONS ©2001 MANITOWOC 0901-777-PG-US-E-10000

Grove Manitowoc National Crane Potain

# MD 485 B M20

## Data Sheet

FEM 1.001-A3
ASCE 7-10

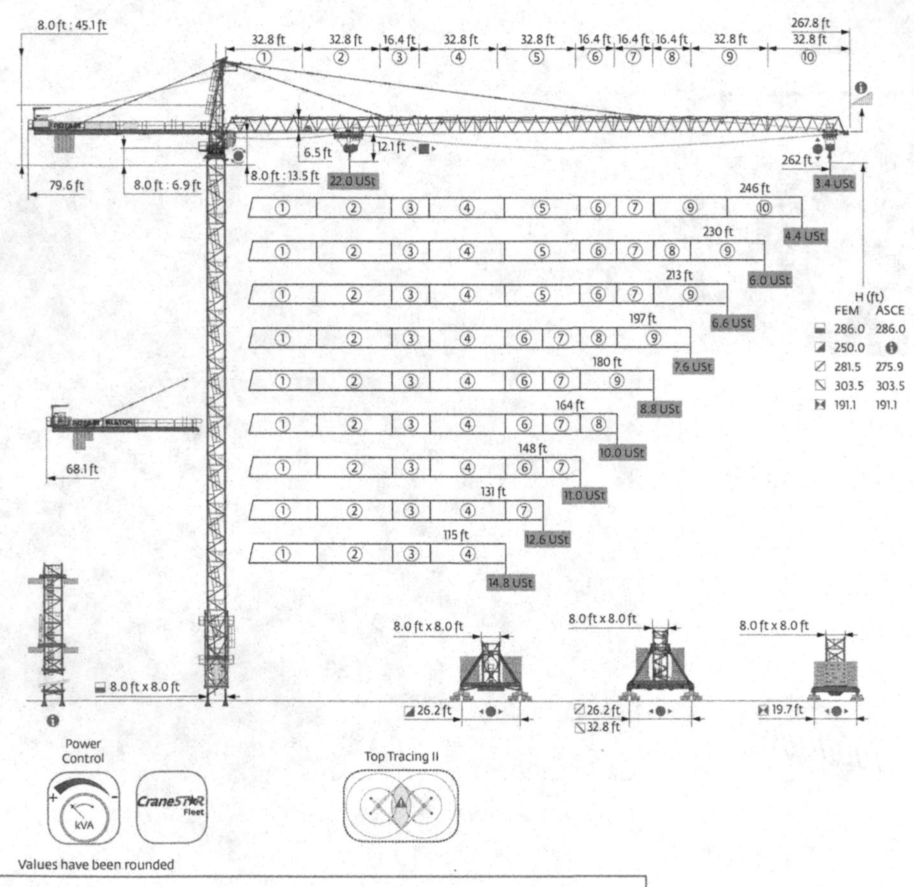


Values have been rounded

# Mast

8.0 ft

115 ft → 262 ft

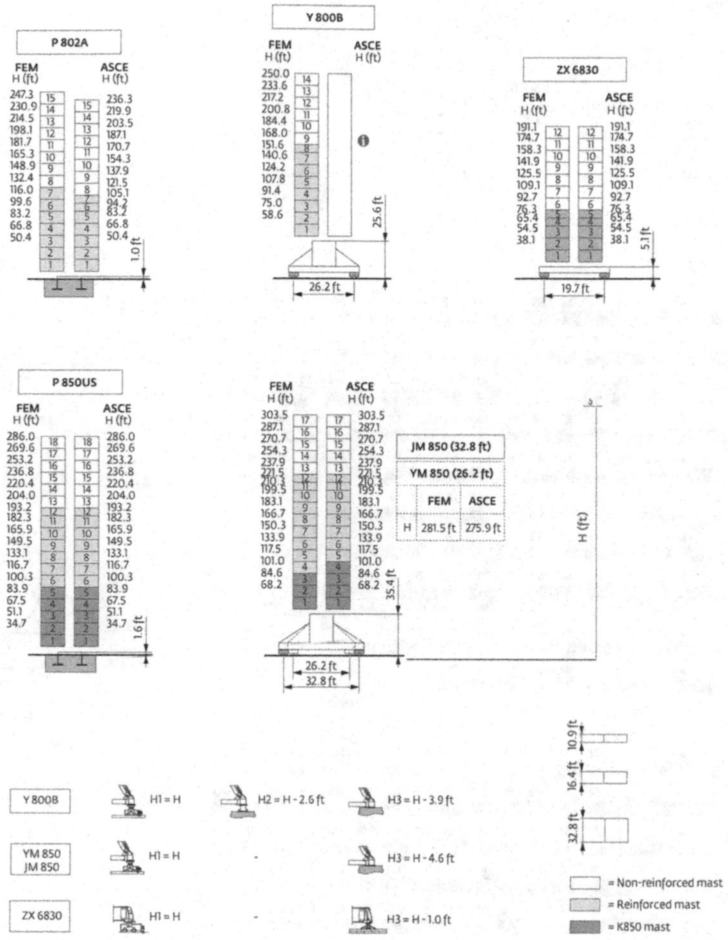


*Note: When "ASCE" is noted in this data sheet it is referring to 115 mph Wind Zone, Exposure B, Design Wind Speed =98 mph. See back cover for design wind speed calculations.*

## Anchorages (Consult us for ASCE 7-10 values)

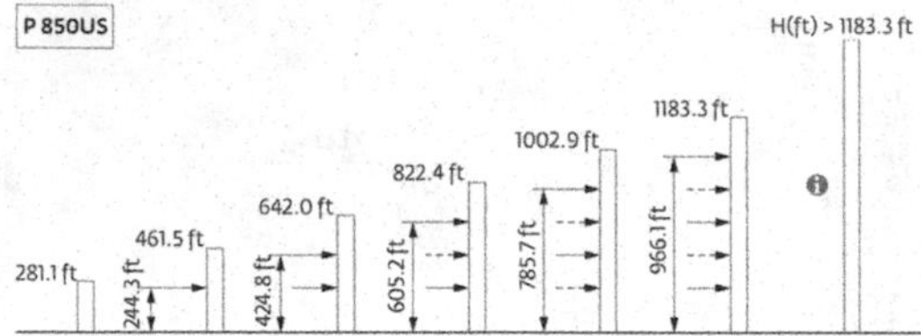


## Load charts

| 262 ft | 12 | ▶ | 57 | 66 | 72 | 82 | 89 | 98 | 99 | 108 | 115 | 121 | 131 | 148 | 164 | 180 | 197 | 213 | 230 | 246 | 262 | ft |
|---|---|---|---|---|---|---|---|---|---|---|---|---|---|---|---|---|---|---|---|---|---|---|
| | | | 22.0 | 18.5 | 16.4 | 14 | 12.7 | 11.1 | 11.0 | 11.0 | 10.3 | 9.6 | 8.7 | 7.5 | 6.6 | 5.8 | 5.2 | 4.6 | 4.2 | 3.7 | 3.4 | USt |
| 246 ft | 12 | ▶ | 62 | 66 | 72 | 82 | 89 | 98 | 105 | 109 | 119 | 121 | 131 | 148 | 164 | 180 | 197 | 213 | 230 | 246 | ft | |
| | | | 22.0 | 20.6 | 18.4 | 15.8 | 14.3 | 12.6 | 11.6 | 11.0 | 11.0 | 10.8 | 9.8 | 8.5 | 7.5 | 6.6 | 6.0 | 5.4 | 4.9 | 4.4 | USt | |
| 230 ft | 12 | ▶ | | 72 | 72 | 82 | 89 | 98 | 105 | 115 | 121 | 127 | 139 | 148 | 164 | 180 | 197 | 213 | 230 | ft | | |
| | | | | 22.0 | 21.9 | 18.8 | 17.2 | 15.2 | 14 | 12.6 | 11.7 | 11.0 | 11.0 | 10.3 | 9.0 | 8.2 | 7.3 | 6.6 | 6.0 | USt | | |
| 213 ft | 12 | ▶ | | | 72 | 82 | 89 | 98 | 105 | 115 | 121 | 127 | 139 | 148 | 164 | 180 | 197 | 213 | ft | | | |
| | | | | | 22.0 | 19.0 | 17.3 | 15.2 | 14.1 | 12.6 | 11.7 | 11.0 | 11.0 | 10.4 | 9.1 | 8.2 | 7.3 | 6.6 | USt | | | |
| 197 ft | 12 | ▶ | | | 74 | 82 | 89 | 98 | 105 | 115 | 121 | 132 | 144 | 148 | 164 | 180 | 197 | ft | | | | |
| | | | | | 22.0 | 19.6 | 18.0 | 15.9 | 14.7 | 13.1 | 12.2 | 11.0 | 11.0 | 10.7 | 9.5 | 8.5 | 7.6 | USt | | | | |
| 180 ft | 12 | ▶ | | | 77 | 82 | 89 | 98 | 105 | 115 | 121 | 131 | 136 | 149 | 164 | 180 | ft | | | | | |
| | | | | | 22.0 | 20.5 | 18.7 | 16.5 | 15.2 | 13.7 | 12.8 | 11.6 | 11.0 | 11.0 | 9.9 | 8.8 | USt | | | | | |
| 164 ft | 12 | ▶ | | | 78 | 82 | 89 | 98 | 105 | 115 | 121 | 131 | 138 | 151 | 164 | ft | | | | | | |
| | | | | | 22.0 | 20.8 | 19.0 | 16.8 | 15.5 | 13.9 | 13.0 | 11.8 | 11.0 | 11.0 | 10.0 | USt | | | | | | |
| 148 ft | 12 | ▶ | | | 79 | 82 | 89 | 98 | 105 | 115 | 121 | 131 | 138 | 148 | ft | | | | | | | |
| | | | | | 22.0 | 21.1 | 19.2 | 17.0 | 15.7 | 14.1 | 13.1 | 11.9 | 11.2 | 10.3 | USt | | | | | | | |
| | | | | | | | | | | | | | | 11.0 | USt | | | | | | | |
| 131 ft | 12 | ▶ | | | 79 | 82 | 89 | 98 | 105 | 115 | 121 | 131 | ft | | | | | | | | | |
| | | | | | 22.0 | 21.2 | 19.4 | 17.1 | 15.8 | 14.2 | 13.2 | 12.0 | USt | | | | | | | | | |
| 115 ft | 12 | ▶ | | | 80 | 82 | 89 | 98 | 105 | 115 | ft | | | | | | | | | | | |
| | | | | | 22.0 | 21.3 | 19.5 | 17.2 | 15.9 | 14.2 | USt | | | | | | | | | | | |

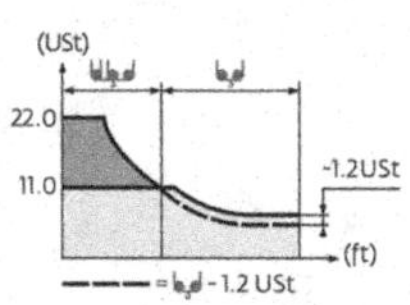


| 262 ft | 8.2 | ▶ | 58 | 66 | 72 | 82 | 89 | 98 | 103 | 105 | 115 | 121 | 131 | 148 | 164 | 180 | 197 | 213 | 230 | 246 | 262 | ft |
|---|---|---|---|---|---|---|---|---|---|---|---|---|---|---|---|---|---|---|---|---|---|---|
| | | | 22.0 | 19.1 | 17.1 | 14.7 | 13.3 | 11.7 | 11.0 | 11.0 | 9.9 | 9.3 | 8.4 | 7.2 | 6.2 | 5.4 | 4.7 | 4.3 | 3.7 | 3.4 | 3.0 | USt |
| 246 ft | 8.2 | ▶ | | 64 | 72 | 82 | 89 | 98 | 105 | 114 | 116 | 121 | 131 | 148 | 164 | 180 | 197 | 213 | 230 | 246 | ft | |
| | | | | 22.0 | 19.0 | 16.3 | 15.0 | 13.1 | 13.7 | 11.0 | 11.0 | 10.4 | 9.5 | 8.2 | 7.2 | 6.3 | 5.5 | 5.0 | 4.4 | 4.1 | USt | |
| 230 ft | 8.2 | ▶ | | | 73 | 82 | 89 | 98 | 105 | 115 | 121 | 133 | 135 | 148 | 164 | 180 | 197 | 213 | 230 | ft | | |
| | | | | | 22.0 | 19.5 | 17.9 | 15.8 | 14.7 | 13.1 | 12.3 | 11.0 | 11.0 | 9.9 | 8.7 | 7.7 | 6.9 | 6.2 | 5.6 | USt | | |
| 213 ft | 8.2 | ▶ | | | 74 | 82 | 89 | 98 | 105 | 115 | 121 | 133 | 135 | 148 | 164 | 180 | 197 | 213 | ft | | | |
| | | | | | 22.0 | 19.5 | 17.9 | 15.8 | 14.7 | 13.1 | 12.3 | 11.0 | 11.0 | 9.9 | 8.7 | 7.7 | 6.9 | 6.3 | USt | | | |
| 197 ft | 8.2 | ▶ | | | 76 | 82 | 89 | 98 | 105 | 115 | 121 | 131 | 137 | 140 | 164 | 180 | 197 | ft | | | | |
| | | | | | 22.0 | 20.3 | 18.5 | 16.4 | 15.2 | 13.7 | 12.8 | 11.7 | 11.0 | 11.0 | 9.0 | 8.0 | 7.3 | USt | | | | |
| 180 ft | 8.2 | ▶ | | | 79 | 82 | 89 | 98 | 105 | 115 | 121 | 131 | 142 | 145 | 164 | 180 | ft | | | | | |
| | | | | | 22.0 | 21.1 | 19.3 | 17.1 | 15.9 | 14.3 | 13.3 | 12.1 | 11.0 | 11.0 | 9.5 | 8.5 | USt | | | | | |
| 164 ft | 8.2 | ▶ | | | 80 | 82 | 89 | 98 | 105 | 115 | 121 | 131 | 144 | 147 | 164 | ft | | | | | | |
| | | | | | 22.0 | 21.4 | 19.6 | 17.4 | 16.1 | 14.6 | 13.6 | 12.3 | 11.0 | 11.0 | 9.7 | USt | | | | | | |
| 148 ft | 8.2 | ▶ | | | 81 | 82 | 89 | 98 | 105 | 115 | 121 | 131 | 145 | 148 | ft | | | | | | | |
| | | | | | 22.0 | 21.6 | 19.8 | 17.5 | 16.3 | 14.7 | 13.8 | 12.5 | 11.0 | 11.0 | USt | | | | | | | |
| 131 ft | 8.2 | ▶ | | | 81 | 82 | 89 | 98 | 105 | 115 | 121 | 131 | ft | | | | | | | | | |
| | | | | | 22.0 | 21.8 | 20.0 | 17.6 | 16.4 | 14.8 | 13.8 | 12.6 | USt | | | | | | | | | |
| 115 ft | 8.2 | ▶ | | | 82 | 82 | 89 | 98 | 105 | 115 | ft | | | | | | | | | | | |
| | | | | | 22.0 | 21.9 | 20.1 | 17.7 | 16.4 | 14.8 | USt | | | | | | | | | | | |

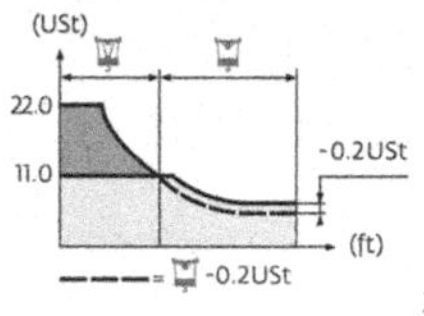

## Base ballast

| | | | | | | | | | | | | | | | | | | |
|---|---|---|---|---|---|---|---|---|---|---|---|---|---|---|---|---|---|---|
| 8.0 ft | Y 800B | H (ft) | 250.0 | 233.6 | 217.2 | 200.8 | 184.4 | 168.0 | 157.1 | 140.6 | 124.2 | 107.8 | 91.4 | 75.0 | 58.6 | | | |
| | FEM | (USt) | 145.5 | 119.0 | 79.4 | 66.1 | 66.1 | 66.1 | 66.1 | 66.1 | 66.1 | 66.1 | 66.1 | 66.1 | 66.1 | | | |
| | ASCE | (USt) | | | | | | | i | | | | | | | | | |
| | YM 850 | H (ft) | 281.5 | 275.9 | 265.1 | 259.5 | 248.7 | 243.1 | 232.3 | 226.7 | 215.9 | 210.3 | 199.5 | 193.9 | 183.1 | 177.5 | 166.7 | |
| | FEM | (USt) | 224.9 | - | 185.2 | - | 145.5 | - | 119.0 | - | 79.4 | - | 66.1 | - | 52.9 | - | 52.9 | |
| | ASCE | (USt) | - | 211.6 | | | | | | | i | | | | | | | |
| | JM 850 | H (ft) | 303.5 | 287.1 | 270.7 | 254.3 | 237.9 | 221.5 | 210.3 | 199.5 | 183.1 | 166.7 | 150.3 | 133.9 | 117.5 | 101.0 | 84.6 | 68.2 |
| | FEM | (USt) | 198.4 | 158.7 | 119.0 | 92.6 | 66.1 | 52.9 | 52.9 | 52.9 | 52.9 | 52.9 | 52.9 | 52.9 | 52.9 | 52.9 | 52.9 | 52.9 |
| | ASCE | (USt) | 198.4 | | | | | | | | i | | | | | | | |
| | ZX 6830 | H (ft) | 191.1 | 174.7 | 158.3 | 141.9 | 125.5 | 109.1 | 92.7 | 76.3 | 65.4 | 54.5 | 38.1 | | | | | |
| | FEM | (USt) | 155.4 | 155.4 | 155.4 | 155.4 | 155.4 | 155.4 | 155.4 | 155.4 | 155.4 | 155.4 | 155.4 | | | | | |
| | ASCE | (USt) | 155.4 | | | | | | i | | | | | | | | | |

## Counter-jib ballast

| | (lb) (+/- 5%) | | | 100 LVF - 150 LCC | | | 150 LVF GH | | |
|---|---|---|---|---|---|---|---|---|---|
| | | | | 13,228 lb | 8818 lb | (lb) | 13,228 lb | 8818 lb | (lb) |
| **262 ft** | 41,877 | 40,929 | 42,825 | 5 | 0 | 66,139 | 3 | 2 | 57,320 |
| **246 ft** | 40,113 | 39,165 | 41,061 | 4 | 1 | 61,729 | 3 | 1 | 48,502 |
| **230 ft** | 39,661 | 38,713 | 40,609 | 4 | 1 | 61,729 | 3 | 1 | 48,502 |
| **213 ft** | 38,118 | 37,170 | 39,066 | 3 | 2 | 57,320 | 2 | 2 | 44,092 |
| **197 ft** | 35,836 | 34,888 | 36,784 | 3 | 1 | 48,502 | 2 | 1 | 35,274 |
| **180 ft** | 34,293 | 33,345 | 35,241 | 2 | 2 | 44,092 | 1 | 2 | 30,865 |
| **164 ft** | 33,235 | 32,287 | 34,183 | 3 | 2 | 57,320 | 2 | 2 | 44,092 |
| **148 ft** | 31,802 | 30,854 | 32,750 | 3 | 1 | 48,502 | 2 | 1 | 35,274 |
| **131 ft** | 29,895 | 28,947 | 30,843 | 2 | 2 | 44,092 | 1 | 2 | 30,865 |
| **115 ft** | 24,659 | 23,711 | 25,607 | 2 | 1 | 35,274 | 1 | 1 | 22,046 |

CBC - 13,228 lb

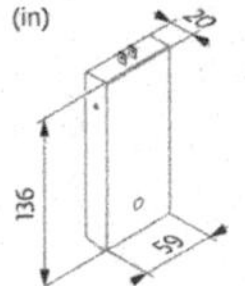


CBD - 8818 lb

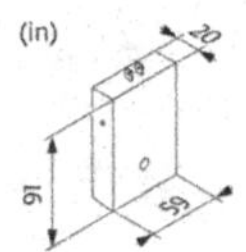

## Component weights

Crane upper : 262 ft - 100 LVF

x 13

x 9

| | | L (ft) | W (ft) | H (ft) | lb (+/- 5%) |
|---|---|---|---|---|---|
| Counter-jib | | 35.4 | 10.2 | 5.6 | 8300 |
| | | 12.1 | 6.2 | 5.6 | 2172 |
| | | 26.9 | 6.2 | 5.6 | 4575 |
| Towerhead | | 13.8 | 7.3 | 38.7 | 16,546 |
| Cab | UltraView | 16.5 | 7.3 | 8.2 | 3704 |
| Pivot | 8.0 ft | 12.5 | 14.0 | 9.7 | 20,349 |
| Hoisting winch (+ rope) | 100 LVF | 10.4 | 5.2 | 6.2 | 9822 |
| | 180 LVF GH | 14.0 | 6.6 | 7.7 | 20,349 |
| | 150 LCC | 12.3 | 5.4 | 6.2 | 12,357 |
| Jib section | ① | 33.5 | 6.6 | 7.8 | 7081 |
| | ② 10 DVF | 34.0 | 6.2 | 7.4 | 8278 |
| | ④ | 33.9 | 6.2 | 7.3 | 4641 |
| | ⑤ | 34.3 | 6.2 | 7.3 | 4023 |
| | ⑨ | 33.3 | 6.2 | 6.5 | 2800 |
| | ⑩ | 33.2 | 6.2 | 6.4 | 1973 |
| Jib section | ③ | 17.6 | 6.2 | 7.4 | 3186 |
| | ⑥ | 17.6 | 6.2 | 7.3 | 2238 |
| | ⑦ | 17.5 | 6.2 | 7.4 | 2535 |
| | ⑧ | 17.1 | 6.2 | 6.6 | 1676 |
| Trolley | 22.0 USt | 5.9 | 7.3 | 5.3 | 1455 |
| Hook block | 22.0 USt | 3.9 | 1.4 | 7.4 | 1940 |
| Trolley | 22.0 USt | 13.5 | 7.2 | 3.8 | 2635 |
| Trolley | 11.0 USt | 7.0 | 7.2 | 3.8 | 1422 |
| Hook block | 22.0 USt | 6.0 | 1.1 | 7.3 | 1951 |
| | 11.0 USt | 3.8 | 0.7 | 5.8 | 981 |

5

# Component weights

| | | | L (ft) | W (ft) | H (ft) | lb (+/- 5%) |
|---|---|---|---|---|---|---|
| Climbing cage | 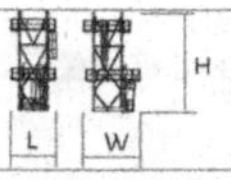 | ◫8.0 ft | 15.2 | 19.0 | 33.6 | 28,484 |
| K 850/KR 849B<br>KRMT 849A<br>K 849A<br>K 850/KR 849A<br>KMT 850.10A<br>KRMT 849C | 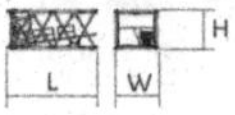 | ◫8.0 ft | 33.6<br>17.2<br>17.2<br>17.2<br>17.5<br>11.7 | 8.3<br>8.4<br>8.3<br>8.3<br>8.3<br>8.4 | 8.2<br>8.3<br>8.2<br>8.2<br>8.2<br>8.3 | 20,878<br>9017<br>7496<br>12,291<br>12,015<br>7066 |
| Fixing angles | 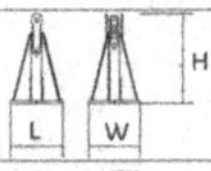 | P 802A | 2.5 | 2.5 | 4.2 | 1193 |
| Fixing angles | 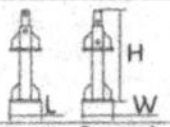 | P 850US | 2.3 | 2.3 | 5.5 | 2127 |
| Chassis mast | 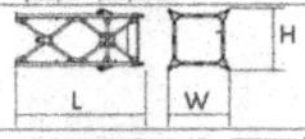 | Y 800B | 19.8 | 9.6 | 9.6 | 19,004 |
| Struts | 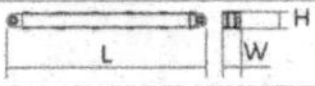 | Y 800B | 18.1 | 1.6 | 1.5 | 2447 |
| 1/2 Side member | 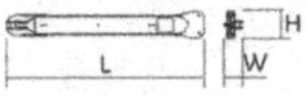 | Y 800B | 18.6 | 4.1 | 2.4 | 3351 |
| Side member | 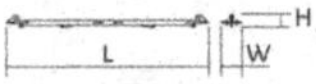 | Y 800B | 39.4 | 4.1 | 2.4 | 6724 |
| Ballast support | 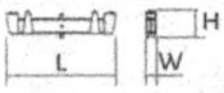 | Y 800B | 12.3 | 1.2 | 3.0 | 2392 |
| Chassis beam | 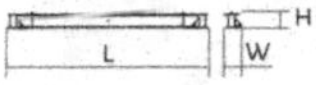 | Y 800B | 28.5 | 2.7 | 2.4 | 4938 |
| Central cross (transport position) | 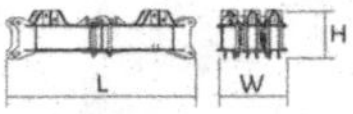 | YM 850<br>JM 850 | 17.1 | 5.6 | 4.9 | 14,771 |
| Chassis mast | 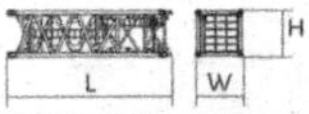 | YM 850<br>JM 850 | 28.7 | 8.2 | 8.2 | 32,187 |
| Chassis girder | 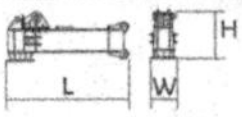 | YM 850<br>JM 850 | 12.5<br>17.1 | 3.0<br>3.0 | 5.1<br>5.1 | 6173<br>7055 |
| Chassis ties | 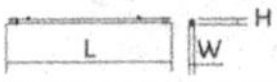 | YM 850<br>JM 850 | 23.6 | 0.8 | 1.1 | 551 |
| Struts | 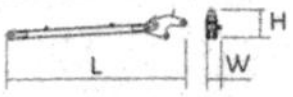 | YM 850<br>JM 850 | 24.6<br>26.9 | 2.5<br>2.5 | 4.3<br>4.3 | 4630<br>5071 |
| Cross girder | 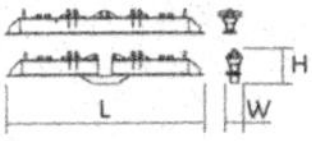 | ZX 6830 | 29.9<br>29.9 | 3.7<br>2.5 | 3.6<br>4.9 | 11,607<br>12,004 |

# Mechanisms

| 480 V - 60 Hz | | | (single fall hoisting) | | | | | (double fall hoisting) | | | | | hp | kW | (cable) |
|---|---|---|---|---|---|---|---|---|---|---|---|---|---|---|---|
| (hoist) | 100 LVF 50 Optima | fpm | 118 | 151 | 207 | 308 | | 59 | 75 | 105 | 154 | | 100 | 75 | 3340 ft |
| | | USt | 11.0 | 8.3 | 5.5 | 2.8 | | 22.0 | 16.5 | 11.0 | 5.5 | | | | |
| | 180 LVF 50 GH Optima | fpm | 210 | 266 | 361 | 561 | 804 | 112 | 144 | 210 | 344 | 400 | 180 | 132 | 3937 ft |
| | | USt | 11.0 | 8.3 | 5.5 | 2.8 | 0.9 | 22.0 | 16.5 | 11.0 | 5.5 | 3.5 | | | |
| | 150 LCC 50 | fpm | 190 | 223 | 282 | 374 | 453 | 95 | 112 | 141 | 187 | 226 | 150 | 110 | 2579 ft |
| | | USt | 11.0 | 8.3 | 5.5 | 2.8 | 1.4 | 22.0 | 16.5 | 11.0 | 5.5 | 2.8 | | | |
| (trolley) | 10 DVF 10 | fpm | 0 → 262 (22.0 USt) 0 → 328 (13.2 USt) 0 → 361 (6.6 USt) | | | | | | | | | | 10 | 7.4 | |
| (slewing) | RVF 183 Optima+ | rpm | 0 → 0.8 | | | | | | | | | | 3 x 12 | 3 x 9 | |
| Y 800B (travel) | RT 584 A1 - 2V | fpm | 28 - 56 | | | | | | | | | | 8 x 8.4 | 8 x 6.2 | |
| YM 850 JM 850 (travel) | | | (i) | | | | | | | | | | | | |
| ZX 6830 (travel) | RT 664 A2B - 2V | fpm | 62 - 125 | | | | | | | | | | 6 x 8.4 | 6 x 6.2 | |

| IEC 60204-32 | kVA |
|---|---|
| 480 V (+6% -10%) 60 Hz | 100 LVF : 106 kVA<br>150 LCC : 170 kVA<br>180 LVF GH : 170 → 98 kVA |

**100 LVF 50 Optima**

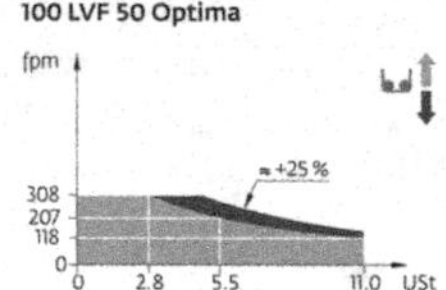

## Key

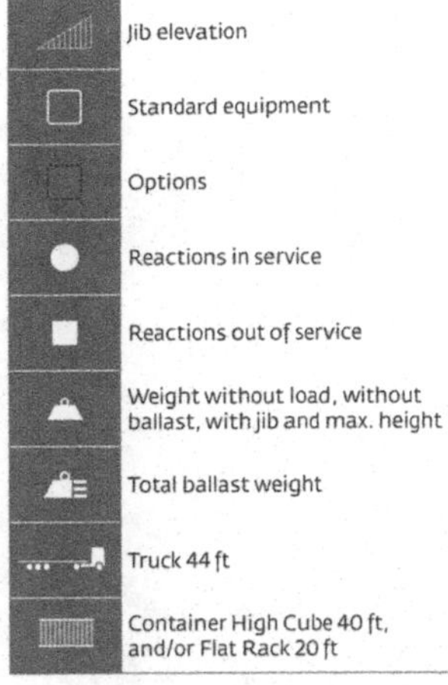


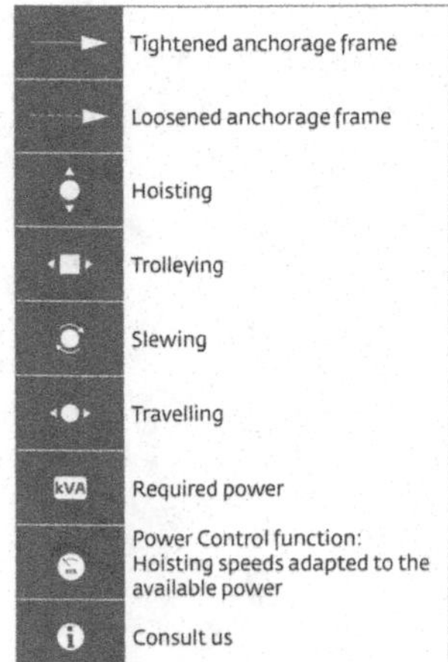


Note: These mast combinations meet the EN 14439 and ASME B30.3-2012 specifications for "out of service" wind conditions, provided the illustrated wind speed matches required design wind for the location of the tower crane. The "out of service" design wind speed was determined in accordance with ASCE 7-10, Figure 26.5-A. The wind velocity, used for this configuration was 98 mph (158 kph), which represents a nominal design 3-second wind gust at 33 ft (10 m) above ground for Exposure B category A. Factor of 0.85 was applied to the 50-year ultimate design wind speed of 115 mph (185 kph), per ASCE 37-02, with the assumption that this crane is considered a temporary structure used during a construction period of 2 years or less.

Constant improvement and engineering progress make it necessary that we reserve the right to make specification, equipment and price changes without notice. Illustrations shown may include optional equipment and accessories, and may not include all standard equipment.

www.manitowoccranes.com


**Manitowoc Cranes**

**Regional Headquarters**

**Americas**
Manitowoc, Wisconsin, USA
Tel: +1 920 684 6621
Fax: +1 920 683 6277

Shady Grove, Pennsylvania, USA
Tel: +1 717 597 8121
Fax: +1 717 597 4062

**Europe, Middle East, Africa**
Dardilly, France
Tel: +33 472 18 2020
Fax: +33 472 18 2000

**China**
Shanghai, China
Tel: +86 21 6457 0066
Fax: +86 21 6457 4955

**Greater Asia-Pacific**
Singapore
Tel: +65 6264 1188
Fax: +65 6862 4040


Potain MD 485 B M20
Code 04-026-.25M-0714

# Appendix C

## *Estimating Equipment Costs Using Simulations and Stochastic Methods*

### INTRODUCTION

Risk analysis involves the quantification of uncertainty, ambiguity, and variability. When estimating equipment ownership costs, sustained production rates, and other outputs essential to modern equipment fleet management decisions, each parameter is bounded by the limits found in historical data. If the data is available, it will always contain the lowest and highest observed values, as well as all other values in between. This allows the equipment manager to utilize analytical statistical methods to reduce the variability to a range of possible values; thereby reducing the uncertainty associated with the output. The data can be used as input to either a deterministic model or a stochastic model, which in turn provides the desired output to make equipment management decisions.

### MODELING EQUIPMENT DECISIONS

Determining the appropriate approach to modeling equipment costs involves an assessment of how much detail is both appropriate and possible. Intuitively, one would think that adding more detail would result in a more accurate cost estimate. However, that is only true if the added detail comes from accurate specific inputs. Generally, estimates are based on assumptions, which actually add noise to the analysis and reduce its ultimate accuracy. For example, in the previous chapters, a number of common assumptions were discussed for estimating equipment depreciation. However, those are authorized methodologies used for tax purposes and do not reflect the actual book value of the equipment at any given point in its useful life. Its salvage value is the price that the owner receives when it sells the equipment and not the value calculated using the straight line or double declining balance depreciation methods, which are merely assumptions. This aspect delineates the major shortcoming of deterministic methods of calculating equipment costs and other parameters: giving an input variable a single value carries with it a high probability that the given value will be wrong.

### DETERMINISTIC EQUIPMENT COST MODELS

The equipment cost models discussed in Chapter 2 are all deterministic models. While they all have proven useful, they all have the following limitations that must be understood when they are used to inform equipment fleet management decisions:

- Most rely on a mathematical average to characterize most input variable values. Average or mean values are skewed by extreme values in the data. Research has shown that using the mode or the median values from a dataset will provide more accurate estimates.
- By definition, historical data is not current since it comes from the past. Therefore, the assumption is that future changes will operate in the same manner as past changes.

Experience has shown that modeling dynamic systems that involve technology is difficult to do with accuracy by using a static algorithm that relies on deterministic values. An unforeseen technological advance can render a piece of equipment obsolete overnight. Hence, the analyst must take care to interpret the model's results based on past data in light of the best information with regard to future market trends.

- Sensitivity analysis involves rerunning the model using different high and low values for each assumed input variable. This is not only tedious and time-consuming but also provides only a point estimate of the changes induced in the model with no associated probability of occurrence. Hence, professional judgement will determine whether a given variable is sensitive to change.

## STOCHASTIC EQUIPMENT COST MODELS

Stochastic equipment cost models using Monte Carlo simulation permits the analysts to view all the possible outcomes for the decision under analysis and assess the impact of risk, allowing for better decision making under uncertainty. Monte Carlo simulation is a mathematical approach to quantitative account for risk in decision making. Monte Carlo simulation is based on game theory and relies on the development of a probability density function for the possible outcomes of a mathematical model by running a large number of iterations using values selected using a random number generator embedded in commercial computer software. It produces a range of possible outcomes and the probabilities with which they may occur. It also computes extreme outcomes on the high and low ends of the curve along with all possible consequences in between.

Monte Carlo simulation analyzes risk analysis through models that graphically illustrate all possible results. It does so by substituting a range of values, defined by individual probability distributions for each stochastic variable that has inherent uncertainty. It then iterates through the model using a different set of random values from the set of probability functions. Monte Carlo simulation often involves tens of thousands of recalculations and produces a distribution of possible outcome values. The resulting probability distribution allows the analyst to associate a statistical level of confidence with each possible value and informs the final decision by describing uncertainty in variables of a risk analysis. Monte Carlo simulation provides the following advantages over deterministic analyses:

- Probabilistic output: Possible results are provided with a probability of each outcome's occurrence.
- Graphical output: A Monte Carlo simulation generates graphs of different outcomes and their chances of occurrence, providing a visual means to communicate results.
- Sensitivity analysis: Most commercial software packages simultaneously evaluate variable sensitivity and its impact on outcomes. It is often displayed as a tornado diagram, thus again, providing a visual output.
- Scenario analysis: Monte Carlo simulation facilitates the analysis of scenarios by showing which inputs had which values together when certain outcomes occurred.
- Correlation of inputs. Monte Carlo simulation permits the analyst to model interdependent relationships between input variables.

## MONTE CARLO ANALYSIS EXAMPLE

Example C.1 is adapted from Minnesota Department of Transportation Report No. MN/RC 2015–16 *Major Equipment Life-Cycle Cost Analysis* [1] and demonstrates the use of Monte Carlo simulation to analyze equipment ownership cost.

**Example C.1**: The calculation of the equipment life in the example was performed using both deterministic and stochastic input variables. The Peurifoy method, detailed in Chapter 2, was employed to calculate the equipment ownership costs, using the input parameters to formulate the stochastic model consisting of solely cost variables. The costs were analyzed on an annual basis for all the parameters. The stochastic and deterministic life cycle cost analysis (LCCA) models use equations shown in Table 2.9 to determine the operating costs for the equipment.

## ECONOMIC LIFE ANALYSIS

Determining economic life for an equipment fleet is a critical component of the LCCA. The economic life, or the optimal time to sell a piece of equipment, requires the usage of equivalent uniform annual cost (EUAC) calculations to annualize all costs over the entire life span for a piece of equipment. In most instances, the year in which the lowest EUAC occurs will be the optimal economic life. This value is used for evaluation of the equipment fleet. The stochastic model uses confidence levels associated with the output. Also, the stochastic economic life evaluation uses the same equations as the deterministic method with stochastic inputs. Table C.1 summarizes the stochastic inputs that were applied to the economic life calculations.

The stochastic economic life is determined by a range of confidence levels associated with the EUAC. The level of confidence is taken as 70%–90%. Then a sensitivity analysis is undertaken to measure the change in market value and the repair and maintenance costs. When repair and maintenance costs sensitivity exceeds the sensitivity of the change in market value, a trigger point is reached, which indicates that the equipment may have reached its economic life, as is shown in Figure C.1.

The trigger point is reached when the sensitivity of the repair and maintenance costs intersects with the sensitivity of the change in market value. This occurs at year 6 in Figure 12 where the cost intersects with the sensitivity of the change in market value. Keeping the equipment beyond the trigger point indicates that the repair and maintenance costs are more uncertain at this point in time than the market value. In other words, the probability that cost of retaining the equipment will be greater than its resale value is high.

## DETERMINISTIC ECONOMIC LIFE

The deterministic economic life is calculated and shown in Figure C.2. The example shown in the figure is the deterministic economic life of a 2006 Volvo Loader. The economic life of the loader was found to be 4 years using the lowest EUAC.

**TABLE C.1**
**Stochastic Values for the Inputs Used in the Economic Life Determination**

| Parameter | Range of Values |
|---|---|
| Interest Rate | 3%–16% |
| Tire Cost | Varied by machine |
| R&MC | 35%–80% |
| Change in Market Value | 8%–15% |
| Diesel Fuel Prices | \$3.38/gal–\$4.13/gal |
| Tire Repair Factor | 12%–16% |

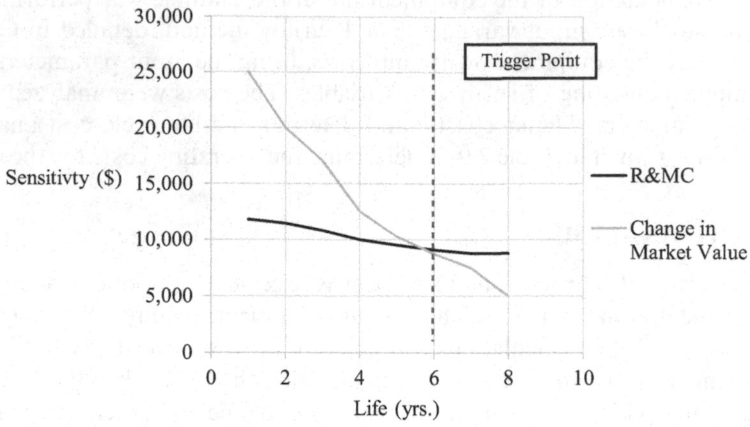


**FIGURE C.1** Trigger Point Determination Using Sensitivity Analysis

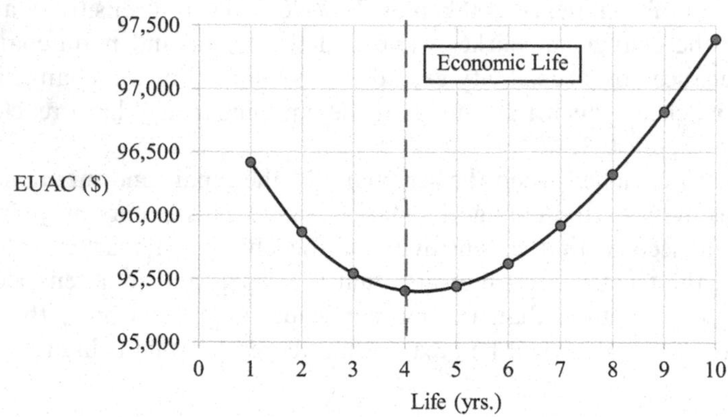


**FIGURE C.2** Deterministic Economic Life of the 2006 Volvo Loader

The stochastic determination of the economic life for the 2006 Volvo loader is depicted in Figure C.3. The confidence levels are shown with the optimal replacement age specified by the lowest EUAC. The economic life for the loader varies from year 5 to 8 depending on the level of confidence.

The stochastic economic life range for the loader supplies more detail than a deterministic determination. Using the range of values for the input parameters provides a more certain calculation of the economic life. Additionally, the range offers the fleet manager options to assess the replacement of equipment.

Monte Carlo simulations also determine the sensitivity of the inputs for the economic life calculation. Based on the sensitivity results of the change in market value and repair and maintenance costs, a trigger point for the machines was established. The sensitivity of each variable is related to the mean of the annual life cycle cost associated with the piece of equipment. The range in the values is represented in dollar amounts. The wider the range the more sensitive the input is to the mean. Figure C.4 displays the results from the sensitivity analysis performed in the seventh year of the 2006 Volvo loader. The results show that the change in market value is more sensitive than the repair and maintenance costs given the year under investigation.

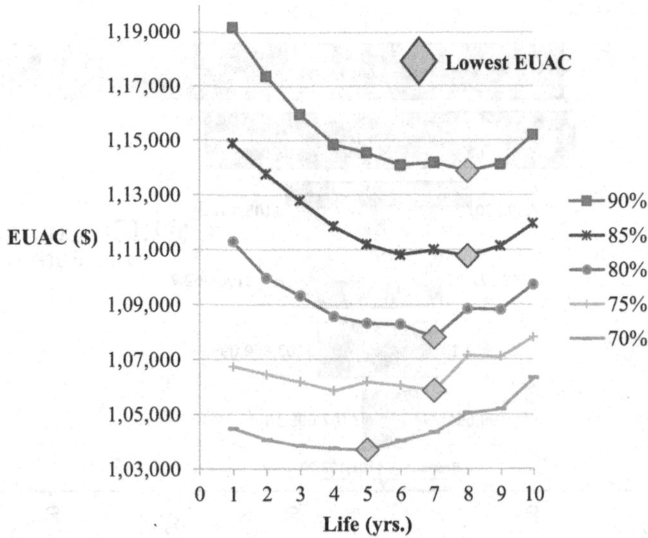


**FIGURE C.3** Stochastic Economic Life of the 2006 Volvo Loader

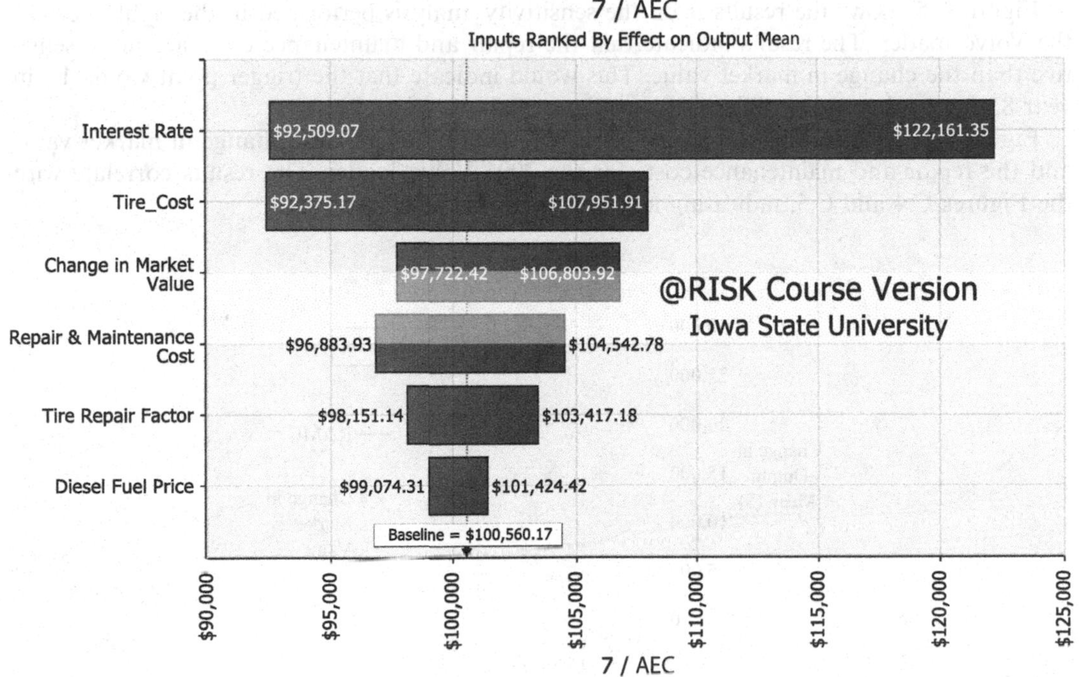


**FIGURE C.4** Sensitivity Analysis for the 2006 Volvo Loader in Year 7

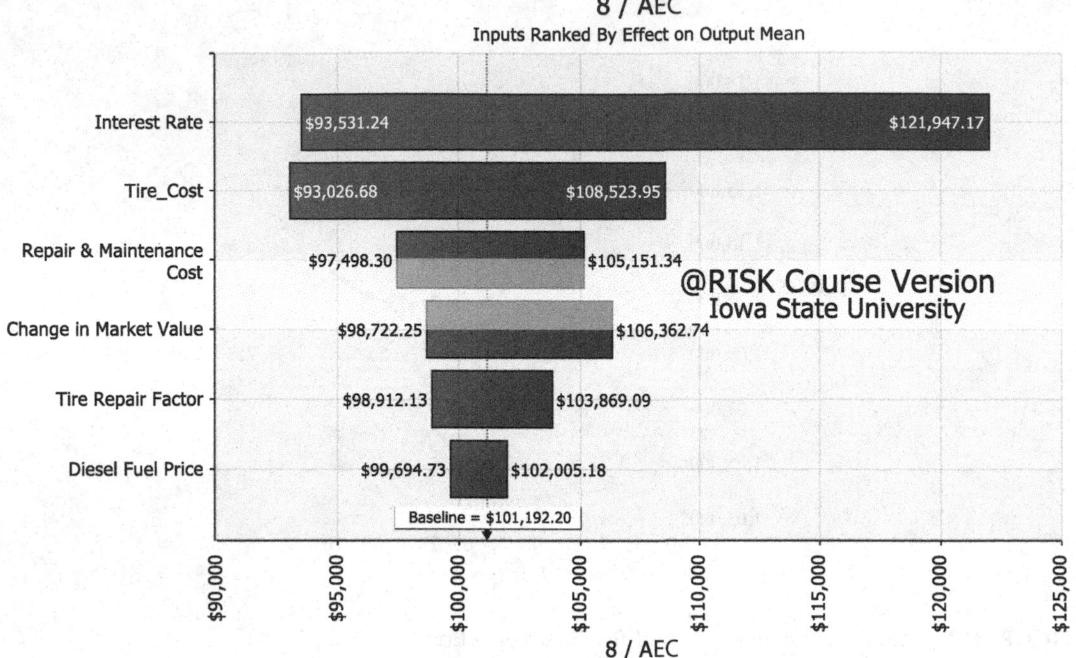


**FIGURE C.5** Sensitivity Analysis for the 2006 Volvo Loader in Year 8

Figure C.5 shows the results from the sensitivity analysis performed in the eighth year of the Volvo loader. The results indicate that the repair and maintenance costs are more sensitive than the change in market value. This would indicate that the trigger point would be in year 8, due to the results differing from Figure C.4.

Figure C.6 contains the plot of the sensitivity fluctuations for the change in market value and the repair and maintenance costs for the 2006 Volvo loader. The results correlate with the Figures C.4 and C.5, indicating a trigger point in the eighth year.

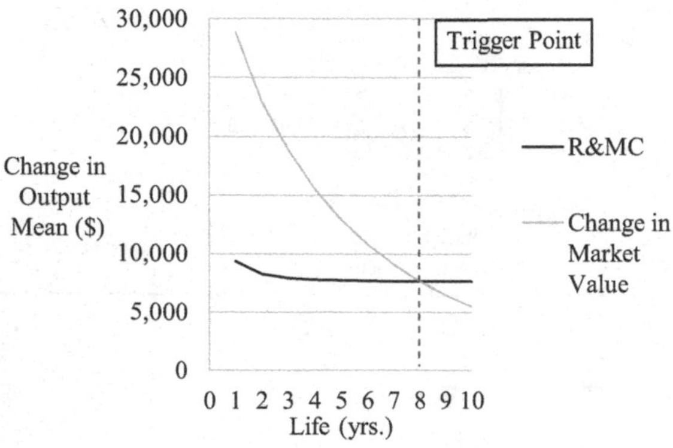


**FIGURE C.6** Change in the Output Mean for 2006 Volvo Loader

The results displayed in Figure C.5 indicate that the sensitivities of the two inputs intersect at year 8, signifying the change in the sensitivity. The intersection of the two parameters is the trigger point for equipment fleet managers.

## REFERENCES

[1] Gransberg, D.D. and O'Connor, E.P. (2015). *Major Equipment Life-Cycle Cost Analysis.* Report No. MN/RC 2015–16. St. Paul, MN: Minnesota Department of Transportation, Research Services and Library.

# Index

Page numbers in *italics* refer to Figures, and page numbers in **bold** refer to Tables.

## C



## D

## G



## H



## I



## J



## L

## M

## N



## O



## P

## T

Printed in the United States
By Bookmasters